WE ARE *HERE*, AND IT IS **NOW**...

THERE IS AN **ART** TO **DESIGN.**

BLEEP

BLOOP

I DESIGN BY **WRITING CODE.**

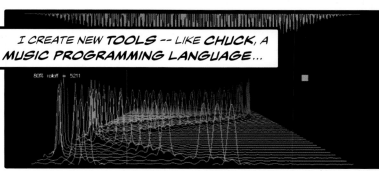

I CREATE NEW **TOOLS** -- LIKE **CHUCK,** A **MUSIC PROGRAMMING LANGUAGE**...

```
// our radius
.99999 => float R;
// our delay order
500 => float L;
// set delay
L::samp => delay.delay;
// set dissipation factor
Math.pow( R, L ) => delay.gain;
// place zero
-1 => lowpass.zero;

// fire excitation
1 => imp.gain;
// for one delay round
L::samp => now;
// cease fire
0 => imp.gain;
```

...TO SYNTHESIZE **NEW SOUNDS** AND EXPERIMENT WITH **MUSICAL INTERACTIONS.**

I DESIGN **SOUND** AND **GRAPHICS** IN TANDEM.

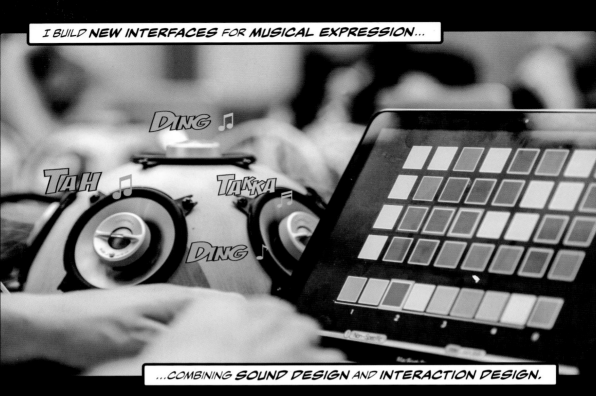

WE USE THESE INSTRUMENTS TO CRAFT PERFORMANCES FOR **LAPTOP ENSEMBLES**...

...LIKE **SLORK**, THE STANFORD LAPTOP ORCHESTRA!

I DESIGN **APPS** TO **TRANSFORM** MOBILE DEVICES...

...INTO **EXPRESSIVE MUSICAL INSTRUMENTS**...

♪ Tooooot

...TO BE PLAYED BY **ANYONE**...

...**NOVICES AND PROS.**

Hot N Cold - Katy Perry
Smule Ocarina, Guitar, Vocal

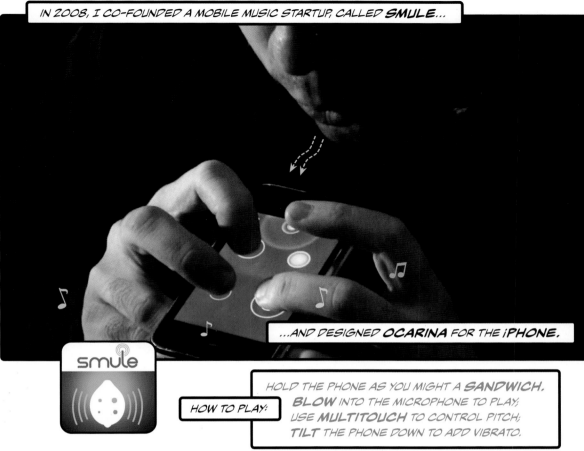

IN 2008, I CO-FOUNDED A MOBILE MUSIC STARTUP, CALLED **SMULE**...

...AND DESIGNED **OCARINA** FOR THE *i***PHONE**.

smule

HOW TO PLAY:

HOLD THE PHONE AS YOU MIGHT A **SANDWICH**.
BLOW INTO THE MICROPHONE TO PLAY;
USE **MULTITOUCH** TO CONTROL PITCH;
TILT THE PHONE DOWN TO ADD VIBRATO.

I SERVED AS **CHIEF CREATIVE OFFICER** AND **CTO** DURING OUR **EARLY YEARS**, UNTIL I STEPPED DOWN IN 2013.

WITHIN THAT TIME, I ALSO DESIGNED **OCARINA 2**, **MAGIC PIANO**, AND OTHER MUSIC-MAKING **ARTIFACTS.**

THESE GAMES, TOYS, INSTRUMENTS HAVE REACHED MORE THAN **200 MILLION** USERS.

THERE IS **ANOTHER DIMENSION** TO ARTFUL DESIGN: A **SOCIAL FABRIC** THAT **CONNECTS** US IN NEW, EXPRESSIVE WAYS.

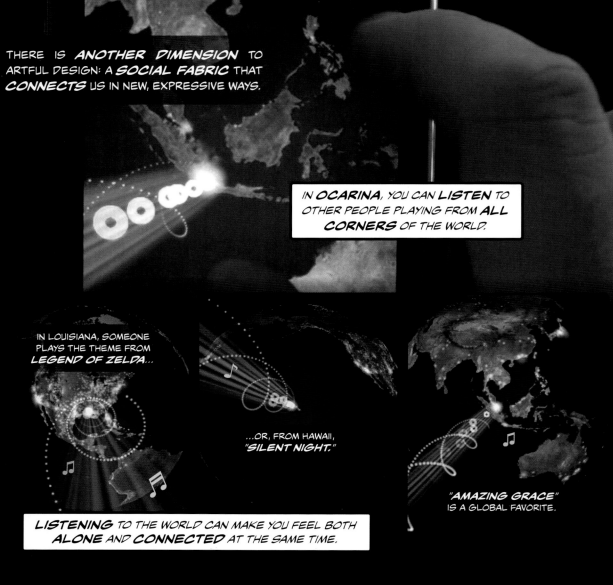

IN **OCARINA**, YOU CAN **LISTEN** TO OTHER PEOPLE PLAYING FROM **ALL CORNERS** OF THE WORLD.

IN LOUISIANA, SOMEONE PLAYS THE THEME FROM **LEGEND OF ZELDA**...

...OR, FROM HAWAII, **"SILENT NIGHT."**

"AMAZING GRACE" IS A GLOBAL FAVORITE.

LISTENING TO THE WORLD CAN MAKE YOU FEEL BOTH **ALONE** AND **CONNECTED** AT THE SAME TIME.

THIS IS DESIGN AS **EXPERIENCE**, **VISCERAL** AND **HUMAN**...

...A SMALL FEELING THAT THERE IS **SOMEONE**, **SOMEWHERE OUT THERE**... AND THAT WE ARE MORE **ALIKE** THAN DIFFERENT.

THIS KIND OF DESIGN WOULD NOT BE POSSIBLE WITHOUT **TECHNOLOGY**...

...YET HOPEFULLY THE USER **NEVER NOTICES** THE TECHNOLOGY.

ARTFUL DESIGN

TECHNOLOGY IN SEARCH OF THE SUBLIME

written and designed by
GE WANG

STANFORD UNIVERSITY PRESS

This work would not be possible, as it is, without support from the **Guggenheim Foundation**, as well as Stanford University's **School of Humanities and Sciences** and **School of Engineering**; nor would it have been as rich without Creative Commons and the folks who shared through it.   Please see Acknowledgments and Image Credits at the conclusion of this book for more information.

STANFORD UNIVERSITY PRESS

Stanford, CA

sup.org

Cataloging-in-Publication Data is available at the Library of Congress.

ISBN 978-1-5036-0052-2 (paper)
ISBN 978-1-5036-0803-0 (ebook)

Printed in Korea.

 OF
CONTENTS

A **ROADMAP** FOR OUR ADVENTURE AHEAD!

PRELUDE

YOU ARE HERE

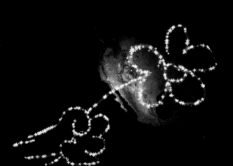

MANIFESTO

In our age of rapidly evolving technology and unyielding human restlessness and discord, design ought to be more than simply functional; it should be expressive, socially meaningful, and humanistic. Design should transcend the purely technological, encompass the human, and strive for the sublime.

Sublime design presents itself, first and last, as a useful thing, but nestled within that window of interaction lies the novel articulation of a thought, an idea, a reflection—an invisible truth that speaks to us, intimate yet universal, purposeful without necessity of purpose, that leaves us playful, understood, elevated. It is a transformation so subtle that it escapes our conscious grasp but that once experienced—like music—we would never want to be without again.

Design should be artful.

CHAPTER 1
DESIGN IS _____

ART WITH A
PURPOSE?

FORMALIZED
INTENT?

PLANNING WITH
AESTHETICS?

HUMANISTIC
ENGINEERING?

POETIC
TECHNOLOGY?

PHILOSOPHY IN
THE MAKING?

DESIGN IS **EVERYWHERE**: THE **ARCHITECTURE** OF THE BUILDINGS AND CITIES WE INHABIT, THE MUSICAL **INSTRUMENTS** AND **GAMES** WE PLAY, THE **SOFTWARE** WE USE, THE **CLOTHES** WE WEAR, THE **ARRANGEMENT** OF OUR LIVING ROOM FURNITURE, AND THE **ORGANIZATION** OF OUR DAILY LIFE...

THE **TRIBE** OF HUMANITY
HAS **ALWAYS DESIGNED.**

WE **ARE**, THEREFORE WE **DESIGN.**

BUT WHAT **IS** DESIGN?
WHAT IS ITS **NATURE**, ITS **PURPOSE**,
ITS **MEANING?** IS IT **ENGINEERING?**
IS IT **ART?** OR IS IT **SOMETHING
ELSE** ENTIRELY?

IN THIS RAPIDLY EVOLVING
TECHNOLOGICAL AGE, WHAT
POSSIBILITIES DOES DESIGN HOLD? HOW
MIGHT WE **UNDERSTAND** ARTFUL DESIGN AS
A CREATIVE DISCIPLINE TO **SHAPE** AND
ENRICH OUR WORLD?

LET'S BEGIN
WITH A **CASE
STUDY**...

THIS **EVERYDAY OBJECT** CAN ILLUSTRATE SOMETHING ABOUT **DESIGN.**

I FOUND IT AT A **SCHOOL SUPPLY STORE:**

...AN ORDINARY-LOOKING **PENCIL BAG**, EXCEPT...

...FOR STARTERS, IT HAS **EYES!**

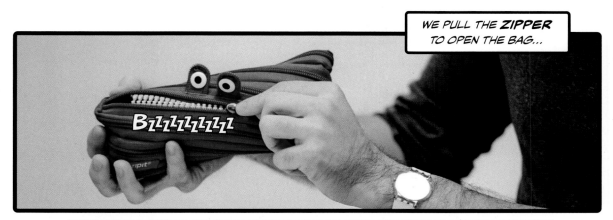

WE REALLY CAN'T "UNSEE" THE FACE!

YEAH, THAT'S RIGHT, BOYS AND GIRLS, YOU GONNA BE SEEIN' *THIS* MUG WELL INTO YOUR DREAMS! ⁒*BWAHAHA!*⁒

THIS IS *DESIGN* PLAYING ON OUR **ANTHROPOMORPHIC** TENDENCIES -- TO DISCERN *HUMANLIKE* FORMS IN INANIMATE OBJECTS.

THE ANTHROPOMORPHIC *FORM* OF THIS OBJECT BORROWS FROM ITS *FUNCTION* AS A CONTAINER WITH A ZIPPER OPENING -- AND INVOKES A SENSE OF *DELIGHT* BY DESIGN.

WHOA... AND I THOUGHT I WAS JUST A PENCIL BAG...

...IT IS ALSO AN EXAMPLE OF AN OBJECT'S *FORM* FOLLOWING ITS *FUNCTION.*

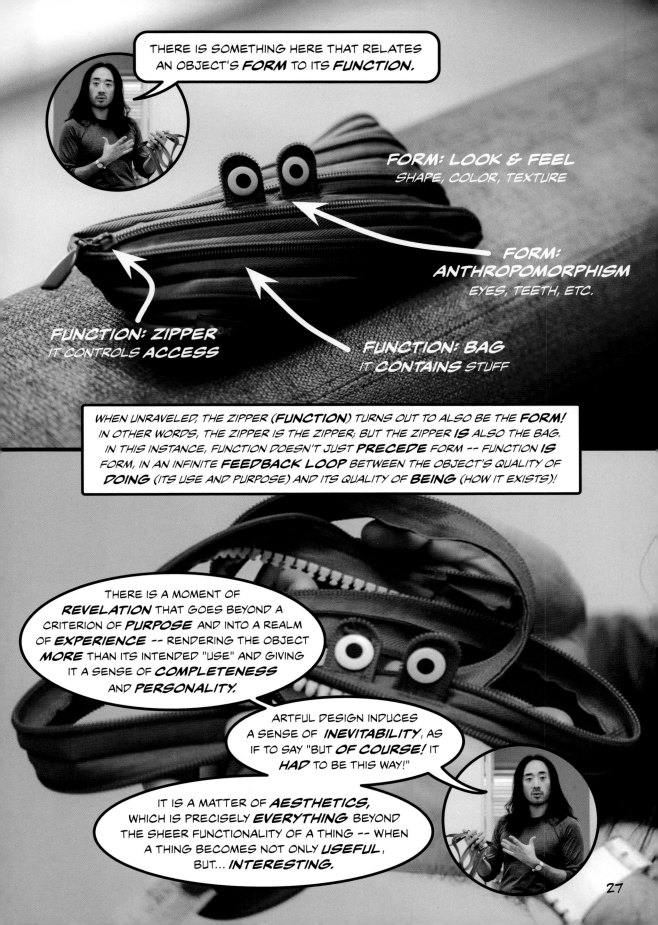

THERE IS SOMETHING HERE THAT RELATES AN OBJECT'S **FORM** TO ITS **FUNCTION.**

FORM: LOOK & FEEL
SHAPE, COLOR, TEXTURE

FORM: ANTHROPOMORPHISM
EYES, TEETH, ETC.

FUNCTION: ZIPPER
IT CONTROLS **ACCESS**

FUNCTION: BAG
IT **CONTAINS** STUFF

WHEN UNRAVELED, THE ZIPPER (**FUNCTION**) TURNS OUT TO ALSO BE THE **FORM!** IN OTHER WORDS, THE ZIPPER IS THE ZIPPER; BUT THE ZIPPER **IS** ALSO THE BAG. IN THIS INSTANCE, FUNCTION DOESN'T JUST **PRECEDE** FORM -- FUNCTION **IS** FORM, IN AN INFINITE **FEEDBACK LOOP** BETWEEN THE OBJECT'S QUALITY OF **DOING** (ITS USE AND PURPOSE) AND ITS QUALITY OF **BEING** (HOW IT EXISTS)!

THERE IS A MOMENT OF **REVELATION** THAT GOES BEYOND A CRITERION OF **PURPOSE** AND INTO A REALM OF **EXPERIENCE** -- RENDERING THE OBJECT **MORE** THAN ITS INTENDED "USE" AND GIVING IT A SENSE OF **COMPLETENESS** AND **PERSONALITY.**

ARTFUL DESIGN INDUCES A SENSE OF **INEVITABILITY**, AS IF TO SAY "BUT **OF COURSE!** IT **HAD** TO BE THIS WAY!"

IT IS A MATTER OF **AESTHETICS**, WHICH IS PRECISELY **EVERYTHING** BEYOND THE SHEER FUNCTIONALITY OF A THING -- WHEN A THING BECOMES NOT ONLY **USEFUL**, BUT... **INTERESTING.**

27

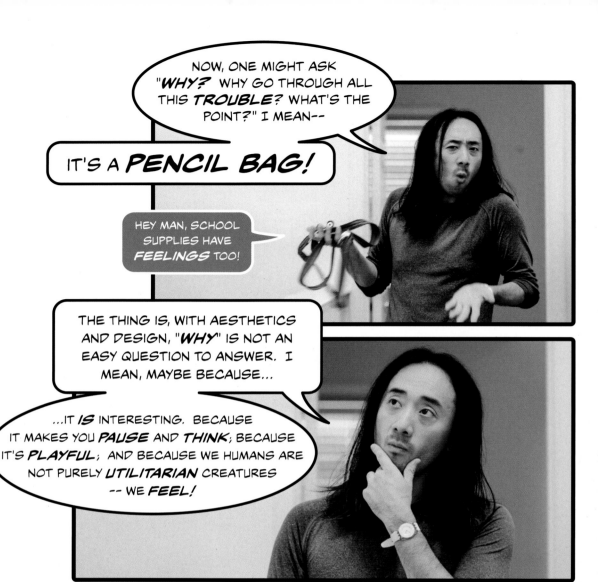

NOW, ONE MIGHT ASK "**WHY?** WHY GO THROUGH ALL THIS **TROUBLE?** WHAT'S THE POINT?" I MEAN--

IT'S A **PENCIL BAG!**

HEY MAN, SCHOOL SUPPLIES HAVE **FEELINGS** TOO!

THE THING IS, WITH AESTHETICS AND DESIGN, "**WHY**" IS NOT AN EASY QUESTION TO ANSWER. I MEAN, MAYBE BECAUSE...

...IT **IS** INTERESTING. BECAUSE IT MAKES YOU **PAUSE** AND **THINK**; BECAUSE IT'S **PLAYFUL**; AND BECAUSE WE HUMANS ARE NOT PURELY **UTILITARIAN** CREATURES -- WE **FEEL!**

"SOMETHING ABOUT IT" APPEALS TO OUR SENSE OF BEAUTY, OF SIMPLICITY -- BUT ALSO REMINDS US THAT **NOT** EVERYTHING WORTHWHILE NEEDS TO SERVE A PRACTICAL PURPOSE (LIKE THE IDEA OF PLAY). THE PENCIL BAG IS BOTH **PURPOSEFUL** AND, AT THE SAME TIME, **FREE** FROM PURPOSE.

I CAN HONESTLY SAY THIS IS AS CLOSE AS I'VE COME TO **FEELING** EMOTION ABOUT A SCHOOL SUPPLY.

PUT ME DOWN, **PUNK!**

AESTHETICS AND **DESIGN**

Aesthetics is how we experience a thing—how it emotionally, intellectually, psychologically, and socially affects us. It is everything **beyond** the thing's function. Yet aesthetics does not usurp or live apart from functionality; instead, it gives context, meaning, and essence to a thing, making it what it is.

In the kitchen of design, aesthetics is not the spice with which you garnish the casserole of functionality: it **is** the casserole, the ingredients you bake into it. If functionality is the casserole's nutrition, then aesthetics is its texture, taste, and quality.

A thing without aesthetics is like food devoid of flavor. If functionality is what a thing does **for** you, aesthetics is what it does **to** you.

Artful design is design with fundamental emphasis on aesthetics.

ALL THIS FOOD TALK IS MAKING ME *HUNGRY.* SAY... YOU GOT ANY WRITING UTENSILS I CAN *MUNCH* ON?

⸰BELCH⸰

FOR EXAMPLE, MY REACTION TO *THIS* PENCIL BAG WAS AN *AESTHETIC* ONE!

WITH THAT, LET'S RETURN TO THE QUESTION: WHAT **IS** DESIGN? THROUGHOUT THE BOOK, WE DEVELOP **PRINCIPLES**, **DEFINITIONS**, AND **IDEAS** THAT AIM TO CAPTURE THE ESSENCE OF **ARTFUL DESIGN**, BUILDING A PRACTICAL PHILOSOPHY OF DESIGN. HERE IS THE FIRST PRINCIPLE (A **WORKING DEFINITION** ACTUALLY), AIMED AT GROUNDING OUR DISCUSSION OF THE NATURE OF DESIGN.

⊕ PRINCIPLE 1.1

A *WORKING DEFINITION* OF *DESIGN*

A *SHAPING* OF OUR WORLD, AIMED AT ADDRESSING AN *INTENDED* PURPOSE...

BY ITSELF, THIS DESCRIBES **ENGINEERING** AND DISTINGUISHES **DESIGN** FROM **PURE ART.**

...CARRIED OUT WITH *AESTHETICS* AND *HUMANITY*...

THIS SECOND TENET DESCRIBES **ART**, AND DISTINGUISHES **DESIGN** FROM **ENGINEERING.**

...WITHIN A *MEDIUM.*

LIKE BOTH ENGINEERING AND ART, DESIGN NEEDS A SPECIFIC **VEHICLE** TO EXPRESS ITSELF. ANYTHING WE CAN REASONABLY WORK WITH CAN BE CONSIDERED A MEDIUM: SOFTWARE, WORDS IN SEQUENCE, CLAY, SOUNDS, IMAGES, ZIPPERS... IN THIS CONTEXT, THE MEDIUM AMOUNTS TO THE **TECHNOLOGY** IN WHICH THE DESIGN EXPRESSES ITSELF.

COROLLARY: ANYTHING WORTH DESIGNING IS WORTH DESIGNING BEAUTIFULLY

DESIGN IS AN ACT OF **ALIGNMENT.** WE DESIGN TO BRING THE WORLD INTO **PRAGMATIC** ALIGNMENT WITH WHAT WE CONSIDER TO BE **USEFUL** AND INTO **AESTHETIC** ALIGNMENT WITH OUR NOTION OF WHAT'S **GOOD** AND **BEAUTIFUL**, OR "THE WAY THINGS OUGHT TO BE." WITHIN THIS **CREATIVE** ENDEAVOR ARE REAL, RICH, EXPRESSIVE OPPORTUNITIES TO SPEAK TO OUR **HUMAN DIMENSION,**

CONTEXTUALIZING ARTFUL DESIGN
WHAT IT IS AND WHAT IT IS NOT

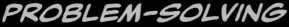

PROBLEM-SOLVING
A *PROBLEM* IN SEARCH OF A *SOLUTION*

PROBLEM-SOLVING IS A *CREATIVE PROCESS* OF DISCOVERY, IDENTIFICATION, AND DEFINITION OF A *PROBLEM* (OR *NEED*), FOLLOWED BY FINDING *SOLUTIONS* TO ADDRESS THAT NEED. PROBLEM-SOLVING IS, BY NATURE, *TOP-DOWN.* THE DISCIPLINE OF *DESIGN THINKING* FITS INTO THIS.

"PEOPLE NEED A CASE TO HOLD THEIR PENCILS!"

VS.

CREATIVE DESIGN
TECHNOLOGY IN SEARCH OF A *PROBLEM*

CREATIVE DESIGN DOES NOT START WITH A PROBLEM BUT WITH AN AVAILABLE *MEDIUM* (A TECHNOLOGY, A TOOL, A TECHNIQUE) AND *TINKERS* IN UNDIRECTED EXPLORATION OUT OF SUSTAINED *CURIOSITY*, AIMING TO DISCOVER *NEW THINGS* AND WHAT THEY MIGHT BE USED FOR. CREATIVE DESIGN IS *BOTTOM-UP* -- IT WORKS BACKWARD FROM THE MEDIUM.

"ZIPPERS ARE COOL; WHAT CAN WE DO WITH THEM?"

EVOLUTION!

ARTFUL DESIGN  DEFINITION 1.2
TECHNOLOGY IN SEARCH OF *THE SUBLIME!*

ARTFUL DESIGN IS AN *EVOLUTION* OF CREATIVE DESIGN IN SYNTHESIS WITH PROBLEM-SOLVING; IT IS *USEFUL* DESIGN *EMPHASIZING AESTHETICS*; IT DERIVES *NOT ONLY* FROM *NEEDS* BUT ALSO FROM THEIR UNDERLYING *HUMAN VALUES.*

IT IS *REDUNDANT* TO SAY "ARTFUL DESIGN" (ALL GOOD DESIGN IS ARTFUL), AND BY DEFAULT, "DESIGN" IN THIS BOOK IMPLIES ARTFUL DESIGN. BUT IT'S ESSENTIAL HERE TO EMPHASIZE THE "ARTFUL," TO REMIND US TO DESIGN NOT SOLELY FOR APPARENT NEEDS, BUT TO ALSO STRIVE FOR A TYPE OF *EVERYDAY POETRY.*

"WHOA, WHAT IF THE PENCIL BAG IS THE ZIPPER?!"

THE SUBLIME?

THE SUBLIME DESCRIBES A *DEEPLY HUMAN AESTHETIC EXPERIENCE*, ONE WE WILL EXPLORE THROUGHOUT THIS BOOK. FOR NOW, THINK OF STANDING AT THE EDGE OF AN *ABYSS*, LOOKING OUT INTO *INFINITY.* THERE ARE TWO OVERRIDING, CONTRASTING EMOTIONS: A SENSE OF *AWE*, AND A TWINGE OF *FEAR* -- AND THE COMBINATION OF THE TWO BRINGS A MOMENT OF *CLARITY* AND *WONDER.* IT IS *TRANSCENDENCE* OF A MEDIUM -- A GLIMPSE INTO *TRUTH* AND *BEAUTY* BEYOND OUR FULL COMPREHENSION. WHILE SUBJECTIVE TO EACH INDIVIDUAL, IT IS ALSO A *UNIVERSAL* HUMAN EXPERIENCE. (I AM NOT SURE MY ENCOUNTER WITH THE PENCIL BAG WASN'T SUBLIME...)

AESTHETICS OF ARTFUL DESIGN

THE **ABIDING ELEMENTS** OF AESTHETICS IN ARTFUL DESIGN, ROOTED IN AN INTERPLAY OF **SENSE** AND **COGNITION**, OF **REASON** AND **SENTIMENT**

SONIC   VISUAL

TACTILE   ## MATERIAL   OLFACTORY

THE DIRECTLY **PERCEIVED**

HUH. IS THERE LIKE A QUIZ ON THIS LATER?

STRUCTURAL

HOW SOMETHING IS **PUT TOGETHER**; THE RELATIONSHIP BETWEEN ITS PARTS

INTERACTIVE

ACTION, RESULT, MAPPING, AGENCY; MATERIALITY MEETS FUNCTIONALITY; **HUMAN** MEETS **TECHNOLOGY**

DIRECT, PHYSICAL, SENSORY

EMOTIONAL / PSYCHOLOGICAL

EMOTIONAL ENGAGEMENT, MEANING, POETRY, PATHOS; SATISFACTION IN THE FULFILLMENT OF PURPOSE; INTERFACE OF PERCEPTION AND REASON

SOCIAL

SPEAKING TO OUR SOCIAL INSTINCTS, OUR IMPERATIVE TO PARTICIPATE, TO BELONG

CONCEPTUAL, INTANGIBLE, INVISIBLE

MORAL-ETHICAL

HUMANIST DIMENSION ("DOES IT **DO GOOD?**"), ETHOS, THE **CONSCIENCE** OF THE DESIGN

YA KNOW, DESIGN HAS A WAY OF ESCAPING CONCRETE *FORMALIZATION*, YET IT IS *MORE* FORMALIZABLE THAN "IT IS IF YOU SAY IT IS." WE *CAN* MAKE *REASONED ARGUMENTS* ABOUT THE NATURE OF DESIGN AND ARTICULATE *NOTIONS* TO CAPTURE ITS *ESSENCE* -- NOT AS A COMPLETE "GENETIC MAP," NOR AS RECIPES, BUT AS *IDEAS* AND THINGS TO *THINK* WITH.

ARTFUL DESIGN IS ESSENTIAL IN AN AGE OF EVER-ACCELERATING TECHNOLOGY AND, I BELIEVE, RICHLY FERTILE AS A *CONSCIOUS PRACTICE* AND ELEVATED *DISCIPLINE*, WORTHY OF *INTELLECTUAL* AND *ARTISTIC* PURSUIT! SO, HOW *DO* WE THINK ABOUT IT?

GOING BACK IN TIME TO AN *IDEA* FROM *ARISTOTLE...*

⇒YO⇐ WE OUGHT TO EXPECT *NO MORE* PRECISION THAN A SUBJECT *NATURALLY AFFORDS!*

FOR EXAMPLE, WHILE THE MATHEMATICIAN AND THE ETHICIST BOTH MUST MAKE *RIGOROUS ARGUMENTS* FOUNDED ON LOGIC, IT WOULD BE FOLLY TO EXPECT THE *SAME* TYPE OF *PRECISION* FROM THE TWO!

NICOMACHEAN ETHICS: ARISTOTLE'S BOOK ON ETHICS, HAPPINESS, VIRTUE; POSSIBLY THE ORIGINAL SELF-HELP BOOK

EXPECT NO MORE *PRECISION* THAN A SUBJECT NATURALLY AFFORDS!

🜨 META-PRINCIPLE 1.4

DESIGN IS BOTH *PRECISE* AND *TACIT*, EMBODYING ASPECTS OF *ENGINEERING* AND SOMETHING MORE AKIN TO *ART.* *PRESENT* BUT *INVISIBLE*, DESIGN IS NESTLED BETWEEN FUNCTION AND FORM, PRAGMATICS AND AESTHETICS. WITH THIS *DUALITY* IN MIND, LET'S CONTINUE WITH A FUNDAMENTAL TENET OF DESIGN...

DESIGN HAS EVERYTHING TO DO WITH **PURPOSE**, AND IT TAKES PLACE BETWEEN THE **TWO** KINDS OF **REASONS** WE DO **ANYTHING** IN LIFE: THE **MEANS-TO-AN-END** AND THE **END-IN-ITSELF**. BALANCING THESE TWO ELEMENTS IS A **CENTRAL TENET** OF ARTFUL DESIGN.

PRINCIPLE 1.5

DESIGN IS MEANS VS. ENDS!

MEANS-TO-AN-END

THAT WHICH SERVES AN **EXTERNAL PURPOSE, USE,** OR **FUNCTION**

END-IN-ITSELF

SOMETHING **GOOD IN ITSELF**, WHOSE VALUE LIES PRIMARILY IN ITS **INTRINSIC** WORTH

VS.

AS HUMANS, WE VALUE THINGS FROM **BOTH** SIDES...

MEANS-TO-AN-END	END-IN-ITSELF
PRAGMATICS	AESTHETICS
FUNCTION	FORM
PRACTICAL NEEDS	VALUES & BELIEFS
TECHNOLOGY	ART
TOOLS	FRIENDSHIP
WORK	PLAY
DOING	BEING
KNOWLEDGE (FOR AN EXTERNAL PURPOSE)	KNOWLEDGE (FOR ITS OWN SAKE)
SOLUTIONS	EXPERIENCE

LET'S **EXAMINE** SOME OF THESE FURTHER!

THE CLASSIC **DICHOTOMY** OF FORM VS. FUNCTION FROM MODERN ARCHITECTURAL DESIGN MAY SEEM OUT OF PLACE AND OUT OF DATE IN REALMS OF SOFTWARE, THE VIRTUAL, THE EXPERIENTIAL, THE TINY, AND THE INVISIBLE. BUT AS LONG AS HUMANS REMAIN INTERESTED IN **MAKING USEFUL THINGS** AND CAPABLE OF **AESTHETIC JUDGMENT**, FORM-AND-FUNCTION, I ASSURE YOU, IS **NOT DEAD:** IT **RE-CONTEXTUALIZES** ITSELF IN NEW DOMAINS.

PRINCIPLE 1.6

DESIGN IS AN **INTERPLAY** BETWEEN **FUNCTION** AND **FORM**

FUNCTION
WHAT A THING **DOES**

VS.

FORM
HOW A THING **IS**

- ### **WHAT** IS IT FOR?

  WHAT **PURPOSE** DOES THE OBJECT SERVE? WHAT MUST IT **DO?** WHAT MUST IT **NOT** DO? IS IT A TOOL? MUSICAL INSTRUMENT? GAME? TOY? FOR EDUCATION? ENTERTAINMENT?

- ### **WHO** IS IT FOR?

  WHO IS THE INTENDED USER OF THE DESIGN? EXPERTS? NOVICES? KIDS? GROWN-UPS? COFFEE DRINKERS? MUSICIANS? ENGINEERS? STUDENTS? TEACHERS? CATS?

I DIDN'T KNOW I WAS **SO COMPLEX...**

- ### **HOW IT WORKS**

  WHAT IS THE OBJECT'S **INTERFACE?** HOW **WELL** DOES IT WORK? HOW DOES IT **RESPOND**, OR **RECOVER** FROM OPERATOR ERROR? HOW EASY IS IT FOR NOVICES, AND WHAT IS THE **CEILING** OF **EXPRESSIVENESS** FOR EXPERTS?

- ### HOW IT **LOOKS**

  SHAPE, COLOR, PATTERN, TEXTURE; HOW THESE ELEMENTS **CHANGE** OVER TIME, HOW SOMETHING MOVES, MORPHS; HOW IT DIRECTS THE EYE; PROPORTION AND BALANCE; WHAT HITS YOUR SENSES

- ### HOW IT **SOUNDS**

  TIMBRE, TEXTURE, DENSITY, DYNAMICS; ARTICULATION, CHANGE OVER TIME; HOW IT ENGAGES OUR EARS

- ### HOW IT **FEELS**

  MATERIALITY, TACTILITY, THE WAY A THING BEHAVES, RESPONDS, **HANDLES**; THE EXPERIENCE OF INTERACTION; THE WAY IN WHICH A DESIGN IS PUT TOGETHER TO REVEAL **MEANING**

- ### HOW IT **MAKES YOU FEEL**

  THE EMOTIONAL **CONSEQUENCE** OF A THING AS EXPERIENCED, ITS "THING-ALITY," ITS OVERALL AESTHETICS; DOES IT MAKE YOU **FEEL** SOMETHING? WHAT IT **MEANS** TO YOU

FORM ??? FUNCTION

WHAT **VERBS** GO IN THE BLANK? LET'S **TRY** SOME.

FORM EVER **FOLLOWS** FUNCTION

THIS DESIGN **MANTRA** (ATTRIBUTED TO AMERICAN ARCHITECT **LOUIS SULLIVAN**) SAYS THE **FORM** OF AN OBJECT (E.G., THE SHAPE OF A TEAPOT) SHOULD BE COGNIZANT OF AND BASED ON ITS **INTENDED PURPOSE** (E.G., CONTAINING AND POURING TEA). IT DOES **NOT** MEAN FORM IS LESS IMPORTANT, OR THAT THE FUNCTION MUST BE CONSIDERED BEFORE FORM; GOOD DESIGN CONSIDERS **BOTH** FROM THE GET-GO AND LETS ONE INFLUENCE THE OTHER. THE ART OF DESIGN ISN'T JUST HOW BEAUTIFUL AN OBJECT IS OR HOW WELL IT WORKS, BUT HOW FORM AND FUNCTION CONTRIBUTE TO A SINGULAR, **COHESIVE**, FUNCTIONAL-AESTHETIC ENTITY.

FORM **EXPRESSES** FUNCTION

FUNCTION IS **GENERAL**, WHEREAS **FORM** HAS TO DO WITH THE **SPECIFIC** WAYS IN WHICH A DESIGN ACTUALLY **EXPRESSES** ITS PURPOSE TO ITS USER. SAYING "THIS PENCIL BAG HOLDS PENCILS" IS PROBABLY THE LEAST INTERESTING THING WE CAN SAY ABOUT A PENCIL BAG, BECAUSE WE CAN ALREADY SEE THAT'S WHAT IT DOES. FORM PROVIDES **SPECIFICITY** ON THE FOUNDATION OF GENERAL FUNCTIONALITY, ADDRESSING NOT JUST **WHAT** BUT **HOW** IT IS EXPRESSED.

BUTTON (PRESS)   HANDLE (PULL)   WHEEL (ROTATE)   SCREW (TURN)   PIANO (LINEAR BUTTONS)   STRANGE PIANO (CIRCULAR BUTTONS)

FORM **MODULATES** FUNCTION

AN OBJECT'S FORM **CONTEXTUALIZES** ITS FUNCTION, **UNFOLDS** IT, IMBUES IT WITH **MEANING**, APPEALING TO OUR EMOTIONS AND PLAYFULNESS. FORM CAN MAKE A STATEMENT, OR INVITE US TO SEE THINGS DIFFERENTLY. OUR RESPONSE CAN BE **BEHAVIORAL** (WE DO SOMETHING) OR **AESTHETIC** (WE FEEL SOMETHING). HOW A SPORTS CAR HANDLES IS PART OF BOTH ITS FUNCTION AND AESTHETIC. I.M.PEI'S **PYRAMID ENTRANCE** TO THE LOUVRE EVOKES POWERFUL SHAPES; ITS TRANSPARENCY ILLUMINATES THE ATRIUM BELOW WHILE GRACEFULLY ACKNOWLEDGING THE SURROUNDING ARCHITECTURE.

MORE THAN FUNCTION,
DESIGN IS
EXPERIENCE.

IF FUNCTION DIDN'T MATTER...

WE'D BE USING THINGS THAT WORK POORLY, INEFFICIENTLY, OR NOT AT ALL. IF FUNCTION DIDN'T MATTER, THINGS WOULD FALL APART, PRESENT PHYSICAL DANGERS, CREATE FRUSTRATION, AND ULTIMATELY, WE'D HAVE CHAOS!

FUNCTION!

FORM!

BOTH ARE **NECESSARY!**

NEITHER **ALONE** IS **SUFFICIENT!**

"THERE SHOULD BE NO SUCH THING AS **ART** DIVORCED FROM **LIFE**, WITH **BEAUTIFUL** THINGS TO LOOK AT AND **HIDEOUS** THINGS TO USE." -- BRUNO MUNARI, **DESIGN AS ART**

IF FORM DIDN'T MATTER...

WE'D ALL BE **CONTENT** TO LIVE IN **WAREHOUSES**, WEAR THE SAME DULL CLOTHES, FEED ON THE SAME TASTELESS FOOD, HAVE NO **PREFERENCE** FOR WHAT'S BEAUTIFUL -- NOR WOULD WE CARE THAT SOMETHING IS HIDEOUS OR DISTASTEFUL. THE STUFF OF DYSTOPIAS, LIFE WOULD BE **EXISTENCE WITHOUT FLAVOR.**

DESIGN IS ALL ABOUT **MAKING CHOICES**, WHICH DO MORE THAN FULFILL AN ESTABLISHED PURPOSE -- THEY **ARTICULATE PREFERENCES**, IN PARTICULAR THOSE OF THE USER, AS EXPRESSED BY THE DESIGNER. DESIGN BEGINS WITH A **GENERALITY** OF PURPOSE, BUT IS A PROCESS OF FINDING EVER-GREATER **SPECIFICITY.**

PRINCIPLE 1.7 DESIGN IS AN

ARTICULATION OF PREFERENCE

AS HUMANS, WE HAVE **PREFERENCES** FOR BOTH THE **PRAGMATIC** AND THE **AESTHETIC.** WE SET OUT TO SOLVE A PROBLEM, BUT IN AN **AESTHETIC** WAY.

ARTFUL & HUMANISTIC

SOMETHING CREATED THAT **REFLECTS** OURSELVES, OUR EXISTENCE AS HUMANS

SATISFYING

SOMETHING WE FIND **EFFECTIVE** AND **PLEASING** IN ITS USE

BEAUTIFUL

SOMETHING WE FIND **MEANINGFUL** AND **TRUTHFUL**, SIMPLY FOR BEING THE WAY IT IS

MUTUALLY INFUSED WITH

PRAGMATICS * AESTHETICS

PURPOSEFUL

SOMETHING **USEFUL** ENABLED BY THE DESIGN

PERSONALITY

SOMETHING THAT IMBUES A DESIGN WITH RICHNESS AND CHARACTER, CONGRUENT WITH ITS PURPOSE AND NATURE

INTERACTIVE

SOMETHING THAT **MEDIATES** US AND THE WORLD AROUND US; INPUT TO OUTPUT, ACTION TO CONSEQUENCE

EASE OF USE VS. EXPRESSIVENESS

SKILL **FLOOR**: HOW EASY TO LEARN
SKILL **CEILING**: HOW DEEP CAN EXPERTS GO

IN A WAY, *DESIGN* IS WHAT WE *DO* WHEN OUR *INTENTIONALITY* EXCEEDS KNOWN *METHODOLOGY.*

IN OTHER WORDS, WE WANT TO BE ABLE TO DO SOMETHING IN A CERTAIN WAY, BUT DON'T KNOW *HOW*, SO WE CREATIVELY TAKE STEPS TO TRY TO ADDRESS THAT ESTABLISHED AIM.

FOR EXAMPLE: "I WANT A *TOOL* TO *TRANSFORM* EVERYDAY SOUNDS INTO MUSIC"...

...OR: "I WANT A CUSTOM-ORDER HAMBURGER DELIVERED TO MY MOUTH AUTOMATICALLY JUST BY *THINKING* ABOUT IT. UH... WHAT WOULD THAT TAKE?"

DESIGN IS INTENTIONALITY EXCEEDING METHODOLOGY

PRINCIPLE 1.8

DESIGN *BEGINS* WHEN WE HAVE AN *INTENDED OUTCOME* WITHOUT KNOWING *HOW* TO ACHIEVE IT, OR IF IT'S EVEN *POSSIBLE* AS ENVISIONED.

THIS IS WHY THERE IS *NO* SPECIFIC *PLAYBOOK* FOR DESIGN. IF THERE WERE, IT WOULD *NOT* BE DESIGN, WHICH ALWAYS ENTAILS THE *SEARCH* FOR SOME YET *UNKNOWN* WAY OF HOW THE PIECES *FIT* TOGETHER, IN A SPECIFIC *CONTEXT*, WITH A SPECIFIC SET OF *VISIBLE AIMS* AND *SUBTLE PREFERENCES.* WHICH IS TO SAY...

PRINCIPLE 1.9

DESIGN IS ADDITION

WHEN *INTENTIONALITY EXCEEDS METHODOLOGY*, WE *INVENT.*

DESIGN IS A *CREATIVE* ENDEAVOR. IT *IMAGINES* AND *CRAFTS* SOMETHING THAT PREVIOUSLY DID NOT EXIST IN THE WORLD. IT'S MORE THAN *PLANNING* OR *THINKING* ALONE; IT IS THE ACT OF *INSTANTIATION.*

DESIGN IS SUBTRACTION

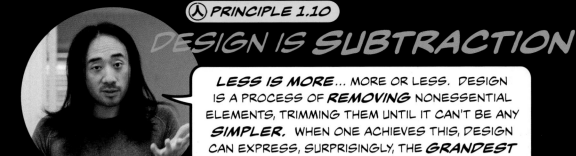

LESS IS MORE... MORE OR LESS. DESIGN IS A PROCESS OF **REMOVING** NONESSENTIAL ELEMENTS, TRIMMING THEM UNTIL IT CAN'T BE ANY **SIMPLER.** WHEN ONE ACHIEVES THIS, DESIGN CAN EXPRESS, SURPRISINGLY, THE **GRANDEST** IDEAS. AS LEONARDO DA VINCI SAID, "SIMPLICITY IS THE ULTIMATE SOPHISTICATION."

THE **HANGING MOBILES** OF ALEXANDER **CALDER** ARE **KINETIC SCULPTURES.** SIMPLE FORMS HANG FROM THE CEILING AND RESPOND TO THE ENVIRONMENT!

THE COMPONENTS ARE METICULOUSLY WEIGHTED SUCH THAT EACH LEVEL OF THE SCULPTURE IS **BALANCED** AND FREE TO ROTATE. (COMPUTER SCIENTISTS MIGHT RECOGNIZE THIS AS A **BALANCED TREE,** WHICH IS ALSO A HIERARCHICAL **DATA STRUCTURE.**)

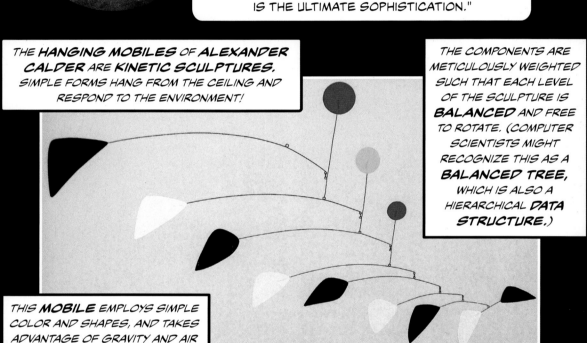

THIS **MOBILE** EMPLOYS SIMPLE COLOR AND SHAPES, AND TAKES ADVANTAGE OF GRAVITY AND AIR CURRENT, EFFORTLESSLY CAPTURING A SENSE OF **FLIGHT,** WEIGHTLESSNESS, MOTION, **RECURSION**... A SYSTEM IN DELICATE BALANCE.

"DESIGN IS SO **SIMPLE,** THAT'S **WHY** IT IS SO **COMPLICATED**..."
-- PAUL RAND

SIMPLICITY IS **NOT SIMPLE** TO ACHIEVE. IT TAKES TIME AND EFFORT TO MAKE SOMETHING **SIMPLE** BUT **NOT SIMPLISTIC.** DESIGN AIMS TO REDUCE THE PARALYZINGLY LIMITLESS STARTING SPACE TO A CONCISE, COGENT FOCAL POINT FOR BOTH DESIGNER AND USER. THIS IS SOMETIMES DONE THROUGH THE DEFINITION OF **CREATIVE CONSTRAINTS.** (FOR EXAMPLE: MAKE A HANGING SCULPTURE THAT IS **BALANCED** AND **FREE TO MOVE** AT EACH LEVEL.)

SEAGRAM BUILDING
NEW YORK CITY
COMPLETED IN 1958

ARCHITECT *LUDWIG MIES VAN DER ROHE* EMBRACED MINIMALIST GEOMETRIES AND UNDERSTATED FACADES...

...AN EXAMPLE OF *"LESS IS MORE"* IN ACTION, IT USHERED IN A NEW ERA OF SIMPLE, ELEGANT SKYSCRAPERS AND AN *EVOLVED* THINKING ABOUT DESIGN AND SIMPLICITY!

MIES IS REGARDED AS ONE OF THE PIONEERING MASTERS OF *MODERN ARCHITECTURAL DESIGN.* HE SMOKED *CIGARS* AND SAID *EPIC* THINGS...

GOD IS IN THE DETAILS.

...AND...

I DON'T WANT TO BE *INTERESTING*, I WANT TO BE *GOOD.*

41

SIMPLICITY IN DESIGN IS ROOTED IN TWO AIMS: **EFFECTIVENESS** AND **ELEGANCE.** EFFECTIVENESS IS THE PERCEIVED FUNCTIONAL **EASE-OF-USE,** WHEREAS ELEGANCE PERTAINS TO THE **GRACE** WITH WHICH THE DESIGN FULFILLS ITS INTENT.

IT'S SIMILAR TO MATHEMATICS, WHERE THE **ELEGANT** SOLUTION TENDS TO BE THE **MOST DIRECT,** THE **CLEANEST,** THE ONE WITH THE **FEWEST STEPS** IN ILLUMINATING A GNARLY PROBLEM. SUCH SOLUTIONS ARE SEEN AS **BEAUTIFUL** AND EVEN **SUBLIME.** AESTHETIC BEAUTY, TO MANY, IS THE VERY QUEST OF MATHEMATICS.

THE GREAT MATHEMATICIAN **LEONHARD EULER** DERIVED WHAT WE NOW CALL **EULER'S FORMULA,** WHICH DESCRIBES TAKING A NUMBER TO AN **IMAGINARY EXPONENT** (E.G., i IS AN IMAGINARY NUMBER EQUAL TO THE SQUARE ROOT OF -1) AS THE SUM OF TWO ORTHOGONAL **SINUSOIDS** -- ONE REAL AND ONE IMAGINARY. (THIS, BY THE WAY, IS SUPREMELY USEFUL IN COMPUTER MUSIC.)

EULER'S FORMULA:
$$e^{i\theta} = \cos(\theta) + i*\sin(\theta)$$

SETTING θ TO π YIELDS
$$e^{i\pi} = \cos(\pi) + i*\sin(\pi)$$

WHICH REDUCES TO
$$e^{i\pi} = -1 + 0$$

NOW ADD 1 TO BOTH SIDES
$$e^{i\pi} + 1 = 0$$

SUPREMELY **BEAUTIFUL** AND **TERRIFYINGLY PURE...**

AS A SPECIAL CASE OF EULER'S FORMULA, **EULER'S IDENTITY** RELATES **FIVE** OF THE MOST **FUNDAMENTAL NUMBERS** IN THE **UNIVERSE,** IN A **SUBLIMELY SIMPLE** RELATIONSHIP. PONDERING THIS FORMULA, ONE BEARS WITNESS TO A FUNDAMENTAL RELATIONAL **TRUTH** OR -- IF YOU WILL -- **DESIGN** OF THE UNIVERSE WE INHABIT. IT'S BEAUTIFUL IN ITS **UTTER SIMPLICITY** AND MIND-BLOWING TO REALIZE THAT THESE NUMBERS **FIT TOGETHER** SO **ELEGANTLY.**

DESIGN IS **CONSTRAINTS,**
WHICH GIVE RISE TO INTERACTIONS AND, IN TURN, *AESTHETICS*

AESTHETICS

AN *ARTFUL DESIGN ICEBERG!*

WHAT IS ULTIMATELY **EXPERIENCED**

COGNITION

I'M A USER!

SENSE

USERS

⸮SPLASH⸮ ⸮SPLASH⸮

INTERACTIONS

WHERE THE USER ENGAGES THE DESIGN

FORM

DESIGNERS

FUNCTION

I'M NOT ABOVE USING METAPHORICAL *ICEBERGS* TO ILLUSTRATE A POINT!

CONSTRAINTS

UNDERLYING *RULES* THAT GOVERN THE *USE* OF THE DESIGN (AND THE ONLY PART THAT DESIGNERS HAVE *DIRECT CONTROL* OVER)

WE ARE DESIGNERS!

TECHNOLOGY

PURPOSE / IDEA

BATTERED HULLS OF FAILED DESIGNS

MURKY DEPTHS OF UNTENABILITY

ABYSS OF ILL-DEFINITION

DIFFERENT FROM **ENGINEERING** AND **PURE ART**, **DESIGN** INCORPORATES **BOTH** IN BALANCE, CAPABLE OF USEFULNESS AND OF BEING **MORE THAN ITSELF**. A SYNTHESIS OF ART AND ENGINEERING, DESIGN IS ITS OWN CRAFT.

DESIGN IS ARTFUL ENGINEERING

PRINCIPLE 1.12

ENGINEERING

"THIS IS THE WAY IT WORKS"

EVALUATION: **OBJECTIVE**

ABOUT: OUR **PHYSICAL WORLD**

CORE VALUES: **KNOWLEDGE** AND **USEFULNESS**

PURE ART

"THERE'S SOMETHING BEAUTIFUL ABOUT IT"

EVALUATION: **SUBJECTIVE**

ABOUT: OUR **HUMANITY, TRANSCENDENCE**

CORE VALUES: **TRUTH, BEAUTY, JUSTICE**

THE ART OF DESIGN

"THERE'S SOMETHING BEAUTIFUL ABOUT THE WAY IT WORKS"

EVALUATION: **NOT** EXCLUSIVELY **OBJECTIVE** OR **SUBJECTIVE**

ABOUT: USEFUL SHAPING OF THE WORLD THAT INHERENTLY **REFLECTS** AESTHETIC AND HUMANISTIC VALUES

CORE VALUES: **TRUTH** IN **USEFULNESS**

WHEN ART, ENGINEERING, AND HUMANITIES ARE UNDERSTOOD IN SYNTHESIS, THE NEED FOR DISTINCTION **DISAPPEARS**; THEY NATURALLY MELD INTO A SINGULAR PURSUIT, DRAWING AS NEEDED FROM ITS CONSTITUENT DISCIPLINES.

LEONARDO DA VINCI EMBRACED PAINTING, SCULPTING, ARCHITECTURE, SCIENCE, MUSIC, ENGINEERING, MATHEMATICS, INVENTION, WRITING, ASTRONOMY, CARTOGRAPHY.

HE ALSO DESIGNED **FLYING MACHINES**...

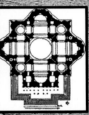

HIS **VITRUVIAN MAN** RELATES PROPORTIONS (FORM) AND ANATOMY (FUNCTION) OF THE HUMAN BODY. IT REMAINS A TESTAMENT TO THE BLEND OF ART AND SCIENCE DURING THE RENAISSANCE PERIOD.

...AND **PAINTED**.

MICHELANGELO'S **DAVID** STANDS OVER 4 METERS TALL; NOTE THE DISPROPORTIONATELY LARGE HEAD AND HANDS, PERHAPS TO ACCENTUATE THESE INTRICATE PARTS OF THE SCULPTURE (WHILE POINTING OUT THE TWO FACULTIES -- MIND AND HANDS -- THAT DEFEATED **GOLIATH**). MICHELANGELO ALSO PAINTED THE CEILING OF THE **SISTINE CHAPEL** AND CONTRIBUTED TO THE ARCHITECTURAL DESIGN FOR **ST. PETER'S BASILICA** IN VATICAN CITY. HE SAW **NO CLEAR DIVISION** BETWEEN SCULPTURE, PAINTING, ARCHITECTURE, AND ENGINEERING.

ROMANTIC ERA GERMAN COMPOSER, OPERA DIRECTOR, AND CONDUCTOR **RICHARD WAGNER** REVOLUTIONIZED MUSIC AND THEATER THROUGH HIS EMBRACING OF **GESAMTKUNSTWERK** (TOTAL-ART-WORK), THROUGH WHICH HE SOUGHT TO SYNTHESIZE POETIC, VISUAL, MUSICAL, AND DRAMATIC ART FORMS.

350 YEARS LATER...

WEIMAR, GERMANY, 1919

GESAMTKUNSTWERK
(TOTAL **WORK** OF ART)

ARTISTS AND ARCHITECTS MUST ALSO BE **CRAFTSPEOPLE!** WE SHOULD EXPERIENCE WORKING WITH MANY DIFFERENT ARTISTIC MEDIUMS, INCLUDING **INDUSTRIAL DESIGN, CLOTHING DESIGN, MUSIC,** AND **THEATER!**

MODERN ARCHITECTURE PIONEER **WALTER GROPIUS**, FOUNDER OF THE **BAUHAUS** SCHOOL

DESIGN SHOULD DO **MORE** THAN FULFILL A **SPECIFIED FUNCTION**; IT SHOULD **INSPIRE** US, **ALTER** SOMETHING IN US, MAKE OUR LIVES FEEL **RICH** AND **CAPACIOUS.** DESIGN PLAYS INTO WHAT PEOPLE **WANT**, BUT IT ALSO CAN SPEAK TO SOMETHING MORE **INVISIBLE**...

...OUR HOPES, OUR FEARS, AND OUR **CAPACITY** FOR PLAYFULNESS, COMPASSION, AND BEAUTY.

GOOD DESIGN **ENABLES** US; GREAT DESIGN **UNDERSTANDS** US.

(LIKE A **PENCIL BAG** THAT **UNDERSTANDS** WE ARE **PLAYFUL** CREATURES...)

DESIGN IS THE ART OF HUMANIZING TECHNOLOGY

🅧 PRINCIPLE 1.14

DESIGN IS NEVER COMPLETE UNTIL IT ACCOUNTS FOR THE **PERSON** EXPERIENCING IT. THE **CHOICES** WE MAKE IN DESIGN CAN COMPEL A USER TO TAKE **ACTION** OR INFLUENCE THE USER'S **THINKING.** IT IS THE ART OF MAKING **USEFUL** THINGS THAT ALSO MAKE US **FEEL**, AND **FEEL HUMAN.**

ANYTHING CAN BE DESIGNED: A BOOK, FILM, POEM, SPEECH, PIECE OF MUSIC -- THEY ARE DESIGNED TO **DO SOMETHING** TO THE AUDIENCE.

THIS HOLDS IF YOU ARE DESIGNING FOR **ONE** PERSON...

SPEAKING OF PEOPLE: THE ICON FOR THIS BOOK IS TAKEN FROM THE CHINESE PICTOGRAPH FOR "**HUMAN**," "**PERSON**," OR "**PEOPLE.**" IT IS PRONOUNCED "**REN.**"

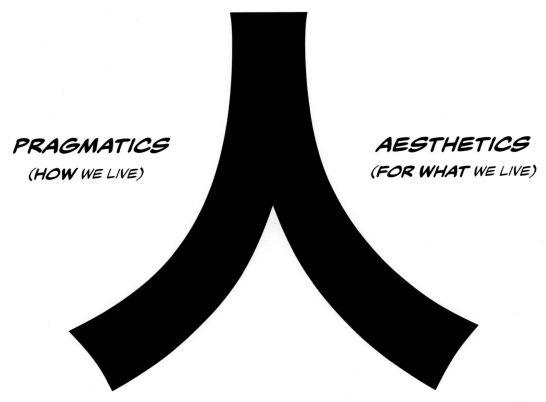

PRAGMATICS
(**HOW** WE LIVE)

AESTHETICS
(FOR **WHAT** WE LIVE)

ONE OF THE SIMPLEST PICTOGRAPHS IN CHINESE, IT ENCODES AN IMPLICIT **SOCIAL CONTRACT:** PEOPLE NEED TO **LEAN** ON ONE ANOTHER IN ORDER TO SURVIVE.

THE TWO LEGS MUTUALLY **SUPPORT EACH OTHER,** RESULTING IN A SINGLE UPRIGHT ENTITY. NEITHER CAN STAND ALONE...

...LIKE **DESIGN ITSELF.**

47

THIS BRINGS US TO A CORE **ETHOS**
IN THE MANIFESTO OF ARTFUL DESIGN...

 PRINCIPLE 1.15

DESIGN NOT ONLY FROM *NEEDS* -- BUT FROM THE *VALUES* BEHIND THEM

WE ARE OFTEN TAUGHT TO *DESIGN* FROM *PRACTICAL NEEDS.* WHILE NEEDS INDEED *GIVE RISE* TO DESIGN, THE MOST POWERFUL DESIGNS ARE ALSO *GROUNDED* IN *VALUES*, DEEPLY-HELD BELIEFS, AND AN *UNDERSTANDING* OF HUMAN BEINGS. DESIGN RESIDES IN A PLACE WHERE THE BEAUTIFUL, PLAYFUL, AND MEANINGFUL INTERACT WITH THE PRACTICAL AND USEFUL TO CONSTITUTE A *FULLNESS* OF *EXPERIENCE.*

WHAT DO YOU *MEAN* I WASN'T DESIGNED OUT OF A *NEED* FOR A PENCIL BAG THAT TRANSFORMS INTO ITS OWN ZIPPER?!

THE PENCIL BAG BEING ITS OWN ZIPPER ISN'T JUST A *LITTLE* SUPERFLUOUS -- IT IS *UTTERLY* SUPERFLUOUS (I.E., IT SERVES NO PRACTICAL PURPOSE)! YET THEREIN LIES ITS *CHARM* AND *ESSENCE.* IT MAKES THIS PENCIL BAG *INTERESTING*, BECAUSE IT SEEMS TO *UNDERSTAND* SOMETHING ABOUT US, THAT WE ARE PLAYFUL CREATURES. YOU MIGHT SAY IT WAS DESIGNED FROM THE *NEEDS* OF A PENCIL BAG, AND THE HUMAN *VALUES* OF *DELIGHT* AND *PLAYFULNESS* -- ACCOMPLISHED THROUGH A *FUNCTIONAL-AESTHETIC LEAP* THAT COMBINES NEEDS, VALUES, AND THE UNDERLYING MEDIUM (*"WHAT IF* THE ENTIRE BAG WERE MADE OF ITS OWN ZIPPER?!").

THE **ENDS-IN-THEMSELVES** CAN OFTEN SEEM **SUPERFLUOUS**, BECAUSE FUNCTIONALLY SPEAKING, THEY **ARE.** BUT THAT'S PRECISELY **WHY** WE ENGAGE IN THEM -- FOR THEIR **INTRINSIC VALUE.** IN AN INEXTRICABLY HUMAN WAY, ENDS-IN-THEMSELVES ARE **ESSENTIAL** -- TO THE EXTENT THAT SOMETIMES THE **EXPERIENCE** OF A DESIGN IS MORE **VALUABLE** AND **MEANINGFUL** THAN THE OSTENSIBLE **FUNCTION** IT SERVES, WHERE THE AESTHETICS **TRANSCENDS** THE PURPOSE.

THAT DOESN'T MEAN WE CAN **DISCARD** OR DEEMPHASIZE **NEEDS**, WHICH ARE, AFTER ALL, **NECESSARY** TO HUMAN LIFE. THIS SETS UP A **DIALECTIC** OF **OPPOSING FORCES** BETWEEN THE MEANS-TO-ENDS AND THE ENDS-IN-THEMSELVES, ONCE AGAIN WITH PRAGMATICS ON ONE SIDE AND AESTHETICS ON THE OTHER.

PRAGMATICS   **VS.**   AESTHETICS

FUNCTION   FORM

NEEDS   VALUES

WORK   PLAY

MEANS-TO-ENDS 人 **ENDS-IN-THEMSELVES**

ARTFUL DESIGN

*DESIGN IS THE **RADICAL SYNTHESIS** OF MEANS AND ENDS INTO A **THIRD** TYPE OF A THING -- **BOTH** USEFUL AND BEAUTIFUL*

人 **PRINCIPLE 1.16**

THE TWO SIDES ARE NOT FULLY COMPARTMENTALIZABLE, NOR ARE THEY MUTUALLY INSOLUBLE; THERE **CAN** BE A **UNITY**, A **SYNTHESIS**, AND THAT SYNTHESIS **IS** DESIGN.

DESIGN SHOULD DO **MORE** THAN FULFILL PRACTICAL **NEEDS**; IT IS ITSELF AN **EXPERIENCE** WORTH CONSIDERING.

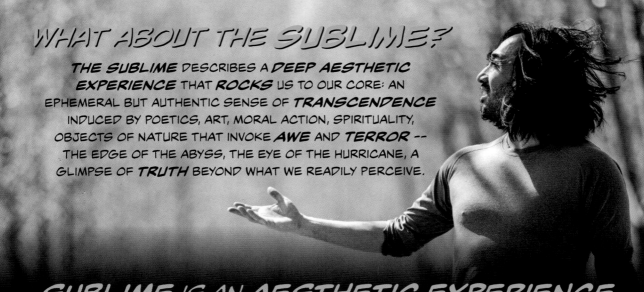

WHAT ABOUT THE SUBLIME?

THE SUBLIME DESCRIBES A **DEEP AESTHETIC EXPERIENCE** THAT **ROCKS** US TO OUR CORE: AN EPHEMERAL BUT AUTHENTIC SENSE OF **TRANSCENDENCE** INDUCED BY POETICS, ART, MORAL ACTION, SPIRITUALITY, OBJECTS OF NATURE THAT INVOKE **AWE** AND **TERROR** -- THE EDGE OF THE ABYSS, THE EYE OF THE HURRICANE, A GLIMPSE OF **TRUTH** BEYOND WHAT WE READILY PERCEIVE.

SUBLIME IS AN AESTHETIC EXPERIENCE

IT IS OFTEN ACCOMPANIED BY **INTENSE**, **CONTRASTING** EMOTIONS, LIKE AWE AND TERROR -- NOT THE KIND THAT MAKE YOU FLEE, BUT THOSE THAT MAKE YOU **STOP IN WONDER.** A CONSEQUENCE OF BOTH **SENSE** AND **REASON**, THE SUBLIME IS A MOMENT OF **CLARITY**, A FLEETING GLIMPSE OF THINGS AS THEY REALLY ARE, AN INTERNAL REALIZATION OF THE IMMENSITY OF EVERYTHING AND OUR PART IN IT, HOWEVER MINUSCULE. THE SUBLIME IS AN **INDIVIDUAL** REACTION, BUT ALSO REFLECTS SOMETHING **UNIVERSAL** IN ALL OF US -- A DESIRE TO TOUCH THE **TRANSCENDENT**, TO SEE AND KNOW THINGS **AS THEY REALLY ARE.** WHAT DOES IT MEAN TO **DESIGN** SUBLIMELY?

THE SUBLIME IS **NOT A FEATURE** TO PUT INTO A PRODUCT, BUT A **CONSEQUENCE** OF **EXPERIENCE.** TO DESIGN **ARTFULLY** IS TO DESIGN WITH **AUTHENTICITY**, TO SHAPE THINGS THAT STRIVE FOR **DEEP BEAUTY**: A HARMONY OF FORM AND FUNCTION IN SEARCH OF **TRUTH**, CLARITY, AN IDEAL, AND OUR **COMMON HUMANITY.** TRANSCENDING SHEER UTILITY, SUBLIME DESIGN STRIVES TO **UNDERSTAND** WHO WE **ARE** AND WHO WE **WANT TO BE.**

> DESIGN SHOULD SEEK TO **ELEVATE** US, NOT IN STATUS, BUT IN **AWARENESS**, MAKE US MORE THOUGHTFUL, WITTY, EMPATHETIC, VULNERABLE... MORE **HUMAN.**

⚖ PRINCIPLE 1.17

DESIGN SHOULD UNDERSTAND US

DESIGN IS A **SEARCH** FOR **WHO WE ARE** -- A **MIRROR** TO LIFE ITSELF; NOT UNLIKE **ART**, IT SHOULD MAKE US **FEEL**, AND FEEL **HUMAN.**

DESIGN IS **CREATED**

DESIGN IS A REALIZATION OF **INTENTION.** WHETHER IT'S WITH WORDS, CODE, PAINT, NUMBERS, FABRIC, CONCRETE AND STEEL, IMAGES, SOUNDS, MOVING PARTS, EXPERIENCES, OR THE RULES OF A SOCIAL SYSTEM, THE **STYLE** SHAPES THE **CONTENT** (**HOW** WE SAY SOMETHING IS AS IMPORTANT AS **WHAT** WE SAY), THE **FORM** MODULATES THE **FUNCTION** AND GIVES IT SIGNIFICANCE BEYOND PURE UTILITY. IN SHORT, DESIGN IS A PROCESS BY WHICH THE **MEDIUM BECOMES THE MESSAGE.**

DESIGN IS **EXPERIENCED**

TO THE **USER**, DESIGN IS MORE THAN THE **FACILITATION** OF A MEANS-TO-AN-END. IT **TRANSFORMS** AN EXPRESSION OF **PURPOSE** INTO A REALIZED **EXPERIENCE** THAT CAN **NO LONGER** BE BROKEN DOWN INTO ITS CONSTITUENT PARTS OR BE FACTORED CLEANLY INTO FORM OR FUNCTION. IT IS THROUGH **AESTHETICS** THAT WE **UNFOLD** THE INVISIBLE MEANING OF WHAT WE DO, WHAT WE MAKE; AND IT IS THROUGH THE AESTHETIC DIMENSION THAT WE HAVE AN AUTHENTIC HOPE OF BETTER **UNDERSTANDING OURSELVES.**

THIS PENCIL BAG **DESIGNS ME**, AND MY EXPERIENCE WITH IT!

"WE SHAPE OUR BUILDINGS, THEREAFTER THEY SHAPE US."
-- WINSTON CHURCHILL

PRINCIPLE 1.18

DESIGN IS A **THING IN MOTION**

DESIGN IS CHOICES, ACTIONS, AND CONSEQUENCES; IT'S SOMETHING WE **DO** TOWARD AN ESTABLISHED AIM, BUT ALSO TO ENGAGE US, MOVE US, AND HELP US UNDERSTAND OURSELVES. THE THINGS WE DESIGN, IN TURN, **DESIGN US.**

IT IS **REMARKABLE** HOW DESIGN MUST EMBRACE SO MANY SEEMINGLY DISPARATE DISCIPLINES AND SCHOOLS OF **IDEAS.** IN OUR EXPLORATION OF ARTFUL DESIGN, WE WILL NEED TO **MELD** COMPUTER SCIENCE, AESTHETICS, HUMAN-COMPUTER INTERACTION, TOY AND GAME DESIGN, THE SOCIAL, AND THE PHILOSOPHICAL WITH **EVERYDAY LIFE.**

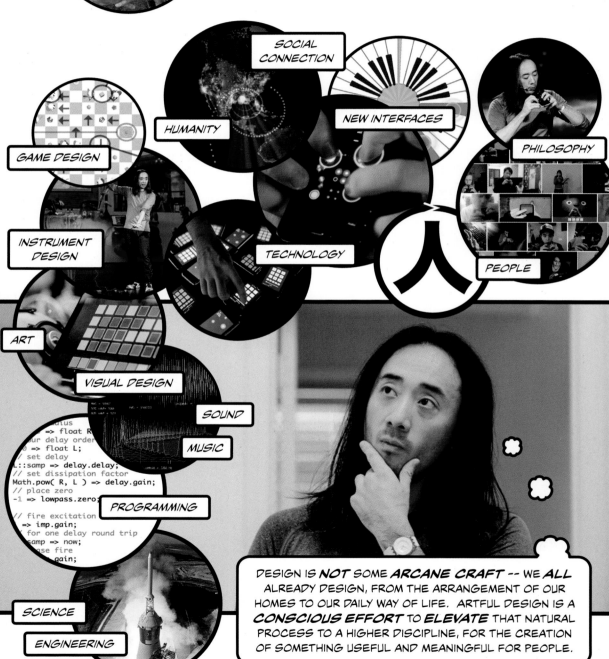

SOCIAL CONNECTION

HUMANITY

NEW INTERFACES

PHILOSOPHY

GAME DESIGN

INSTRUMENT DESIGN

TECHNOLOGY

PEOPLE

ART

VISUAL DESIGN

SOUND

MUSIC

PROGRAMMING

SCIENCE

ENGINEERING

DESIGN IS **NOT** SOME **ARCANE CRAFT** -- WE **ALL** ALREADY DESIGN, FROM THE ARRANGEMENT OF OUR HOMES TO OUR DAILY WAY OF LIFE. ARTFUL DESIGN IS A **CONSCIOUS EFFORT** TO **ELEVATE** THAT NATURAL PROCESS TO A HIGHER DISCIPLINE, FOR THE CREATION OF SOMETHING USEFUL AND MEANINGFUL FOR PEOPLE.

AND SO WE HAVE REACHED THE END OF **CHAPTER 1.**
ALONG THE WAY WE HAVE LAID OUT A NUMBER OF **FIRST
PRINCIPLES -- LENSES** THAT WE WILL LOOK
THROUGH IN OUR EXPLORATION OF **ARTFUL DESIGN.**

WHERE TO NEXT?

CHAPTER 2 IS A STORY OF DESIGNING
MUSICAL EXPRESSION WITH EVERYDAY
TECHNOLOGY, AND OUR JOURNEY FORWARD
EXPLORES THE DESIGN OF SIGHT, SOUND,
INTERFACES, TOYS, GAMES, AND SOCIAL
EXPERIENCES. WE CONTINUE TO DEVELOP
OUR **PRINCIPLES** -- ALL CONTRIBUTING
TO AN OVERALL **PHILOSOPHY** AND
MANIFESTO OF **ARTFUL DESIGN!**

AT THE END OF EACH CHAPTER, THERE ARE
WHAT I CALL **DESIGN ETUDES**, OR STUDIES
AND EXERCISES DESIGNED TO FURTHER
EXPLORE THE IDEAS IN THE CHAPTER.

THERE IS ALSO A LIST OF RELATED
MATERIALS -- READINGS, MOVIES, VIDEO
GAMES, RESOURCES -- IN THE **ANNOTATED
BIBLIOGRAPHY** AT THE END OF THE BOOK.

SUPPLEMENTAL MATERIALS,
COMMENTARIES, AND RESOURCES FOR
DESIGN ETUDES, CAN BE FOUND HERE:
HTTPS://ARTFUL.DESIGN/

CHAPTER 1
DESIGN ETUDE

> TRY THEM AND GET OTHERS TO DO THEM!

DESIGNERS SHOULD BE ABLE TO **ARTICULATE** THEIR DESIGNS AND SAY **HOW** AND **WHY** THE CHOICES WERE MADE.

• PART 1: TAKING **NOTICE**

TAKE NOTE OF **THREE THINGS** IN YOUR DAY YOU FIND **BEAUTIFUL** AND THAT YOU RECOGNIZE TO BE **DESIGN**.

EXAMPLES

- AN EVERYDAY OBJECT
- A TOY OR A GAME
- A KITCHEN APPLIANCE
- THE LAYOUT OF A ROOM
- THE PLOT (TWIST) OF A MOVIE
- A SOCIAL INTERACTION
- A TOOL
- A BUILDING

• PART 2: **MEANS** AND **ENDS**

FOR EACH THING YOU NOTED IN PART 1, PERFORM A **FUNCTIONAL-AESTHETIC** ANALYSIS: THINK ABOUT AND ARTICULATE WHY YOU FIND IT BEAUTIFUL. DOES IT HAVE TO DO WITH ITS FORM, OR FUNCTION, OR A SURPRISING INTERPLAY OF BOTH? IS IT SATISFYING? HOW SO? DO YOU FIND IT ELEGANT? WHAT EMOTIONAL RESPONSES DOES IT ELICIT? IN WHAT WAY IS IT MEANINGFUL TO YOU? CHARACTERIZE ITS FUNCTION/PURPOSE VS. ITS FORM/AESTHETICS. WHAT **MEANS-TO-AN-END** DOES IT SERVE, AND WHAT **END-IN-ITSELF** DOES IT SPEAK TO? WRITE A FEW SENTENCES OR MAKE A DIAGRAM OF YOUR ANALYSIS.

• PART 3: **GUERRILLA** DESIGN

INFILTRATE YOUR DAILY LIFE WITH DESIGN. ADD AESTHETICS TO SOMETHING THAT DOESN'T SEEM TO NEED IT (THIS IS OFTEN THE BEST PLACE TO EXPERIMENT WITH AESTHETICS). IF YOU DRAW SOMEONE A MAP TO YOUR HOME, DO IT WITH AESTHETICS, GIVE IT PERSONALITY. IF YOU COMPOSE A MESSAGE, EMPLOY A POETIC STRUCTURE THAT FITS THE MESSAGE! WHATEVER YOU DO, **DO IT WITH AESTHETICS** -- EXPERIMENT, AND HAVE FUN!

TECHNOLOGY AND ART

Our own human nature impels us to **lean forward**, to hurtle ourselves into ever more advanced technological futures. Driven by curiosity, desire, and aspiration, but unchecked by wisdom that we do not yet (and may never fully) possess, we find ourselves in a precarious unbalance—one that could lead to decay, destruction, and even extinction, just as it could lead to some unimagined human enlightenment. From where we stand today, the former seems more likely.

But, all is not lost, for by the same human nature, we are moved by beauty, justice, kindness. Perhaps that is why art exists, both as a conscious act of truth and as humanity's unconscious endeavor to balance our perpetual drive forward. Technology must be balanced by beauty, for it is in the sublime that we have hope of finding salvation from our ever-restless selves, to transcend mere existence into a life richly lived.

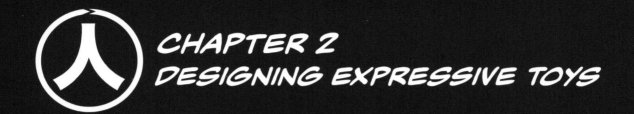

CHAPTER 2
DESIGNING EXPRESSIVE TOYS

THERE IS A MOMENT, IN THE **RISE** OF EVERY *UBIQUITOUS TECHNOLOGY*, WHEN IT MOVES FROM A *VISIBLE* ENTITY TO AN *INVISIBLE* REALM OF *POSSIBILITIES*...

...IT *EMBEDS* ITSELF INTO *EVERYDAY LIFE*, RECEDING FROM CONSCIOUS NOTICE AS A TECHNOLOGY AND EMERGING AS A *MEDIUM*, WHICH ASKS OF US TWO QUESTIONS:

WHAT CAN WE *DO* WITH IT?

HOW WILL IT *CHANGE* US?

HERE IS A STORY ABOUT USING SUCH A UBIQUITOUS TECHNOLOGY TO BUILD *EXPRESSIVE MUSICAL TOYS,* PLAYABLE BY *ANYONE,* AIMED TO *CONNECT* PEOPLE TO MUSICAL EXPRESSION...

Astounding

MUSIC SCIENCE

Reg. U. S. Pat. Off.

OCTOBER 1948
25 CENTS

IN THIS ISSUE

UNLEASHING THE MOBILE: THE MOST **UBIQUITOUS** COMPUTING TECHNOLOGY IN HISTORY, FOR MUSIC!

AFTER THE SECOND WORLD WAR, COMIC BOOKS AND SCIENCE FICTION THRIVED IN A **GOLDEN AGE** -- USHERING IN A NEW ERA OF **IMAGINATION** AND **STORYTELLING!**

WHILE COMIC **SUPERHEROES** RACED TO SAVE HUMANITY FROM ALL MANNER OF INTERGALACTIC DESTRUCTION, SCIENCE FICTION AUTHORS LIKE **ISAAC ASIMOV** AND **ARTHUR C. CLARKE** PONDERED HUMANITY'S **FUTURE**, LIKE THE GREAT PHILOSOPHERS BEFORE THEM, THEY EXPLORED OUR ROLE AND PLACE IN THE **UNIVERSE.**

ASIMOV'S **FOUNDATION** SERIES CHRONICLES A **FUTURE HISTORY** OF HUMANITY REACHING THE STARS, FORMING A **GALACTIC EMPIRE** AND A NEW BRANCH OF SCIENCE (CALLED **PSYCHOHISTORY**) THAT **PREDICTS** THE COLLECTIVE ACTIONS OF MASS POPULATIONS OF PEOPLE.

SCIENCE FICTION AND DESIGN HAVE SOMETHING IN **COMMON,** AT THEIR CORE, THEY AREN'T REALLY ABOUT SCIENCE OR TECHNOLOGY, BUT ARE ABOUT THE **QUESTIONS** TO WHICH HUMANITY DOESN'T YET HAVE THE ANSWERS OR **EVEN KNOW TO ASK,** OUR STORY BEGINS IN ANCIENT TIMES WITH **MUSIC** AND **TECHNOLOGY**...

MUSIC IS **ESSENTIAL** TO OUR **HUMAN EXPERIENCE.** IN ALL OF HISTORY, THERE IS **NO** CIVILIZATION OR CULTURE THAT DID NOT HAVE SOME KIND OF **MUSIC.** IT IS ALSO IMPORTANT TO NOTE THAT HUMANS **MAKE USE** OF **TECHNOLOGY** IN ORDER TO MAKE MUSIC, AND HAVE DONE SO **THROUGHOUT** HISTORY.

40,000 YEARS!

THIS **BONE FLUTE** -- AN EXAMPLE OF **ANCIENT MUSIC TECHNOLOGY** -- WAS MADE FROM A BEAR FEMUR. IT WAS FOUND IN NORTHWESTERN SLOVENIA AND IS MORE THAN 40,000 YEARS OLD. **TODAY,** MORE THAN EVER, WE STILL DESIGN NEW INSTRUMENTS, AS **NEW TECHNOLOGIES** GIVE RISE TO NEW MUSICAL **ARTIFACTS,** AND IN TURN, NEW MUSICAL **EXPERIENCES.** THROUGH ALL THIS, THE CONSTANT IS THAT **MUSIC** AND **TECHNOLOGY** HAVE ALWAYS **CO-EVOLVED.**

UNTIL THE **INDUSTRIAL REVOLUTION,** MUSICAL INSTRUMENTS SUCH AS THE **PIPE ORGAN** RANKED AMONG THE **MOST** TECHNOLOGICALLY COMPLEX MACHINES IN THE WORLD.

THIS ORGAN IS LOCATED IN ST. STEPHEN'S CATHEDRAL IN PASSAU, GERMANY. IT HAS **17,774 PIPES** AND **223 REGISTERS,** WHICH ARE OPERATED BY THE ORGANIST USING **STOPS** TO ACHIEVE DIFFERENT TIMBRES.

DUE TO MECHANICAL LINKAGE, PNEUMATICS, AND DISTANCE FROM THE PIPES, **DELAY** BETWEEN KEYPRESS AND CORRESPONDING SOUND CAN TAKE **HUNDREDS** OF **MILLISECONDS!** ORGANISTS **LEARN** TO PLAY WITH THIS LATENCY.

PIPE ORGAN CONTROLS ARE **COMPLEX**

VS.

PIPE ORGAN CONSOLE

SUBMARINE CONSOLE

THE **COMPUTER** IS **THE** TECHNOLOGY OF OUR TIME. **UNIQUE** DUE TO ITS **PROGRAMMABILITY**, THE COMPUTER IS A **META-TECHNOLOGY** THAT CAN BE FASHIONED INTO **OTHER** TECHNOLOGIES!

THE COMPUTER ENABLED NEARLY **EVERY** ARTIFACT EXAMINED IN THIS BOOK: AUDIOVISUAL SOFTWARE, PROGRAMMING LANGUAGES, TOYS, GAMES, SOCIAL SYSTEMS. THIS **BOOK** ITSELF WAS CREATED WITH THE AID OF COMPUTERS.

ALRIGHT! LET'S MAKE SOME **COMPUTER MUSIC!**

YAH!

MUSICALLY SPEAKING, THE COMPUTER HAS BEEN ATTRACTIVE TO RESEARCHERS AND COMPOSERS DUE TO ITS POTENTIAL FOR ENTIRELY **NEW SOUNDS** AND **FANTASTICAL AUTOMATIONS.**

IBM SYSTEM/360
(CIRCA 1965)

HOWEVER, THIS IS 1965, AND IT'S GONNA TAKE LIKE AN **HOUR** TO COMPUTE **EVERY SECOND** OF DIGITAL AUDIO... THEN WE GOTTA **CONVERT** IT INTO **ANALOG** TAPE SO WE CAN ACTUALLY LISTEN TO IT...

OKAY, I'LL ORDER **CHINESE FOOD!**

SPEAKING OF **FOOD**...

YUM. COOKING IS QUITE AN **ADVENTURE.** IT CAN BE DELICIOUS OR END IN GASTRONOMICAL DISASTER...

...OR **BOTH.**

A FITTING **ANALOGY** FOR MAKING COMPUTER MUSIC (AND PROGRAMMING IN GENERAL) IS **COOKING** FOR YOURSELF -- SURVEYING THE PANTRY, FIRING UP THE STOVE, AND EXPERIMENTING WITH **RAW INGREDIENTS** TO CONCOCT NEW DISHES OF YOUR INVENTION! THERE IS SOMETHING **AUTHENTIC** AND **ADVENTUROUS** TO THE ACTIVITY, A FEELING THAT THE WORLD IS FILLED WITH **POSSIBILITY.**

IN THIS SENSE, THE COMPUTER IS MORE THAN A TOOL -- IT'S A **KITCHEN!** BACK AT PRINCETON, WHERE I DID MY PH.D. IN COMPUTER SCIENCE, WE HAD THE **PRINCETON SOUND KITCHEN** -- A COLLECTIVE OF COMPUTER MUSICIANS **COOKING** WITH COMPUTERS!

BUT I DIGRESS (WHICH TENDS TO HAPPEN AROUND FOOD). WHAT I AM **TRYING** TO SAY IS...

...**MUSIC TECHNOLOGY** IS A DYNAMIC, COMPLEX, AND CULTURAL ENTITY. ITS PRESENT EXPLORATIONS (E.G., IN THE FORM OF COMPUTER MUSIC) CAN BE TRACED BACK TO **ANCIENT** ROOTS THROUGH AN ORGANIC **EVOLUTION** THAT CONTINUES TODAY. IT DRAWS UPON **MANY DISCIPLINES** -- COMPUTER SCIENCE, HUMAN-COMPUTER INTERACTION, MUSIC THEORY, PERFORMANCE, HUMAN PERCEPTION AND COGNITION, ART, PHILOSOPHY, AND **MORE**...

...IT'S SOMETHING OF A METAPHORICAL **KITCHEN SINK** (WHILE WE ARE ON THE SUBJECT OF KITCHENS): AN EVER-EXPANDING HODGEPODGE OF BOLDLY NEW, NOSTALGICALLY OLD, AND AT TIMES **CONFLICTING** SETS OF IDEAS, TECHNIQUES, AESTHETICS, AND CURIOSITIES.

THE TRUE **COMMONALITY** UNDERLYING ALL THIS IS FOUND IN THIS INEXTRICABLY **HUMAN** THING WE CALL **MUSIC** -- AND IN THE FACT THAT ANYTHING **NEW** WE INTEND TO DO MUST BE **DESIGNED.** AS ARISTOTLE ONCE SAID...

⸮AHEM⸮ IF THE ART OF SHIPBUILDING WERE **INHERENT** IN WOOD, WE HAVE ONLY BUT TO LEAVE WOOD LAYING ABOUT...

SO YEAH. YOU GOTTA ACTUALLY **DESIGN** STUFF.

LET'S GROUND THIS EXPLORATION IN A MORE **CONCRETE EXAMPLE.** WE WILL LOOK AT THE DESIGN OF MUSICAL TOYS FOR A **SPECIFIC** MEDIUM -- THE **MOBILE PHONE.** BEFORE IT BECAME A UBIQUITOUS TECHNOLOGY, IT WAS THE STUFF OF **SCIENCE FICTION.**

SPACE CADET TO **GROUND CONTROL**: ENGAGING PORTABLE PHONE... COMMENCING **SPACE SELFIES...**

1948

1962

"THE TIME WILL COME WHEN WE WILL BE ABLE TO CALL A PERSON ANYWHERE ON EARTH, MERELY BY DIALING A NUMBER. HE WILL BE LOCATED AUTOMATICALLY, WHETHER HE IS IN MID-OCEAN, IN THE HEART OF A GREAT CITY, OR CROSSING THE SAHARA. THIS DEVICE ALONE MAY CHANGE THE PATTERNS OF SOCIETY AND COMMERCE AS GREATLY AS THE TELEPHONE, ITS PRIMITIVE ANCESTOR, HAS ALREADY DONE. ITS PERILS AND DISADVANTAGES ARE OBVIOUS; THERE ARE NO WHOLLY BENEFICIAL INVENTIONS. YET THINK OF THE COUNTLESS LIVES IT WOULD SAVE, THE TRAGEDIES AND HEARTBREAKS IT WOULD AVERT. NO ONE NEED EVER AGAIN BE LOST, FOR A SIMPLE POSITION- AND DIRECTION-FINDING DEVICE COULD BE INCORPORATED IN THE RECEIVER ON THE PRINCIPLE OF TODAY'S RADAR NAVIGATION AIDS. AND IN CASE OF DANGER OR ACCIDENT, HELP COULD BE SUMMONED MERELY BY PRESSING AN 'EMERGENCY' BUTTON."

-- ARTHUR C. CLARKE

PROFILES OF THE FUTURE: AN INQUIRY INTO THE LIMITS OF THE POSSIBLE

THAT'S **REMARKABLE!**

CLARKE EFFECTIVELY **PREDICTED** NOT ONLY PERSONAL MOBILE PHONES (WITH GLOBAL POSITIONING) BUT THAT SUCH A SINGLE TECHNOLOGY MIGHT BECOME **PERVASIVE** IN OUR WORLD AND CHANGE THE **PATTERNS** OF LIFE AND SOCIETY!

NEW YORK CITY, 6TH AVENUE. ON THE SIDEWALKS OUTSIDE THE HILTON HOTEL, THE FIRST **PUBLIC MOBILE PHONE CALL** WAS PLACED BY **MARTIN COOPER** OF MOTOROLA. IT WAS MADE WITH A PROTOTYPE THE SIZE AND HEFT OF A **BRICK!**

THE UNIT WEIGHED 3 POUNDS, COMMUNICATED WITH A REMOTE COMPUTERIZED TELEPHONE EXCHANGE, HAD POWER ENOUGH FOR 36 MINUTES OF TALK TIME AND 12 HOURS OF STANDBY. IT WAS ENVISIONED THAT THE SERVICE WOULD COST SUBSCRIBERS $60 TO $100 PER MONTH.

HELLO! I AM CALLING FROM A REAL, HAND-HELD PORTABLE CELLULAR PHONE! AND I AM **WALKING** DOWN THE STREET IN MANHATTAN!

WHOA! AND I ALMOST GOT RUN OVER BY A CAR! THIS THING CAN BE DANGEROUS!

THIS IS THE **LIFE! FISHING**, A COLD **DRINK**, FLOATING ON THE WATER, BEING OUT IN **NATURE**, AND TALKING ON MY 3-POUND **TAKE-ALONG TELEPHONE!**

POPULAR SCIENCE JULY 1973: "NEW TAKE-ALONG TELEPHONES"

1983

THE UNITED STATES FEDERAL COMMUNICATIONS COMMISSION APPROVED THE **MOTOROLA DYNATAC 8000X**, THE FIRST COMMERCIALLY AVAILABLE CELLULAR PORTABLE PHONE. IT WEIGHED 28 OUNCES AND COST $3995!

OF WORLDWIDE MOBILE SUBSCRIBERS (1983): 0

THUS BEGAN THE **AGE** OF THE **MOBILE PHONE**, MADE POSSIBLE BY MANY ENGINEERS AND INVENTORS: MARTIN'S TEAM AT MOTOROLA; HIS BOSS JOHN MITCHELL WHO CHAMPIONED THE TECHNOLOGY; AMOS JOEL JR.'S TEAM AT AT&T/BELL LABS WHO CREATED THE SIGNAL SWITCHING SYSTEM THAT ENABLED A PHONE TO SEAMLESSLY MIGRATE FROM ONE CELL ZONE TO ANOTHER; ALONG WITH COUNTLESS OTHERS WORKING IN SIGNALS, SYSTEMS, AND TELECOMMUNICATIONS.

WHILE MOBILE SUBSCRIPTIONS INCREASED WORLDWIDE, RESEARCHERS AT **XEROX'S PALO ALTO RESEARCH CENTER** (PARC) EXPLORED A **VISION** OF COMPUTING THAT WOULD **PERVADE** EVERYDAY LIFE, WHERE INTERACTIONS WITH COMPUTERS CAN HAPPEN **ANYTIME**, **ANYWHERE**, AND, OPTIMALLY, WITH **MINIMAL AWARENESS** OF THE TECHNOLOGY.

1991 PARC DIRECTOR OF RESEARCH **MARK WEISER** PUBLISHED **"THE COMPUTER FOR THE 21ST CENTURY"** IN **SCIENTIFIC AMERICAN**, PROCLAIMING AN AGE OF **UBIQUITOUS COMPUTING** -- "SOON TO BE UPON US."

THE MOST PROFOUND TECHNOLOGIES ARE THOSE THAT **DISAPPEAR.** THEY WEAVE THEMSELVES INTO THE **FABRIC OF EVERYDAY LIFE** UNTIL THEY ARE INDISTINGUISHABLE FROM IT.

UBIQUITOUS COMPUTING NAMES THE THIRD WAVE IN COMPUTING, JUST NOW BEGINNING. FIRST WERE MAINFRAME COMPUTERS, EACH SHARED BY LOTS OF PEOPLE. THEN IT WAS THE PERSONAL COMPUTING ERA, PERSON AND MACHINE STARING UNEASILY AT EACH OTHER ACROSS THE DESKTOP. NEXT COMES UBIQUITOUS COMPUTING, OR THE AGE OF **CALM TECHNOLOGY**, WHEN TECHNOLOGY RECEDES INTO THE BACKGROUND OF OUR LIVES.

THE PRINCIPLES OF UBIQUITOUS COMPUTING CAN BE DISTILLED INTO THE FOLLOWING:

THE **PURPOSE** OF A COMPUTER IS TO HELP YOU DO **SOMETHING ELSE!**

THE BEST COMPUTER IS A QUIET INVISIBLE SERVANT; IT SHOULD EXTEND **YOUR UNCONSCIOUSNESS.**

TECHNOLOGY SHOULD **CREATE CALM.**

OF WORLDWIDE MOBILE SUBSCRIBERS (1991): 16 MILLION
WORLDWIDE PENETRATION: 3%

COMPUTERS IN WEISER'S VISION ARE **OMNIPRESENT** IN THE WORLD, BUT OPERATE IN THE **BACKGROUND** OF OUR CONSCIOUSNESS -- WITH AN AWARENESS OF OUR IDENTITY, LOCATION, AND EVEN INTENT.

THIS IS THE **ERICSSON R380**, THE FIRST DEVICE TO BE MARKETED AS A **SMARTPHONE.** ITS FRONT PANEL FLIPS OPEN, REVEALING A LARGER TOUCHSCREEN, TRANSFORMING THE DEVICE FROM PHONE TO **PERSONAL DIGITAL ASSISTANT** (PDA) WITH WHICH ONE CAN EMAIL, TAKE NOTES, AND PLAY GAMES.

WORLDWIDE MOBILE SUBSCRIBERS (2000): 738 MILLION
WORLDWIDE PENETRATION: 12%

2007 WAS AN INFLECTION POINT IN THE **EVOLUTION** OF COMPUTING. WORLDWIDE MOBILE PENETRATION HAD EXCEEDED HALF THE EARTH'S POPULATION. APPLE RELEASED THE **iPHONE** AND, A YEAR LATER, MADE IT AVAILABLE AS A PLATFORM FOR DEVELOPERS, USHERING IN A NEW MODEL FOR SOFTWARE AND A NEW ERA OF DO-IT-YOURSELF DESIGNERS AND BUILDERS.

2007 WORLDWIDE MOBILE SUBSCRIBERS (2007): 3.7 BILLION
WORLDWIDE PENETRATION: 51%

WHAT MADE THE **iPHONE** APPEALING FOR DESIGN?

CRITICAL MASS: THIS DEVICE WOULD SOON BE USED BY HUNDREDS OF MILLIONS OF PEOPLE.

A POWERFUL **PROCESSOR**, GREAT FOR SYNTHESIZING REAL-TIME AUDIO!

THE **GRAPHICS** PROCESSING UNIT (GPU) AS A SEPARATE PROCESSOR, OPTIMIZED FOR REAL-TIME GRAPHICS AND ANIMATION!

ONBOARD **SENSORS**: MULTITOUCH SCREEN, ACCELEROMETERS, MICROPHONE, CAMERA, AND (LATER) GYROSCOPE AND COMPASS.

DISTRIBUTION: APPLE'S INTRODUCTION OF AN APP STORE MADE IT EASY FOR USERS TO GET SOFTWARE. WITHOUT THE NEED FOR PHYSICAL SHELF SPACE, THE APP STORE ALSO BYPASSES THE MOBILE CARRIERS THAT PREVIOUSLY CONTROLLED THE SOFTWARE. IT'S SUPER EASY TO DOWNLOAD AN APP!

SOFTWARE DEVELOPMENT KIT (**SDK**) OFFERED AN **EASE** OF USE CLOSER TO DESKTOP DEVELOPMENT, SIMPLER AND MORE SOPHISTICATED THAN MOBILE DEVELOPMENT ENVIRONMENTS THAT PRECEDED IT.

LOCATION-AWARE (GPS), WITH **PERSISTENT NETWORK** CONNECTION.

MORE BROADLY, THE APP-BASED SMARTPHONE IS A **PHYSICAL PERSONAL COMPUTER!**

INSTEAD OF GOING TO A DESKTOP, WE CAN COMPUTE WHEREVER AND WHENEVER WE ARE IN THE MIDST OF **EVERYDAY LIFE.** IN SOME WAYS, IT IS AN EMBODIMENT OF MARK WEISER'S VISION FOR UBIQUITOUS COMPUTING (THOUGH PERHAPS NOT NEARLY AS **INVISIBLE** OR UNOBTRUSIVE AS ENVISIONED)!

THE **GROWTH** OF MOBILE HAS BEEN TRULY REMARKABLE! MOBILE SUBSCRIPTIONS WENT FROM ZERO IN 1980 TO MORE THAN 7 BILLION (NOT EVERYONE ON THE PLANET HAS A MOBILE PHONE; SOME PEOPLE HAVE MULTIPLE DEVICES). THIS GROWTH REPRESENTS NEARLY COMPLETE PENETRATION AMONG THE GLOBAL POPULATION, TOUCHING NEARLY ALL.

7.3 BILLION

7.3 BILLION

APPLE IPHONE 5

HUMAN POPULATION

VS.

MOBILE SUBSCRIPTIONS!

SAMSUNG GALAXY S II

6.6 BILLION

APPLE'S INTRODUCTION OF THE *iPAD* (AND SUBSEQUENT MASS ADOPTION OF TABLET COMPUTING) SHOWS THAT **FORM FACTOR** CAN MAKE ALL THE DIFFERENCE!

5.7 BILLION

smule

2008: MOBILE MUSIC STARTUP FOUNDED AS PART OF THE FIRST WAVE OF APP-BASED MOBILE COMPANIES

4.4 BILLION

人

FOR REFERENCE, IN 1000 A.D., **400 MILLION** PEOPLE LIVED ON EARTH

A TURNING POINT: THE *iPHONE* CHANGED THE IDEA OF THE PHONE FROM A COMMUNICATION-ORIENTED DEVICE TO AN **EVERYTHING-ORIENTED** DEVICE -- THROUGH APPS!

3.4 BILLION

THERE IS AN INTERESTING TREND OF **DECREASING** DEVICE SIZE UNTIL 2007 (AS TECHNOLOGY MATURED) AND A SUBSEQUENT **INCREASE** IN SCREEN SIZE SINCE 2007 (AS COMPUTING SHIFTED MOBILE).

MOTOROLA RAZR V3

ERICSSON R380

1984: FIRST CELLULAR PHONE RELEASED!

MOTOROLA DYNATAC 8900-X2

NOKIA 2146

NOKIA 3210

11 MILLION

738 MILLION

0

1980   1990   2000   2007   2015

ONE DOES NOT DO **RESEARCH**, WHETHER SCIENTIFIC OR ARTISTIC, **KNOWING** QUITE WHERE IT WILL LEAD; WE SIMPLY FOLLOW A PATH OF **INTEREST** AND **CURIOSITY.** CHANCE, TIMING, SENSIBILITIES ALL INFLUENCE THE DIRECTION AND RESULT. HERE IS MY UNPLANNED "SCENIC ROUTE" OF RESEARCH:

IT BEGAN WITH THE DESIGN OF **CHUCK**, A MUSIC PROGRAMMING LANGUAGE IN WHICH **TIME** IS, IN A WAY, **PROGRAMMABLE.** IT IS A TOOL FOR DESIGNING MUSICAL SOUNDS AND INTERACTIONS.

2003

CHUCK WAS USED IN THE **LAPTOP ORCHESTRA**, WHICH SERVED AS **ENSEMBLE** AND RESEARCH **LABORATORY** FOR THE DESIGN OF NOVEL INSTRUMENTS AND NEW FORMS OF COMPUTER-MEDIATED PERFORMANCE. SINCE THAT TIME, MORE THAN 75 LAPTOP ENSEMBLES HAVE EMERGED WORLDWIDE!

2005

2007

WITH THE RISE OF POWERFUL SMARTPHONE TECHNOLOGY, A NEW PERSONAL COMPUTING REVOLUTION WAS UPON US. OUR RESEARCH EXPANDED, **MUTATING** INTO THE **MOBILE PHONE ORCHESTRA** IN 2007. IT BROUGHT TWO THINGS INTO FOCUS: FIRST, THE **PERSONAL NATURE** OF MOBILE PHONES MADE FOR A DIFFERENT BREED OF COMPUTERS, ONE THAT LIVES ALONGSIDE US IN **EVERYDAY LIFE.** SECOND, THIS TECHNOLOGY PRESENTED NEW POSSIBILITIES FOR MUSICAL INTERACTIONS.

HOW MIGHT WE TAKE WHAT WE'VE LEARNED FROM DESIGNING COMPUTER-BASED INSTRUMENTS AND BRING THE EXPERIENCE TO **MORE PEOPLE?** **CURIOSITIES** SUCH AS THIS LED JEFF SMITH AND ME TO FOUND THE MOBILE MUSIC STARTUP **SMULE!**

2008

JEFF

TINA

CHRYSSIE

ROB

TURNER

smule

FOR BETTER OR WORSE, THE TASK OF **NAMING** THE STARTUP FELL TO ME. OUR THEN-FLEDGLING COMPANY AMBITIOUSLY AIMED TO BRING MUSIC-MAKING TO **HUGE POPULATIONS** OF PEOPLE, AND IT REMINDED ME OF MY FAVORITE SCIENCE FICTION NOVELS, THE **FOUNDATION** SERIES, BY **ISAAC ASIMOV!**

MY **IMAGINARY CONVERSATION** WITH THE **GOOD DOCTOR** WENT SOMETHING LIKE THIS:

smule

THE **S** STANDS FOR "**SONIC**," AND THE **MULE** COMES FROM YOUR **FOUNDATION TRILOGY**, DR. ASIMOV!

HAH, BUT THE MULE IS A COMPLEX, MALEVOLENT CHARACTER!

INDEED. WHILE I DIDN'T INTEND FOR SMULE TO BE **DIABOLICAL** IN ANY WAY, I DID ENVISION **SMULE** TO REACH AND INFLUENCE **MASS POPULATIONS** OF PEOPLE (KINDA LIKE THE MULE), IN OUR CASE, TO **MAKE MUSIC!**

OH I SEE, I THINK. BUT ⋛**SHHHH**⋚, SAY NO MORE, OR YOU MIGHT GIVE AWAY THE STORY!

MOST DEF. I WOULDN'T **DREAM** OF IT! **EVERYONE** SHOULD DISCOVER THE **FOUNDATION** SERIES FOR THEMSELVES! BACK IN THE DAY, WE GAVE A COPY TO EACH NEW EMPLOYEE!

SWEET! WELL, THANKS FOR THE **AIRTIME!** SO LONG FOR NOW AND **GOOD LUCK,**

THANK **YOU!**

(AND ALSO, UMM... YOU ARE MY HERO.)

A **TOY** IS DESIGNED TO BE LARGELY DEVOID OF **PRACTICAL** USE; ITS PRIMARY "FUNCTION" IS ONE OF **PLAY.** SUCH A **COMMITMENT** TO **LACK OF UTILITY** IS PERHAPS NONE MORE APPARENT THAN IN THE **FIRST** APP I DESIGNED AND BUILT WITH **SMULE. SONIC LIGHTER** IS A SOUND-BASED **ARTIFACT** FOR THE *iPHONE*, AN EXPERIMENT IN **PHYSICAL** INTERACTION DESIGN, USING AVAILABLE SENSORS ON THE DEVICE: MULTITOUCH, ACCELEROMETERS, MICROPHONE. THE **IRONY** IS, OF COURSE, THAT AS A LIGHTER...

...IT IS UTTERLY **USELESS!**

UH... THIS COULD BE... **NONE** MORE USELESS.

FOR ALL THE THINGS IT **CAN** DO, THE **ONE THING** IT **CANNOT** IS TO PHYSICALLY LIGHT SOMETHING AFLAME. IT WAS ONE OF THE **FIRST** LIGHTERS ON THE APP STORE (ACTIVE FROM 2008-2010), AND ITS DESIGN REMAINS UNIQUE FOR SEVERAL REASONS:

THE **AESTHETICS** EMBRACE THE CORE ELEMENTS OF A LIGHTER AND **REJECT** ORNAMENTATION (E.G., NO LIGHTER CASE WAS RENDERED -- THE DEVICE ITSELF IS THE CASE!)

THE FLAME ANIMATES SMOOTHLY, ORGANICALLY, AND MOVES IN ACCORDANCE WITH **TILT** (USING ACCELEROMETERS) AND **TOUCH** (THE FIRE DANCES UNDER ONE'S FINGERS). IT'S **PHYSICAL.**

HE AESTHETIC OF **PHYSICALITY** GOES FURTHER: THE LAME VIRTUALLY **SINGES** THE SIDE OF THE DEVICE WHEN OTATED ON ITS SIDE, ACCOMPANIED BY VIRTUAL SMOKE AND SUBTLE CRACKLING SOUND. THIS IS A STATEMENT THAT HE DEVICE ISN'T **SIMULATING** A LIGHTER -- IT **IS** THE GHTER... AND ALL THE WHILE **DOUBLING DOWN** ON THE BSURDITY OF A LIGHTER THAT **IS NOT** A LIGHTER.

SIZZLE

THE MAIN **AESTHETIC LEAP** OF SONIC LIGHTER, HOWEVER, IS THIS:

YOU PHYSICALLY **BLOW** INTO THE LIGHTER TO MAKE THE FLAME DANCE...

...AND IF YOU BLOW HARD ENOUGH, YOU **BLOW OUT** THE FLAME!

HIS DESIGN GESTURE STRETCHES HE **LIMITS** OF A LIGHTER THAT ISN'T, Y CONNECTING THE **VIRTUAL** TO THE HYSICAL, AND TAKING A **GAG TOY** O ITS **ILLOGICAL EXTREME!**

BLOW

EMBEDDED WITHIN THE APP, A SMALL **CHUCK PROGRAM** TRACKS THE STRENGTH OF THE SOUND RESULTING FROM BLOWING INTO THE **MICROPHONE.**

TRACKING **BREATH** ISN'T THE ONLY USE OF **SOUND INPUT** IN SONIC LIGHTER...

SONIC LIGHTER CAN INITIATE A SIMPLE **NEAR FIELD COMMUNICATION** BY EMITTING A SPECIAL **SONIC SIGNAL.**

THE SIGNAL IS **BARELY AUDIBLE,** BUT IF YOU LISTEN CAREFULLY, YOU CAN HEAR A **PAIR** OF **SINE WAVES** AT SPECIFIC FREQUENCIES.

ALL DORMANT SONIC LIGHTERS "LISTEN" FOR SOUND AT THESE FREQUENCIES. BY TRACKING THE OUTPUT OF TWO **RESONANCE FILTERS** TUNED TO THESE FREQUENCIES, A SONIC LIGHTER CAN RELIABLY KNOW ANOTHER IS TRYING TO IGNITE IT.

RIIIIIING

WHEN EMITTING THE SIGNAL, THE LIGHTER GOES INTO **BLOW TORCH** MODE!

WITH SOUND, ONE CAN **IGNITE** OTHER SONIC LIGHTERS SIMPLY BY PLACING DEVICES IN CLOSE PROXIMITY...

...IN FACT, ONE CAN IGNITE **ALL** PHONES IN THE VICINITY THAT CAN "HEAR" THE SIGNAL!

SO YES, IT **CAN** TECHNICALLY LIGHT SOMETHING... ANOTHER SONIC LIGHTER!

(AND NO, STILL NOT USEFUL -- BUT HEY, IT'S KINDA COOL)

IGNITION

IGNITION

IGNITION

IGNITION

RIIIIIING

IGNITION

IGNITION

IGNITION

IF YOU PLUG THE PHONE INTO SPEAKERS (OR EMBED THE SIGNAL IN OTHER SOUNDS), YOU CAN, IN THEORY, SIMULTANEOUSLY IGNITE **THOUSANDS** OF SONIC LIGHTERS! (AN EARLY DESIGN MOTIVATION WAS TO USE IT AT **ROCK CONCERTS.**)

71

TO OUR KNOWLEDGE, **SONIC LIGHTER** WAS THE **FIRST APP** TO INCLUDE A **GLOBE VISUALIZATION** OF ITS USERS. IN AN ARTIFACT THAT IS ALREADY **LOW** ON ACTUAL UTILITY, THIS FEATURE SEEMED LIKE THE **PERFECT THING** TO ADD. BECAUSE, YOU KNOW, IT'S **KINDA COOL** TO SEE **OTHER PEOPLE** AROUND THE WORLD IN STRANGE **SOLIDARITY** IGNITING THEIR FAKE LIGHTERS.

INTERNALLY, THIS USES THE DEVICE'S **GLOBAL POSITIONING SYSTEM** (GPS), AS WELL AS ITS PERSISTENT CONNECTION TO THE INTERNET, UPLOADING GEOGRAPHICAL INFORMATION TO A CENTRAL **SERVER** AND DATABASE, WHICH CAN BE QUERIED FOR RECENT **GLOBAL IGNITIONS.**

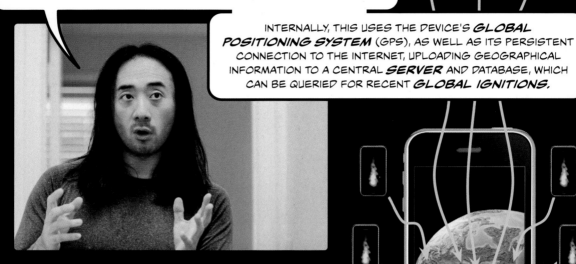

PERHAPS AS A DESPERATE, SUBCONSCIOUS ATTEMPT TO **COMPENSATE** FOR THE APP'S INHERENT USELESSNESS, I WANTED IT TO DO **SOMETHING** "REAL" LIGHTERS **CANNOT.** SO **OKAY**, YOU CAN'T ACTUALLY LIGHT ANYTHING WITH SONIC LIGHTER, **BUT** AT LEAST **NOW** YOU CAN NOT LIGHT ANYTHING **TOGETHER**, WITH STRANGERS AROUND THE WORLD! IT MAY NOT BE MUCH, BUT IT AIN'T **NOTHIN'!**

BEFORE

DAY 1

DAY 2

DAY 3

THE FIRST **SEVEN DAYS** OF THE WORLD LIGHTING UP ON SONIC LIGHTER!

DAY 4

DAY 5

DAY 6

DAY 7

OF IGNITIONS: *250,000*   # OF THINGS ACTUALLY SET ON FIRE: *0*

WHAT IS THE **VALUE** OF SOMETHING LIKE SONIC LIGHTER? DEFINITELY NOT ITS **PRACTICALITY** FUNCTION. BUT THEN AGAIN, THAT WAS NEVER ITS TRUE PURPOSE, WHICH WAS TO BE A **TOY**...

...AND AS A TOY, SONIC LIGHTER MAKES MORE SENSE: IT SPEAKS TO OUR **PLAYFUL** NATURE, **CO-OPTING** TECHNOLOGY TO RECAST SOMETHING **FAMILIAR** INTO AN **UNFAMILIAR** CONTEXT (SUCH **DEFAMILIARIZATION** IN DESIGN IS LIKE TELLING A JOKE: ITS SUCCESS DEPENDS ON THE AUTHENTICITY OF ITS SURPRISE). THIS KIND OF EXPERIENCE CAN DELIGHT US AND FEED OUR IMAGINATION.

PRINCIPLE 2.1

DESIGN FOR PLAY AND DELIGHT

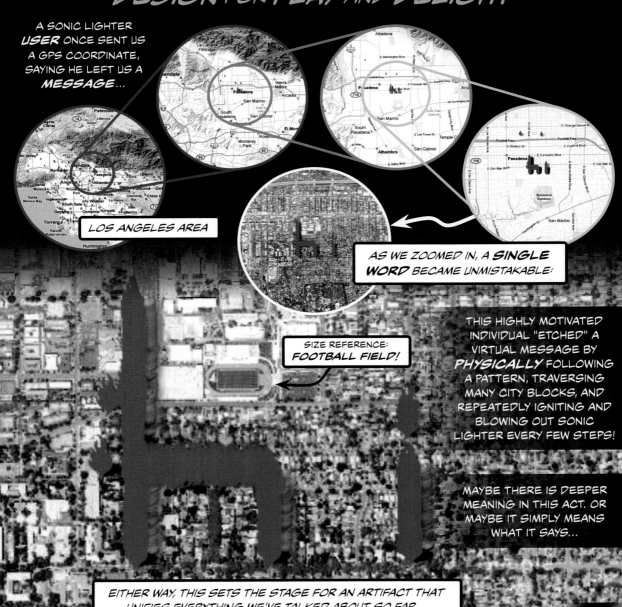

A SONIC LIGHTER **USER** ONCE SENT US A GPS COORDINATE, SAYING HE LEFT US A **MESSAGE**...

LOS ANGELES AREA

AS WE ZOOMED IN, A **SINGLE WORD** BECAME UNMISTAKABLE:

SIZE REFERENCE: **FOOTBALL FIELD!**

THIS HIGHLY MOTIVATED INDIVIDUAL "ETCHED" A VIRTUAL MESSAGE BY **PHYSICALLY** FOLLOWING A PATTERN, TRAVERSING MANY CITY BLOCKS, AND REPEATEDLY IGNITING AND BLOWING OUT SONIC LIGHTER EVERY FEW STEPS!

MAYBE THERE IS DEEPER MEANING IN THIS ACT. OR MAYBE IT SIMPLY MEANS WHAT IT SAYS...

EITHER WAY, THIS SETS THE STAGE FOR AN ARTIFACT THAT UNIFIES EVERYTHING WE'VE TALKED ABOUT SO FAR...

73

DESIGN AND **PHILOSOPHY**
OF AN EXPRESSIVE MUSIC TOY

OCARINA

A TOYFUL RE-ENVISIONING OF AN
ANCIENT MUSICAL INSTRUMENT,
TRANSFORMED IN THE **KILN** OF
MODERN TECHNOLOGY!

Toooooot

Toooooot

RELEASED IN 2008 AND DESIGNED FOR THE
iPHONE, OCARINA WAS ONE OF THE VERY
FIRST MUSICAL INSTRUMENTS IN THE EMERGING
LANDSCAPE OF APP-BASED COMPUTING.

OF ALL THE THINGS I HAVE DESIGNED, I THINK *OCARINA* MOST CONCISELY *EMBODIES* THE *PRINCIPLES* OF ARTFUL DESIGN.

IT TRANSFORMS THE PHONE INTO A *FLUTE-LIKE INSTRUMENT!*

IT'S *PHYSICAL* AND *VIRTUAL*, EXPLORING BOTH *MUSICAL* INTERACTION AND AN EXPRESSIVE *SOCIAL* DIMENSION.

FOR ME, IT BRINGS TOGETHER A NUMBER OF IDEAS IN DESIGN.

FORM AND FUNCTION

HUMAN-COMPUTER INTERACTION

AUDIO

VISUAL

ECONOMY OF DESIGN

MUSICAL EXPRESSION

PLAY

PHYSICALITY

TOY DESIGN

SOCIAL DESIGN

IF THE *GAG* WITH *SONIC LIGHTER* IS THE ABSURDITY OF A *PRACTICAL* THING THAT COULD *NEVER* FULFILL ITS INTENDED PURPOSE, THEN *OCARINA* IS THE *OPPOSITE*: A *WHIMSICAL* THING THAT *CAN* FUNCTION, AS AN ANCIENT FLUTE-LIKE INSTRUMENT, ON A PHONE! THE *JOKE* HERE, NO LESS ABSURD, IS THAT IT *ACTUALLY WORKS!*

75

BLOW INTO THE PHONE TO **ARTICULATE** THE SOUND. **BREATH** IS TRACKED, PROCESSED, AND MAPPED TO **LOUDNESS.**

HOLD THE PHONE AS YOU MIGHT A **SANDWICH**, RESTING THE PHONE ON THE **THUMBS** AND **RING** FINGERS, LEAVING THE **INDEX** AND **MIDDLE** FINGERS FREE TO MANIPULATE THE TOUCHSCREEN!

BLOW

IT'S BEST TO BLOW INTO THE MICROPHONE FROM A FEW INCHES AWAY (DON'T **EAT** THE PHONE!)

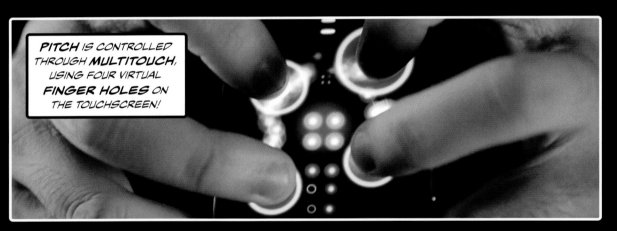

PITCH IS CONTROLLED THROUGH **MULTITOUCH**, USING FOUR VIRTUAL **FINGER HOLES** ON THE TOUCHSCREEN!

THE FINGER HOLES **GLOW** AND **EXPAND** ON TOUCH AND **RETRACT** UPON RELEASE, MAKING THE INTERFACE FEEL **ALIVE** AND **ORGANIC.** FUNCTIONALLY, THIS COMPENSATES FOR A **LACK** OF **TACTILE** FEEDBACK ON A TOUCHSCREEN AND MAKES IT EASIER TO SEE WHAT YOU ARE PRESSING.

TILTING THE PHONE CONTROLS **VIBRATO**, ADDING EXPRESSION TO NOTES (ESPECIALLY THE TAIL OF LONGER ONES) AND AN ADDITIONAL DIMENSION OF **PHYSICALITY** TO THE INTERACTION!

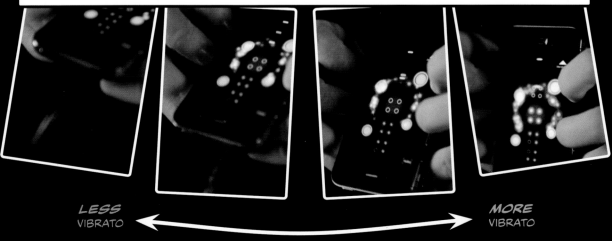

LESS VIBRATO ← → *MORE* VIBRATO

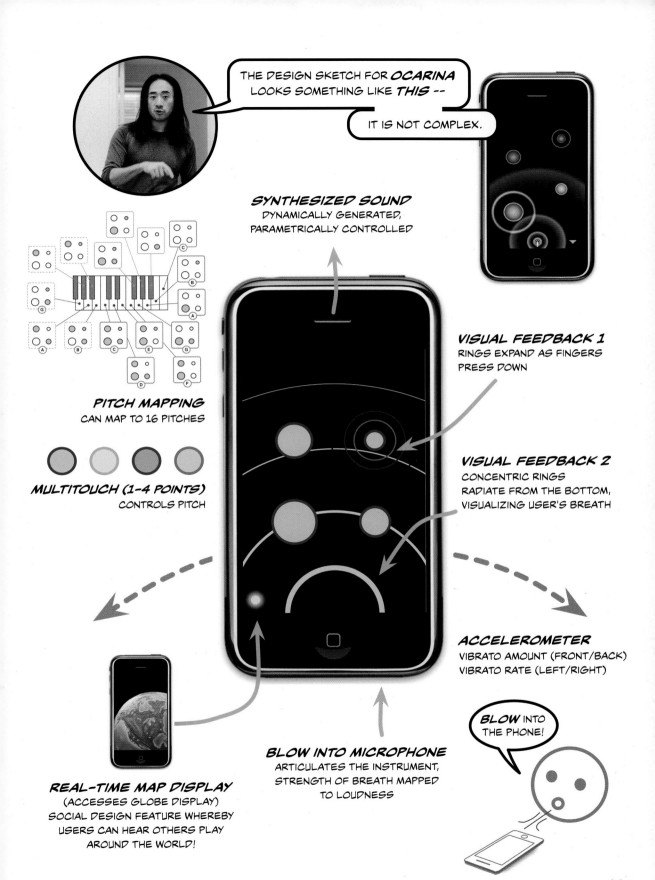

THE DESIGN SKETCH FOR *OCARINA* LOOKS SOMETHING LIKE *THIS* --

IT IS NOT COMPLEX.

SYNTHESIZED SOUND
DYNAMICALLY GENERATED, PARAMETRICALLY CONTROLLED

PITCH MAPPING
CAN MAP TO 16 PITCHES

MULTITOUCH (1-4 POINTS)
CONTROLS PITCH

VISUAL FEEDBACK 1
RINGS EXPAND AS FINGERS PRESS DOWN

VISUAL FEEDBACK 2
CONCENTRIC RINGS RADIATE FROM THE BOTTOM, VISUALIZING USER'S BREATH

ACCELEROMETER
VIBRATO AMOUNT (FRONT/BACK)
VIBRATO RATE (LEFT/RIGHT)

BLOW INTO MICROPHONE
ARTICULATES THE INSTRUMENT, STRENGTH OF BREATH MAPPED TO LOUDNESS

BLOW INTO THE PHONE!

REAL-TIME MAP DISPLAY
(ACCESSES GLOBE DISPLAY)
SOCIAL DESIGN FEATURE WHEREBY USERS CAN HEAR OTHERS PLAY AROUND THE WORLD!

OCARINA'S DESIGN ADHERES TO SOMETHING I CALL **INSIDE-OUT DESIGN,** WHICH WORKS **OUTWARD** FROM AVAILABLE TECHNOLOGICAL INGREDIENTS, TAKING INTO ACCOUNT THEIR **POSSIBILITIES** AND **CONSTRAINTS!**

IN THIS CASE, IT IS ABOUT USING EVERYTHING AVAILABLE ON THE iPHONE TO DESIGN A SINGULAR ARTIFACT.

WE CAN APPLY THIS STRATEGY TO **DISCOVER** AND DETERMINE **WHAT** TO **DESIGN** IN THE FIRST PLACE!

THE CHOICE TO DESIGN AN OCARINA STARTED WITH THE DEVICE ITSELF -- BY CONSIDERING ITS VERY **FORM** AND EMBRACING ITS INHERENT **CAPABILITIES**, "AS IS"!

CASE IN POINT: **WHY** AN **OCARINA?**

(I.E., **WHY NOT** A VIOLIN, GUITAR, PIANO, DRUM, OR SOMETHING ELSE?)

FOR STARTERS, THE **PHYSICAL FORM** AND SIZE OF AN iPHONE IS SIMILAR TO THAT OF A FOUR-HOLE "ENGLISH PENDANT" OCARINA. THE ONBOARD SENSORS (MULTITOUCH SCREEN, ACCELEROMETERS, MICROPHONE) SEEM FITTING FOR THE PHYSICAL INTERACTION OF **OCARINA.**

INSIDE-OUT DESIGN REJECTS **BLUNT TRANSFER** (OR "PORTING") FROM OTHER DOMAINS; INSTEAD IT CHAMPIONS AN **ETHOS** OF DESIGNING FROM THE GROUND UP, EMBRACING THE **MEDIUM** AND ITS CONSTRAINTS, AND THINKING AS BROADLY AS POSSIBLE ABOUT ITS **NEW** POTENTIALS!

PRINCIPLE 2.2 DESIGN INSIDE-OUT

OCARINA WAS **NOT** DESIGNED AS A "MOBILE VERSION" OF AN OCARINA, BUT AS SOMETHING THAT IS ITS OWN EXPERIENCE...

...THERE IS SOMETHING **DISARMING** ABOUT ENGAGING AN EVERYDAY DEVICE IN AN ENTIRELY **DIFFERENT** MANNER...

...AND BY **APPROPRIATING** TECHNOLOGY IN UNCONVENTIONAL WAYS, WE CAN IMBUE A SENSE OF **PLAY** AND **DELIGHT.**

INSIDE-OUT DESIGN POSES AN INTERESTING **TWIST** ON OUR PRINCIPLE OF **FORM** FOLLOWING **FUNCTION. WHEREAS** FORM IS OFTEN TO BE DERIVED NATURALLY FROM FUNCTION, HERE **FORM** (PHONE) INSPIRED THE **FUNCTION** (OCARINA)!

PRINCIPLE 2.3 *SOMETIMES, FUNCTION FOLLOWS FORM*

IN DESIGNING WITH NEW TECHNOLOGY, ESPECIALLY COMMODITY DEVICES WITH MASS ADOPTION, THIS APPROACH OFFERS A **USEFUL CONSTRAINT.** IT'S ABOUT USING PRECISELY THAT WHICH IS **ALREADY THERE -- NOTHING MORE!** IT DIVERTS THE MIND FROM WISHING **"IF ONLY** THERE WERE X..." TO ASKING **"WHAT** CAN WE DO WITH WHAT WE ALREADY **HAVE?"**

   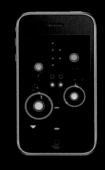

PHONE
2007

OCARINA FORM
"ENGLISH PENDANT"

OCARINA
2008

OCARINA 2
2012

THIS PARTICULAR OCARINA FORM CAN BE TRACED BACK TO **ANCIENT CULTURES,** WHERE OCARINA-LIKE INSTRUMENTS PLAYED AN IMPORTANT ROLE IN TRADITIONS OF SONG AND DANCE.

燻

XŪN

OCARINA IS SIMILAR IN FORM TO THE **XUN*,** ONE OF THE OLDEST CHINESE INSTRUMENTS, DATING BACK **7000 YEARS!** THE MAIN DIFFERENCE IS THAT THE PLAYER BLOWS **ACROSS** THE TOP OF THE XUN, WHEREAS A PLAYER BLOWS **INTO** THE OCARINA. AND THE SOUND OF A XUN IS TYPICALLY MORE **SOLEMN** AND **MOURNFUL** THAN THE OCARINA SOUND.

*PRONOUNCED: "SHEE-YU-EN" ROLLED INTO ONE SYLLABL

AN OPPORTUNITY TO CONSIDER AUDIO, VISUAL, AND PHYSICAL INTERACTION DESIGN IN CONJUNCTION.

[W]ANTED TO CRAFT **OCARINA** AS A WHIMSICAL, **MAGICAL** ARTIFACT -- REFLECTED IN THE [B]REATH MECHANISM, ITS **LOOK** AND **FEEL**, [A]ND IN THE AESTHETIC EQUIVALENCE BETWEEN COMMUNICATION DEVICE AND INSTRUMENT.

[O]NCE AGAIN, THERE IS A COMMITMENT [T]O **NOT ADORN** THE ARTIFACT. THE [A]ESTHETIC STATEMENT IS NOT "THIS [E]MULATES AN OCARINA" BUT RATHER...

...THIS **IS** AN OCARINA!

EVEN THE **NAME** "OCARINA" REFLECTS THIS ETHOS OF PHYSICAL DESIGN AND DELIBERATEL[Y] AVOIDS THE COMMON EARLY NAMING CONVENTIC[N] OF PREPENDING APP NAMES WITH "i" (E.G., iOCARINA). IT IS AN **ARTICULATION** OF THE DIFFERENCE BETWEEN "EMULATES" AND "IS"!

 PRINCIPLE 2.4

TAKE ADVANTAGE OF *PHYSICALITY*
ARTFULLY BLEND THE PHYSICAL WITH THE VIRTUAL

THE **BUTTONS** ARE DESIGNED TO FEEL **RESPONSIVE** AND AS **PHYSICAL** AS POSSIBLE [O]N A FLAT TOUCHSCREEN -- HENCE THEIR ANIMATED EXPANSION: AS IF PRESSING SOMEHOW **FLATTENS** THEM.

[T]RANSLUCENT **GREEN WAVES** SMOOTHLY WASH OVER THE SCREEN IN RESPONSE TO **BREATH** BLOWN TO PLAY THE INSTRUMENT. THEY LOOSELY REPRESENT SOUND WAVES AND [T]HE EXCITATION OF THE INSTRUMENT. THEY ALSO SIGNIFY A **TRANSFORMATION** FROM

THE SOUND OF **OCARINA** IS **GENERATED** IN REAL TIME, USING A SET OF AUDIO SIGNAL PROCESSING ELEMENTS, CONTROLLED FROM INPUT FROM THE MICROPHONE, ACCELEROMETERS, AND TOUCHSCREEN!

THIS IS A **BLUEPRINT PAGE.** YOU'LL FIND OTHERS LIKE IT SPRINKLED THROUGHOUT, CONTAINING CODE AND DOMAIN-SPECIFIC **TECHNICAL** INFORMATION.

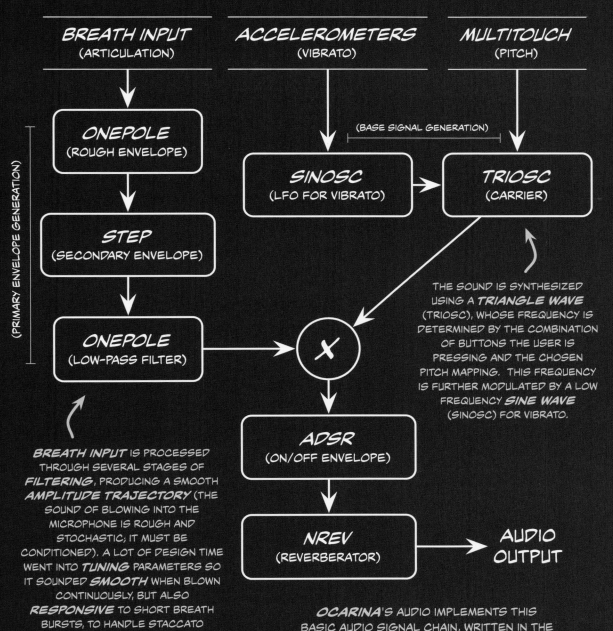

BREATH INPUT (ARTICULATION)

ACCELEROMETERS (VIBRATO)

MULTITOUCH (PITCH)

(PRIMARY ENVELOPE GENERATION)

ONEPOLE (ROUGH ENVELOPE)

STEP (SECONDARY ENVELOPE)

ONEPOLE (LOW-PASS FILTER)

(BASE SIGNAL GENERATION)

SINOSC (LFO FOR VIBRATO)

TRIOSC (CARRIER)

X

ADSR (ON/OFF ENVELOPE)

NREV (REVERBERATOR)

AUDIO OUTPUT

THE SOUND IS SYNTHESIZED USING A **TRIANGLE WAVE** (TRIOSC), WHOSE FREQUENCY IS DETERMINED BY THE COMBINATION OF BUTTONS THE USER IS PRESSING AND THE CHOSEN PITCH MAPPING. THIS FREQUENCY IS FURTHER MODULATED BY A LOW FREQUENCY **SINE WAVE** (SINOSC) FOR VIBRATO.

BREATH INPUT IS PROCESSED THROUGH SEVERAL STAGES OF **FILTERING**, PRODUCING A SMOOTH **AMPLITUDE TRAJECTORY** (THE SOUND OF BLOWING INTO THE MICROPHONE IS ROUGH AND STOCHASTIC; IT MUST BE CONDITIONED). A LOT OF DESIGN TIME WENT INTO **TUNING** PARAMETERS SO IT SOUNDED **SMOOTH** WHEN BLOWN CONTINUOUSLY, BUT ALSO **RESPONSIVE** TO SHORT BREATH BURSTS, TO HANDLE STACCATO ARTICULATIONS.

OCARINA'S AUDIO IMPLEMENTS THIS BASIC AUDIO SIGNAL CHAIN, WRITTEN IN THE **CHUCK** PROGRAMMING LANGUAGE.

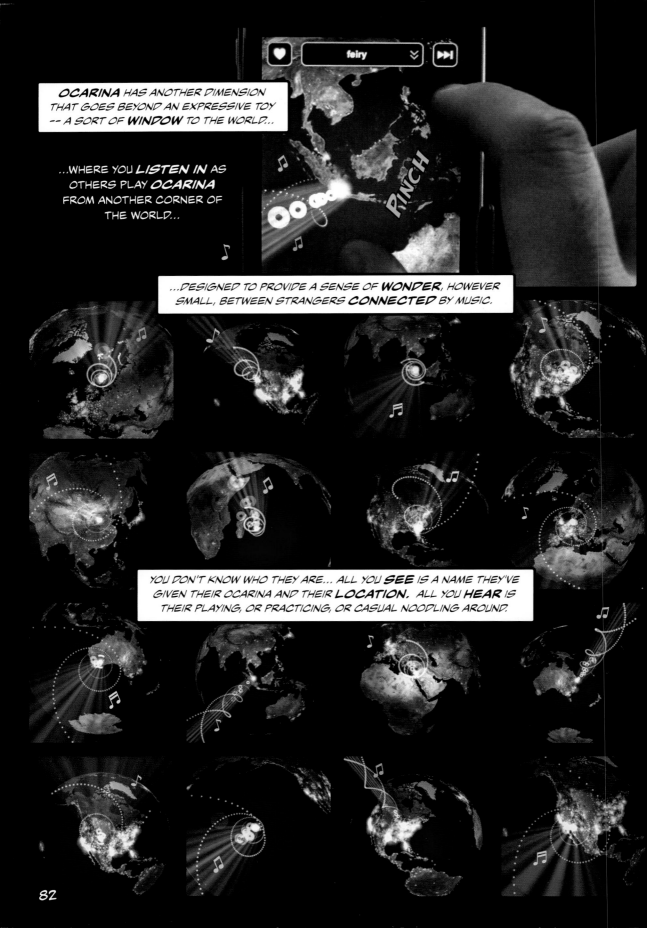

OCARINA HAS ANOTHER DIMENSION THAT GOES BEYOND AN EXPRESSIVE TOY -- A SORT OF **WINDOW** TO THE WORLD...

...WHERE YOU **LISTEN IN** AS OTHERS PLAY **OCARINA** FROM ANOTHER CORNER OF THE WORLD...

...DESIGNED TO PROVIDE A SENSE OF **WONDER**, HOWEVER SMALL, BETWEEN STRANGERS **CONNECTED** BY MUSIC.

YOU DON'T KNOW WHO THEY ARE... ALL YOU **SEE** IS A NAME THEY'VE GIVEN THEIR OCARINA AND THEIR **LOCATION**. ALL YOU **HEAR** IS THEIR PLAYING, OR PRACTICING, OR CASUAL NOODLING AROUND.

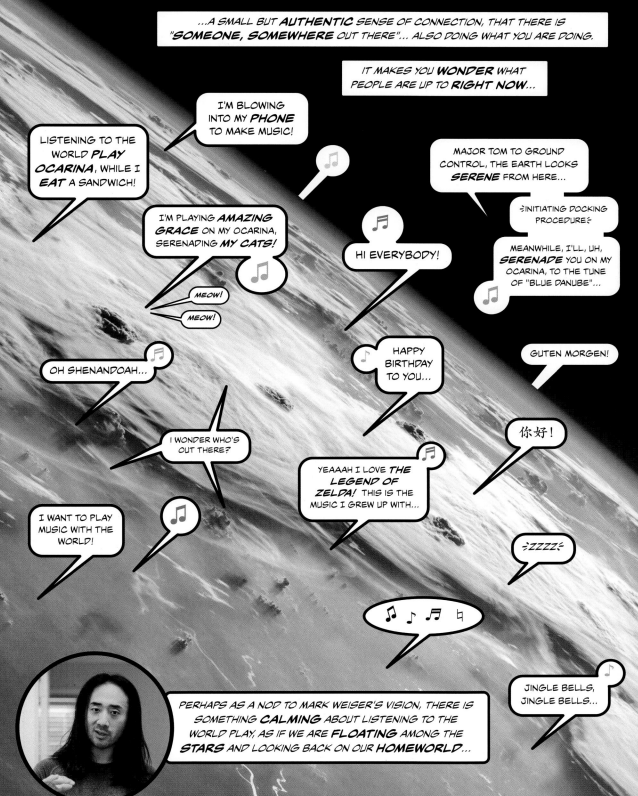

OCARINA MAY BE THE **FIRST** INSTRUMENT THAT LETS ITS PLAYERS LISTEN TO **ONE ANOTHER** PLAYING AROUND THE WORLD...

...A SMALL BUT **AUTHENTIC** SENSE OF CONNECTION, THAT THERE IS **"SOMEONE, SOMEWHERE** OUT THERE"... ALSO DOING WHAT YOU ARE DOING.

IT MAKES YOU **WONDER** WHAT PEOPLE ARE UP TO **RIGHT NOW**...

I'M BLOWING INTO MY **PHONE** TO MAKE MUSIC!

LISTENING TO THE WORLD **PLAY OCARINA**, WHILE I **EAT** A SANDWICH!

MAJOR TOM TO GROUND CONTROL, THE EARTH LOOKS **SERENE** FROM HERE...

⸗INITIATING DOCKING PROCEDURE⸗

I'M PLAYING **AMAZING GRACE** ON MY OCARINA, SERENADING **MY CATS!**

HI EVERYBODY!

MEANWHILE, I'LL, UH, **SERENADE** YOU ON MY OCARINA, TO THE TUNE OF "BLUE DANUBE"...

MEOW!

MEOW!

OH SHENANDOAH...

HAPPY BIRTHDAY TO YOU...

GUTEN MORGEN!

I WONDER WHO'S OUT THERE?

你好!

YEAAAH I LOVE **THE LEGEND OF ZELDA!** THIS IS THE MUSIC I GREW UP WITH...

I WANT TO PLAY MUSIC WITH THE WORLD!

⸗ZZZZ⸗

♫ ♪ ♫ ♮

PERHAPS AS A NOD TO MARK WEISER'S VISION, THERE IS SOMETHING **CALMING** ABOUT LISTENING TO THE WORLD PLAY, AS IF WE ARE **FLOATING** AMONG THE **STARS** AND LOOKING BACK ON OUR **HOMEWORLD**...

JINGLE BELLS, JINGLE BELLS...

83

OCARINA'S GLOBE AIMS FOR A KIND OF **TRANSPORTIVE** EXPERIENCE, SOMETHING BEYOND WHAT A TRADITIONAL OCARINA COULD DO...

THROUGH TECHNOLOGY, **OCARINA** ASPIRES TO SOMETHING **HUMAN** THAT **ISN'T ABOUT** TECHNOLOGY AT ALL.

YOU LISTEN TO THE WORLD, ONE PERSON AT A TIME, THE SOUND REPRODUCED WITH CLARITY, LIKE A SMALL VOICE.

IT COMES FROM A FEW SECONDS AGO, OR ANOTHER TIME IN THE RECENT PAST.

A **LIGHT COLUMN** INDICATES THE **LOCATION** WHERE THE MUSIC COMES FROM.

TWO STREAMS OF BLUE PARTICLES **SPIRAL** OUT OF THE EARTH IN A DNA-LIKE **DOUBLE HELIX.**

EACH NOTE BECOMES A GREEN CIRCLE FLOATING INTO THE STARS...

THERE IS SOMETHING OF THE **UNIVERSAL** IN THE WAY IT MAKES YOU FEEL **LONELY** AND **CONNECTED** AT THE SAME TIME...

PRINCIPLE 2.5

DESIGN **WITH** TECHNOLOGY, TO **TRANSCEND** TECHNOLOGY

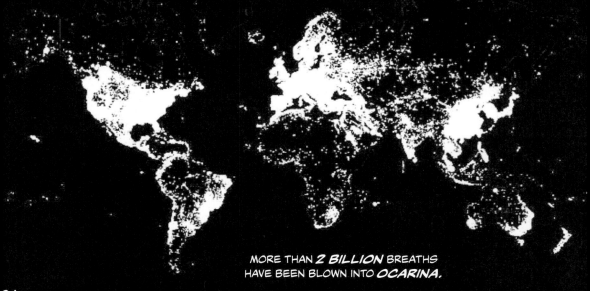

MORE THAN **2 BILLION** BREATHS HAVE BEEN BLOWN INTO **OCARINA.**

*I'D LIKE TO THINK, FOR ALL THE **TECHNOLOGY** THAT MADE A THING LIKE **OCARINA** POSSIBLE, THE RESULT WAS **MORE** THAN THE SUM OF ITS PARTS -- REACHING **TOWARD** THE **HUMAN** AND THE **SUBLIME**...*

FROM AN **OCARINA** USER IN 2009

"THIS IS MY PEACE ON EARTH. I AM CURRENTLY DEPLOYED IN IRAQ, AND HELL ON EARTH IS AN EVERYDAY OCCURRENCE. THE FEW NIGHTS I MAY HAVE OFF I AM DEEPLY ENGAGED IN THIS APP. THE GLOBE FEATURE THAT LETS YOU HEAR EVERYBODY ELSE IN THE WORLD PLAYING IS THE MOST **CALMING** ART I HAVE EVER BEEN INTRODUCED TO. IT BRINGS THE ENTIRE WORLD TOGETHER WITHOUT POLITICS OR WAR. IT IS THE EXACT OPPOSITE OF MY LIFE."

-- DEPLOYED U.S. SOLDIER

PRINCIPLE 2.6 *TECHNOLOGY SHOULD **CREATE CALM***

ON CHINA CENTRAL TELEVISION'S SERIES "**ONE PERSON, ONE WORLD**"...

Y'KNOW, *OCARINA* DEMONSTRATES THAT MUSIC *TRULY* HAS THE POWER TO MOVE PEOPLE, GIVE THEM STRENGTH IN WAYS YOU DON'T EXPECT. AND REGARDLESS OF THE *MEDIUM!*

THAT'S A GOOD QUESTION. I AM TRYING TO FIGURE IT OUT MYSELF...

INDEED!

THESE SOFTWARE *APPS* HAVE TRANSFORMED THE MOBILE PHONE INTO SOMETHING MUCH *MORE* THAN A PHONE. BUT WHAT SHOULD WE MAKE OF SOMETHING LIKE *OCARINA?* IS IT AN *INSTRUMENT?* IS IT A *TOY?*

张国伦
GUOLUN ZHANG
SONGWRITER

陈伟鸿
WEIHONG CHEN
HOST

MAYBE *CLASSIFICATION* ISN'T SO IMPORTANT HERE, BUT I TEND TO THINK OF *OCARINA* AS A TYPE OF *EXPRESSIVE TOY.* LIKE A TOY, IT INVITES *PLAY* AND HAS A LOW BARRIER TO ENTRY. YET IT AFFORDS A KIND OF *EXPRESSIVENESS* ASSOCIATED WITH *INSTRUMENTS!*

I SEE. SO REGARDLESS OF WHAT THESE ARTIFACTS MAY BE, THEY'RE DESIGNED TO ENCOURAGE EVERYDAY PEOPLE TO *MAKE MUSIC*, INCLUDING THOSE WHO OTHERWISE *MIGHT NOT?*

THAT'S THE HOPE. ALTHOUGH I ALSO DESIGN THESE THINGS BECAUSE IT'S FUN AND, AT THE END OF THE DAY, IT'S WHAT I *DO* -- A WAY TO EXPRESS MYSELF.

MAKES SENSE. COMPUTER MUSIC AND DESIGN IS YOUR *ART!*

WILL COMPUTER-BASED INSTRUMENTS SOMEDAY **REPLACE** TRADITIONAL INSTRUMENTS?

I CERTAINLY **HOPE NOT.**

I GET ASKED THAT A LOT! FOR SOME REASON, THAT'S HOW PEOPLE NATURALLY REACT TO THIS SORT OF RESEARCH! IT IS CERTAINLY NOT MY INTENTION TO REPLACE TRADITIONAL INSTRUMENTS! WE ARE **EXPLORERS** -- NOT **DESTROYERS!**

THERE IS A **REASON** WE HAVE SO MANY INSTRUMENTS IN THE WORLD. EACH BRINGS SOMETHING DIFFERENT. FOR EXAMPLE, NOTHING IS GOING TO BE AS GOOD AT BEING A CELLO -- OTHER THAN A CELLO!

A COMPUTER CAN DO A LOT AND AFFORDS NEW SOUNDS AND INTERACTIONS. BUT IT CANNOT DO EVERYTHING! I AM ALWAYS MORE INTERESTED IN THE **HUMAN** IN THE INTERACTION LOOP. IN DESIGNING THESE COMPUTER MUSIC INSTRUMENTS, THE ETHOS IS TO EMBRACE WHAT COMPUTERS ARE GOOD AT DOING, **RECONCILING** IT WITH WHAT PEOPLE ARE GOOD AT DOING!

I WONDER WHY PEOPLE THINK WHAT I DO **THREATENS** TRADITIONAL INSTRUMENTS?

WHAT I AM **TRYING** TO DO WITH **OCARINA** AND COMPUTER MUSIC DESIGN IS TO **ADD** TO THE MUSICAL ECOSYSTEM, TO FIND **RECOMBINANT** WAYS TO RECONCILE FAMILIAR ELEMENTS WITH NEW **EXPERIENCES!** I MEAN, IT'S NOT LIKE THERE IS AN UPPER LIMIT TO HOW MUCH MUSIC CAN BE MADE IN THE WORLD, BEYOND WHICH WE'D SAY "OH, THAT'S TOO MUCH. STOP!" INSTEAD, I BELIEVE WE CAN ALWAYS **MAKE MORE**; MOST CERTAINLY WE ARE **NOT** MAKING ENOUGH...

EXPAND ON THAT!

IF YOU THINK ABOUT IT -- BEFORE COMPUTERS, INTERNET, RADIO, AND RECORDING, PEOPLE **HAD** TO **MAKE MUSIC** WHERE IT WAS HEARD!

IN OTHER WORDS, UNTIL ONLY ABOUT A HUNDRED YEARS AGO, ALL MUSIC WAS MADE **LIVE!**

OKAY, FROM THE TOP!

OOOH I **GOT** THIS!

Tooooooot

IT WASN'T LONG AGO WHEN **FAMILIES** REGULARLY **PLAYED MUSIC** AS A FORM OF ENTERTAINMENT. PEOPLE OFTEN LEARNED TO PLAY INSTRUMENTS OUT OF **INTEREST** AND SO THEY COULD, SAY, PLAY THE LATEST TUNES. IT WAS A FUN **PASTIME.** THEY WEREN'T DOING IT TO "GO PRO" BUT DID IT FOR THEMSELVES, THE PEOPLE AROUND THEM, AND FOR THE JOY OF MAKING MUSIC **SOCIALLY.**

Y'KNOW, IT'S **FASCINATING!** THERE WAS A TIME WHEN THE WORD **AMATEUR** CONNOTED SOMETHING WHOLLY **GOOD!** IT MEANT YOU **LOVED** SOMETHING, LIKE AN INSTRUMENT, ENOUGH TO LEARN IT FOR YOURSELF.

WORD
AMATEUR

= FROM LATIN
AMATOR

= MEANING
ONE WHO LOVES

AMONG OTHER THINGS, AMATEUR MUSICIANSHIP IS ABOUT **PERSONAL ENRICHMENT, ACTIVELY** ENGAGING WITH OUR FAVORITE MUSIC, GETTING OUR HANDS DIRTY TO MAKE MUSIC -- LIKE PLAYING UKULELE IN THE PARK, OR SINGIN' IN THE RAIN!

SADLY, THESE FORMS OF MUSIC-MAKING ARE **VANISHING**... I MEAN, HOW MANY FAMILIES TODAY STILL REGULARLY MAKE MUSIC TOGETHER AFTER DINNER?

THE BIRTH OF **MASS CONSUMPTION** OF MUSIC

SOMEHOW, TECHNOLOGICAL ADVANCEMENTS IN THE 20TH CENTURY CHANGED THE **PERSONAL** AND **SOCIAL** DYNAMICS OF MUSIC-MAKING!

SOUND RECORDING: CAPTURES MUSIC FOR PLAYBACK; PERFORMANCES BECOME **TIMELESS**

EDISON'S WAX CYLINDER RECORDER / PLAYER!

RADIO: BROADCASTS MUSIC; **VAST DISTANCES** NO LONGER AN IMPEDIMENT

PHONOGRAPH

MAGNETIC TAPE

VACUUM TUBE: ENABLES TECHNOLOGY FOR ANALOG ELECTRONICS, LIKE RECEIVERS AND TELEVISIONS

COMPACT DISC

MP3

DIGITAL COMPUTER: OFFERS PRISTINE STORAGE, PROCESSING, TRANSMISSION OF MUSIC

THE CLOUD

?

WHAT'S NEXT?

THE INTERNET: DISTRIBUTES PERSONALIZED MASS MEDIA, SOCIAL NETWORKING; MUSICAL DATA BECOMES PERVASIVE, RANDOM-ACCESS, CENTRALIZED IN COMPUTING CLOUDS!

THESE **INNOVATIONS** HAVE ALTERED THE RELATIONSHIP BETWEEN PEOPLE AND MUSIC -- FOR BETTER AND FOR WORSE -- AS **SIDE EFFECTS** OF THE **EVOLUTION** OF TECHNOLOGY...

WE NOW HAVE MORE **ACCESS** TO MUSIC THAN EVER BEFORE, AS LISTENERS AND CONSUMERS!

YET SOMEHOW I FEEL WE ARE **MAKING LESS MUSIC THAN EVER.**

THAT'S A **SHAME**, BECAUSE WHILE LISTENING TO MUSIC IS WONDERFUL, THERE IS A SUBLIME **JOY IN MAKING MUSIC** -- AN ACTIVITY THAT **ENRICHES** SIMPLY BY HAPPENING AT ALL!

HAS TECHNOLOGY MADE IT SO **EASY** TO CONSUME MUSIC THAT IT NO LONGER SEEMS **NECESSARY** TO **MAKE IT**? HAS SUPER-READY **ACCESS** TO VAST LIBRARIES OF HIGH-QUALITY RECORDINGS OF VIRTUOSI SOMEHOW **INTIMIDATED** OR CURBED OUR DESIRE TO MAKE MUSIC FOR OURSELVES?

MANY PEOPLE'S FIRST REACTION IS THAT BY DELVING INTO COMPUTER-BASED INSTRUMENTS, PEOPLE LIKE ME ARE **THREATENING** TRADITIONAL MUSICAL INSTRUMENTS AND PRACTICES. HOWEVER, THE **INCONVENIENT TRUTH** IS THAT MUSIC-MAKING IS **CONSTANTLY** BEING THREATENED, NOT BY COMPUTER MUSIC RESEARCH BUT RATHER BY THE COUNTLESS "DISTRACTIONS" ENABLED BY MODERN TECHNOLOGY: TELEVISION, STREAMING VIDEO, INTERNET, VIDEO GAMES, ETC. THERE IS NOTHING **INHERENTLY** WRONG WITH THESE ACTIVITIES, BUT THEY DO **ADD UP** AND OCCUPY OUR TIME! IT IS SO MUCH **EASIER** TO **CONSUME** THESE FORMS OF MASS MEDIA THAN TO, SAY, LEARN TO PLAY AN INSTRUMENT!

I WANT TO **CHALLENGE** THIS TREND! IF TECHNOLOGY UNWITTINGLY TOOK AWAY AMATEUR MUSICIANSHIP, PERHAPS WE CAN USE TECHNOLOGY TO **BRING IT BACK,** IN THE CONTEXT OF TODAY'S WORLD.

NICHOLAS COOK, MUSIC FACULTY AT THE UNIVERSITY OF CAMBRIDGE, WRITES...

"MUSIC HAS BECOME PART OF AN AESTHETIC ECONOMY DEFINED BY THE PASSIVE AND INCREASINGLY PRIVATE CONSUMPTION OF COMMODIFIED PRODUCTS RATHER THAN THROUGH THE ACTIVE, SOCIAL PROCESSES OF PARTICIPATORY PERFORMANCE.

IN SHORT, WE SEEM TO HAVE FORGOTTEN THAT MUSIC IS A PERFORMANCE ART AT ALL, AND MORE THAN THAT, WE SEEM TO HAVE CONCEPTUALIZED IT IN SUCH A WAY THAT WE COULD HARDLY THINK OF IT THAT WAY EVEN IF WE WANTED TO..."

-- NICHOLAS COOK

I THINK THERE ARE TWO REASONS *WHY* I DO WHAT I DO. THE FIRST IS TO DESIGN MUSICAL ARTIFACTS, TO TAKE US *BACK* TO A PAST OF *PERSONAL MUSICAL PERFORMANCE* BY TAKING ADVANTAGE OF TECHNOLOGY, AS A *CELEBRATION* OF MUSIC! I WANT FOR US TO *RECLAIM* A SENSE OF PLAYFULNESS IN *MAKING* MUSIC, TO GET PEOPLE TO PLAY MORE MUSIC!

MUSIC-MAKING IS REALLY LIKE THE JOY OF *COOKING* YOUR OWN FOOD. MOST OF US WHO COOK AREN'T DOING IT TO BE PROFESSIONAL CHEFS, BUT WE ENJOY IT NONETHELESS! IF MUSIC IS FOOD FOR THE EAR AND SOUL -- WHY AREN'T WE *COOKING MORE MUSIC* FOR OURSELVES?!

THIS GUY SURE *LOVES* TO TALK ABOUT FOOD... MAKIN' ME *HUNGRY!*

I THINK I SEE WHAT YOU ARE TRYING TO DO!

A *SECOND* GOAL IN MY WORK IS TO LOOK TO THE *FUTURE*, TO DESIGN AND CREATE SOMETHING THAT SIMPLY HAS NOT BEEN POSSIBLE WITHOUT TECHNOLOGY... TO EXPLORE WHAT NEW MUSICAL THINGS AND EXPERIENCES AWAIT DISCOVERY, THAT WE DON'T YET HAVE NAMES FOR, THAT *DEFY CLASSIFICATION.* MIGHT WE CREATE INSTRUMENTS TO BE PLAYED BY A *MILLION STRANGERS* ACROSS THE WORLD? WHAT WOULD THAT SOUND LIKE? HOW WOULD IT *FEEL* TO BE A PART OF *THAT?*

I BECAME A **SONGWRITER** AND **SINGER** BECAUSE I WAS FOLLOWING MY **INTERESTS** IN MUSIC. I SEE YOU HAVE ALSO FOLLOWED YOUR INTERESTS IN **DESIGN.** I WONDER WHERE THESE INTERESTS WILL TAKE US...

YES! AND HOPEFULLY IT WILL BE A FUTURE THAT **EMBRACES** VARIED WAYS OF **MAKING** MUSIC, NEW AND TRADITIONAL ALIKE! TECHNOLOGY **WILL** EVOLVE -- WHAT'S MODERN TODAY SHALL BECOME ANTIQUATED TOMORROW -- BUT THE CORE **HUMAN DESIRE** TO **EXPRESS** WILL STILL BE HERE. THROUGH ARTFUL FASHIONING OF TECHNOLOGY, WE **WILL** SEEK OUT NEW THINGS TO SEE, HEAR, INTERACT WITH -- TO **THINK** AND **FEEL** WITH. THE INSTRUMENT MAY LOOK AND SOUND DIFFERENT, BUT THE SONG REMAINS THE SAME. **MUSIC** IS STILL **MUSIC**, REGARDLESS OF THE MEDIUM. AS THE ANTHEM GOES, "ROCK 'N' ROLL IS HERE TO STAY"!

...I FEEL WE ARE HEADING INTO A **NEW ERA** OF MUSIC AND FUTURE MUSICAL EXPERIENCES! IT SEEMS WE ARE ONLY AT THE **BEGINNING.** MUSIC AND TECHNOLOGY WILL CONTINUE TO **CO-EVOLVE!**

AND IF MUSIC IS -- AS YOU SAY -- **FOOD** FOR THE **SOUL**, HOPEFULLY COMPUTERS WILL ADD TO THE MENU AND PALETTE!

THANK YOU FOR BEING ON "ONE PERSON, ONE WORLD"!

THANK **YOU** FOR HAVING ME!

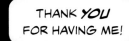

95

AND NOW FOR SOMETHING **COMPLETELY DIFFERENT** (AND **NOT** SO DIFFERENT, IN ITS ROLE AS AN EXPRESSIVE TOY):

I AM T-PAIN!

I AM **AWARE**, MOST OF THE TIME, THAT I AM **NOT** ACTUALLY T-PAIN, BUT THE PURPOSE OF THE **APP** AND NAMING IT "I AM T-PAIN" WAS TO **TRANSFORM** EACH USER INTO T-PAIN WITH HIS SIGNATURE **VOCAL EFFECT**!

⋛PSST⋚ NANCY WHADDYA SAY WE **BUM RUSH** THAT PODIUM 'N' **SING** *"BUY U A DRANK"*

MY FELLOW AMERICANS, IT'S **AUTO-TUNE** FOR YOUR PHONE!

JOE... NOT AGAIN

AUTO-TUNE HAS **PERMEATED** THE VERY FABRIC OF OUR POPULAR CULTURE! FROM ITS INTENDED USE AS A SUBTLE **PITCH CORRECTION** AUDIO EFFECT TO FULL-ON VOCAL EFFECT (AS USED BY ARTISTS FROM **CHER** TO **T-PAIN**), AUTO-TUNE IS **UBIQUITOUS** IN POP MUSIC WORLDWIDE. SOME LOVE IT, OTHERS DO NOT! REGARDLESS, IT SEEMS TO ILLUSTRATE OUR FASCINATION WITH FINDING SOME **STRANGE BALANCE** BETWEEN OUR **NATURAL** VOICE AND AN **UNCANNY** COMPUTERIZED VERSION OF IT!

...AND MAYBE *SOMEDAY* THE *PEOPLE* OF THIS GREAT EARTH WILL *SING* UNABASHEDLY, MERRILY, IN PERFECT HARMONY OR OUT OF TUNE, IN THEIR HOMES, LIVING ROOMS, BATHROOMS, ON THE STREET, IN CARS, INTO A PHONE, INTO THE AIR, WITH OR WITHOUT AUTO-TUNE, WITHOUT JUDGMENT, WITH FRIENDS, FAMILIES, STRANGERS, ENEMIES, PETS... FOR A GOOD REASON, OR NO REASON AT ALL!

HURRAH FOR PEOPLE TAKING TO THE STREETS TO MAKE MUSIC! THE WORLD CAN BE ABUNDANTLY **COLORFUL** WHEN YOU POINT AN EXPRESSIVE LENS AT IT! WHAT MIGHT WE TAKE AWAY FROM THIS? HOW ABOUT...

(⚙) PRINCIPLE 2.7

DESIGN TO **LOWER INHIBITION**

DESIGN IS BEING COGNIZANT OF HOW WE AS HUMANS TEND TO **THINK** AND USING THAT AWARENESS TO ENCOURAGE CERTAIN **BEHAVIORS** THAT, HOPEFULLY, ARE **BENEFICIAL** FOR US!

THE **PSYCHOLOGY** OF AUTO-TUNE ON A PHONE MAY BE GROUNDED IN THE **DISARMING** QUALITY OF MOBILE PHONES AS EXPRESSIVE TOYS: IF IT DOESN'T FEEL SO **SERIOUS**, THEN IT'S SOMEHOW "MORE OKAY" TO EXPRESS YOURSELF! AUTO-TUNE SEEMS TO HIT A **MIDDLE GROUND** BETWEEN ONE'S OWN VOICE AND A **MECHANIZED** VERSION OF IT -- IT'S **YOU** BUT **NOT YOU.** IT'S LIKE GAINING **PLAUSIBLE DENIABILITY** FOR PUTTING ONESELF OUT THERE THROUGH SONG! THAT IS A POWER OF EXPRESSIVE TOYS!

AS FAR AS I KNOW, THE APPLICATION OF AUTO-TUNE HAS NEVER MADE ANY SITUATION **MORE** SERIOUS.

PERHAPS THERE IS SOMETHING ABOUT A **VOCAL EFFECT** THAT LENDS **LEVITY**! THE "IT WAS A JOKE" FACTOR GIVES PEOPLE **PERMISSION** TO EXPRESS!

AUTO-TUNE (KINDA LIKE ALCOHOL) IS A TYPE OF **INHIBITION INHIBITOR!**

INHIBITION INHIBITOR!

SELF-CONSCIOUS...

"WHAT, YOU WANT ME TO **SING?!**"

"POST MYSELF SINGING, TO THE INTERNET? UM, NO ONE NEEDS TO HEAR THAT!"

"I WANT TO EXPRESS SOME DEEP THOUGHTS, BUT I DON'T WANT TO PUT MYSELF OUT THERE... IT'S TOO... VULNERABLE."

AUTO-TUNE!

TOY MENTALITY

OKAY!

"OH, WITH AUTO-TUNE? WHATEVER, SURE!"

"OH HAH, CHECK THIS OUT. I WAS AUTO-TUNE SINGING FOR MY CAT!"

"OOH, AUTO-TUNE SOMEHOW MAKES ME MORE COMFORTAB SINGING MY MIND! IT SIGNALS THAT I AM NOT TAKING MYSEL TOO SERIOUSLY."

BEFORE WE LEAVE THIS CHAPTER...

WELCOME TO **AUTORAP** AND THE **CORNHOLIO STRESS TEST!** IF WE CAN AUTORAP **THIS**, WE CAN AUTORAP **ANYTHING!**

SMULE'S AUTORAP WAS THE BRAINCHILD OF FELLOW RESEARCHERS **PRERNA GUPTA** AND **PARAG CHORDIA!**

STEP ONE: SIMPLY **SPEAK** INTO THE PHONE...

SAY WHAT YOU WANT, SPEAK YOUR MIND, GET CREATIVE!

I AM THE **ALMIGHTY BUNGHOLIO!**

MY BUNGHOLE, IT **SPEAKETH**, IT SAYS...

RAKAKAKAKAKAKAKA TIKKA TIKKA KAKAKAKA!!!

BOING-ING-ING-ING-ING!

MY IMPROVISED **BEAVIS** AND **BUTTHEAD**-INSPIRED MONOLOGUE IS RECORDED AS INPUT INTO AUTORAP.

ARGHGHGH!

STEP TWO: LET THE APP **DO ITS THING...**

100

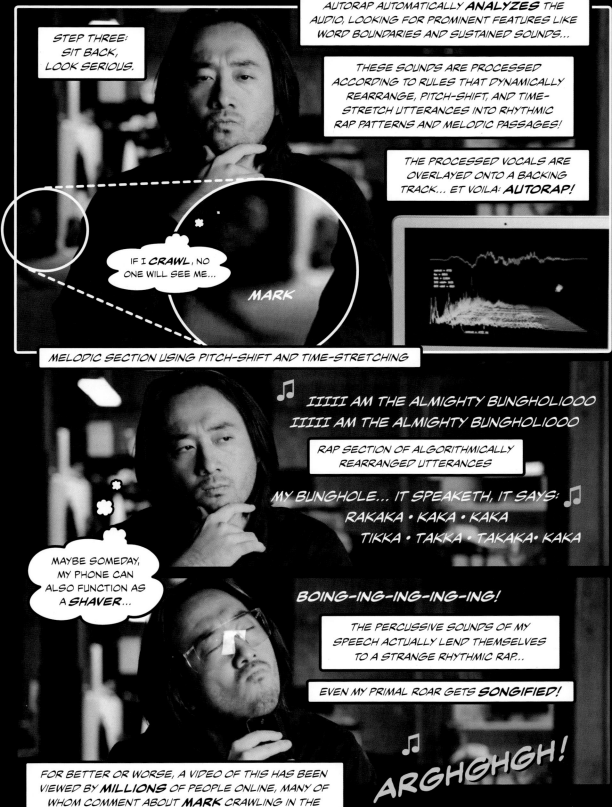

AND SO WE REACH THE END OF CHAPTER 2. WHAT BEGINS AS *SCIENCE FICTION* BECOMES *TECHNOLOGICAL REALITY,* ITS IMPLICATIONS NO LESS FANTASTICAL.

TECHNOLOGY IS A *CATALYST* THAT CONTINUALLY INTERACTS WITH WHO WE ARE AND INFLUENCES OUR EVOLVING WAY OF LIFE. DESIGN IS THE *VEHICLE* TO CRAFT THE DIRECTION AND NUANCE OF THAT EVOLUTION.

CHAPTER 2
DESIGN ETUDE

PART 1: INSIDE-OUT BRAINSTORM

APPLY THE IDEA OF *INSIDE-OUT DESIGN*, AND WORK *BACKWARDS* TO CONCEIVE OF A PHYSICAL EXPRESSIVE TOY. DO THIS BY TAKING ADVANTAGE OF AN *EXISTING EVERYDAY TECHNOLOGY.* WORK RIGOROUSLY WITH THE CONSTRAINTS OF AVAILABLE TECHNOLOGY (NO "IF ONLY WE HAD X") AND CONSIDER AN *AESTHETIC LEAP* THAT *JUSTIFIES* USING THE TECHNOLOGY (SOMETHING YOU COULDN'T DO WITHOUT IT). THE TOY SHOULD AIM TO INVITE THE USER TO BE CREATIVE OR PLAYFULLY EXPRESSIVE IN SOME WAY. SKETCH IDEAS, AND ARTICULATE AS MUCH DETAIL AS POSSIBLE.

PART 2: (OPTIONAL) PROTOTYPE IT!

HOWEVER YOU ARE ABLE, *PROTOTYPE* YOUR DESIGN. *BUILD* THE MINIMAL ESSENTIAL ELEMENTS. THIS IS WHERE THE PROVERBIAL RUBBER MEETS THE ROAD AND WHERE YOU WILL DISCOVER HOW WELL (OR NOT) THE TECHNOLOGIES YOU PLAN TO USE LEND THEMSELVES TO THE TASK. YOU MAY NEED TO *BACKTRACK*, AND PERHAPS EVEN *START OVER* WITH AN ENTIRELY NEW CONCEPT IF YOU GET STUCK. *DO NOT* TRY TO SOLVE HARD TECHNICAL PROBLEMS *UNLESS ABSOLUTELY NECESSARY.* DON'T FIGHT THE TECHNOLOGY. *EMBRACE* ITS STRENGTHS AND LIMITATIONS.

PART 3: ONE MORE THING...

AFTER YOU'VE DESCRIBED YOUR CONCEPT IN AS MUCH DETAIL AS POSSIBLE, THINK OF *ONE FEATURE* TO REALLY *PUSH* THIS *OVER THE EDGE!* FOR EXAMPLE, THE GLOBE IN *OCARINA* WAS A SIMPLE BUT OVER-THE-TOP DESIGN GESTURE. SIMILARLY, YOU MIGHT THINK OF A SPECIFIC *SOCIAL* FEATURE OR A *COLLABORATIVE* USE CASE FOR YOUR DESIGN. GOOD LUCK AND *HAVE FUN!*

EXAMPLE *PHYSICAL LAPTOP + COMPUTER VISION + VISUALIZATION = "LAPTOP ACCORDION"**

(SOMEWHERE BETWEEN AWFUL AND AWESOME) SET LAPTOP *SIDEWAYS* ON LAP AND PLAY BY *OPENING* AND CLOSING THE *SCREEN* (MOTION TRACKED BY COMPUTER VISION, USING FRONTSIDE ONBOARD CAMERA). SOUND IS SYNTHESIZED ON LAPTOP; KEYBOARD IS MAPPED TO PITCH. THE SCREEN ITSELF MIGHT VISUALIZE THE PHYSICAL *OPENING* AND *CLOSING* OF THE INSTRUMENT. AN OVER-THE-TOP VERSION: *SCRUB* A *VIDEO RECORDING* (MAPPED TO SCREEN POSITION) OF SOMEONE'S FACE AS IF THE PERSON IS *TRAPPED* INSIDE THE LAPTOP, AS YOU PLAY IT! LET THE *BAD IDEAS* FLOW, FOR THEY ARE THE *SEEDS* OF GOOD DESIGNS!

*WE ACTUALLY BUILT ONE... COMING IN CHAPTER 5.

AN EXCERPT
FROM **MUSICKING**, BY **CHRISTOPHER SMALL** (1998)

"The fundamental nature and meaning of music lie not in objects, not in musical works at all, but in action, in what people **do**. It is only by understanding what people do as they take part in a musical act that we can hope to understand its nature and the function it fulfills in human life. Whatever that function may be, I am certain, first, that to take part in a musical act is of central importance to our very humanness, as important as taking part in the act of speech, which it so resembles (but from which it also differs in important ways), and second, that everyone, every normally endowed human being, is born with the gift of music no less than with the gift of speech.

If that is so, then our present-day concert life, whether 'classical' or 'popular,' in which the 'talented' few are empowered to produce music for the 'untalented' majority, is based on a falsehood. It means that our powers of making music for ourselves have been hijacked and the majority of people robbed of the musicality that is theirs by right of birth, while a few stars, and their handlers, grow rich and famous through selling us what we have been led to believe we lack."

OHHHHHH **DANG.**

CHAPTER 3
VISUAL DESIGN

THE ARTFUL DESIGNER IS A **PLANNER** AND **BUILDER** WITH **AESTHETIC SENSE**, ABLE TO **SHAPE** TECHNOLOGY WITH THE UNDERSTANDING THAT WE ARE **MULTI-SENSORY**, **MULTI-MODAL** CREATURES WHO EXPERIENCE THE WORLD THROUGH **SIGHT**, **SOUND**, AND **INTERACTION**. WE ARE AWARE OF THIS "MULTI-NESS," MAKE USE OF IT, APPEAL TO IT, AND ULTIMATELY FASHION ENTIRELY **NEW THINGS** OUT OF IT.

THE PREVIOUS CHAPTERS EXPLORED THE **NATURE** OF **ARTFUL DESIGN**, AND A BIT OF THE "**WHY.**" WE NOW DIG DEEPER INTO ITS **BUILDING BLOCKS.** HERE WE EXPLORE **VISUAL DESIGN**: SHAPES, COLORS, TEXTURES, AND WAYS IN WHICH THEY MOVE, CHANGE, EVOLVE. THE **FUNCTION** OF THESE VISUAL ELEMENTS RANGES FROM COMMUNICATION TO INFORMATION AND INTERACTION, WHILE THE **FORM** ENCODES AN ASPECT OF **VISUAL POETRY** THAT HELPS TO **UNFOLD** AND GIVE **MEANING** TO ITS FUNCTION.

WHOA...

THOSE ARE SOME **TRIPPY GRAPHICS** RIGHT THERE...

DUNNO WHY, BUT I'VE ALWAYS BEEN A SUCKER FOR **SPIRALS**...

WE BEGIN FROM THE **SIMPLEST** OF **BEGINNINGS**: NOTHING...

IN THE BEGINNING OF **ALL DESIGN**,
THERE IS ONLY *THE VOID*.

FORMLESS...
AND WITHOUT *ESSENCE.*

DESIGN ONLY FEELS INEVITABLE *AFTER* IT'S FINISHED. WHEN WE ADMIRE AN ELEGANTLY DESIGNED OBJECT, WE FEEL "BUT OF COURSE -- IT *HAD* TO BE THAT WAY!" BUT *ONCE UPON A TIME,* THERE WERE *CHOICES.* DESIGN USUALLY BEGINS, UNPLEASANTLY AND MESSILY, IN A SEA OF PETRIFYING POSSIBILITIES.

PRACTICAL, ARTISTIC, PERSONAL QUESTIONS...

"WHAT *IS* IT?"

"WHAT *MUST* IT DO? WHAT MUST IT NOT?"

"HOW SHOULD IT *LOOK? SOUND? FEEL?*"

"*HOW* WILL IT WORK?"

"EVERY CHOICE IS BUILT UPON SOME PREVIOUS CHOICE. *WHAT IF* I MAKE THE WRONG ONE SOMEWHERE?"

"*WILL* IT WORK?"

BUT IT'S GOTTA BEGIN *SOMEWHERE.*

LET US START... WITH A SIMPLE *EXPLORATION* OF *FORM.*

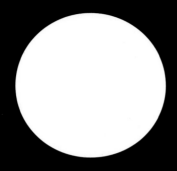

THE *CIRCLE.*

SUBTRACT A SMALLER CIRCLE FROM THE CENTER, AND WE HAVE A *RING.*

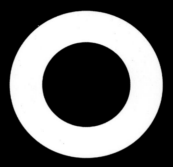

BLUR THE RING.

ADD THE TWO TOGETHER...

NOW WE HAVE AN **ELEMENT** THAT APPEARS TO **GLOW.** MORE IMPORTANTLY, IT'S GOT **PERSONALITY!**

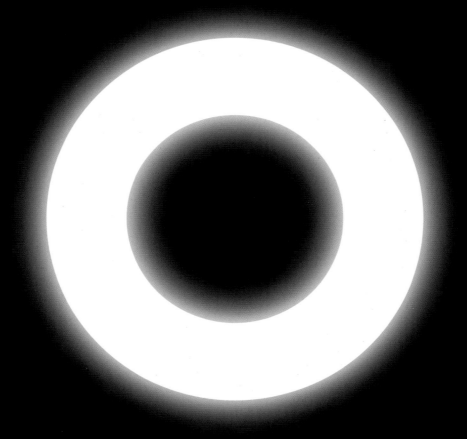

I CALL THIS A **FLARE**
(OR A **HEAVENLY DONUT**).

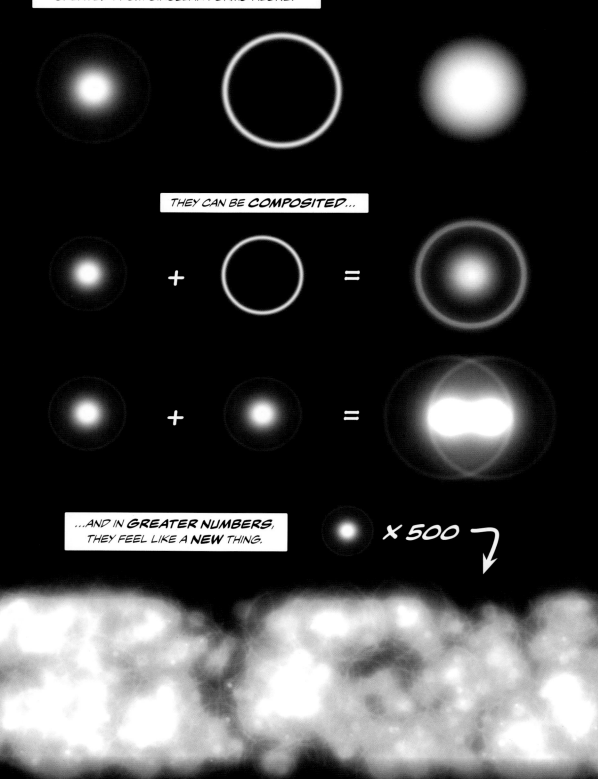

THERE ARE **COUNTLESS** ELEMENTS TO BE CREATED FROM CIRCULAR FORMS ALONE.

THEY CAN BE **COMPOSITED**...

...AND IN **GREATER NUMBERS**, THEY FEEL LIKE A **NEW** THING.

X 500

WE ADD **COLOR.**

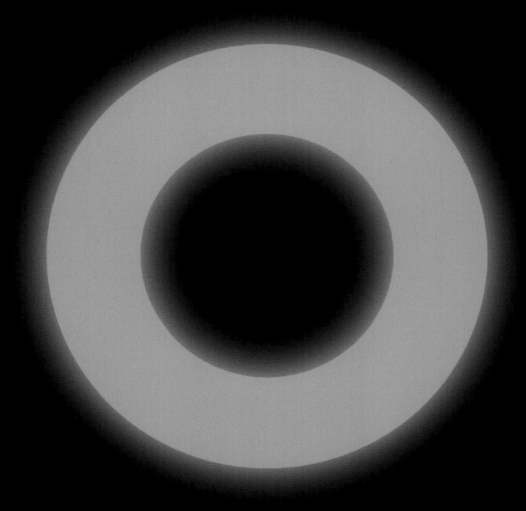

RED: 1.0   *GREEN: 0.5*   *BLUE: 0.0*

WE CAN CREATE 3D **SPHERES** OF LIGHT...

x **12** =

...AND **STRETCH** EACH SPHERE INTO A LIGHT **COLUMN**...

...WHICH WE COLOR

LIKE **LIGHTSABERS.**

WE **ARRANGE** LIGHT COLUMNS IN A CIRCLE: A FLOWER-SHAPED **COLOR WHEEL** OF 12...

SIDE VIEW

TOP VIEW

...IN THE HUES OF A **RAINBOW.**

AS WE MOVE TOWARD THE CENTER, THE GLOWING COLUMNS **BLEND** INTO ONE ANOTHER...

RED: 1.0
GREEN: 0.5
BLUE: 0.0

RED: 1.0
GREEN: 1.0
BLUE: 0.0

RED: 0.5
GREEN: 1.0
BLUE: 0.0

RED: 0.0
GREEN: 1.0
BLUE: 0.0

RED: 0.0
GREEN: 1.0
BLUE: 0.5

RED: 1.0
GREEN: 0.5
BLUE: 0.0

RED: 1.0
GREEN: 0.3
BLUE: 0.0

RED: 1.0
GREEN: 0.0
BLUE: 0.0

COMPRISED OF NOTHING MORE THAN THE SAME SIMPLE BUILDING BLOCKS...

...144 **INSTANCES** OF THE SAME FLARE, 12 IN EACH OF 12 COLUMNS...

RED: 1.0
GREEN: 0.0
BLUE: 0.5

...AN EXAMPLE OF TAKING SIMPLE ELEMENTS TO AN **EXTREME**, FORMING SOMETHING COMPLETELY DIFFERENT.

RED: 1.0
GREEN: 0.0
BLUE: 1.0

RED: 0.5
GREEN: 0.0
BLUE: 1.0

RED: 0.0
GREEN: 0.0
BLUE: 1.0

THE **WHITE** LIGHT COLUMNS IN THE **OCARINA** GLOBE WERE CREATED FROM SIMPLE TEXTURED IMAGES; THE **BLUE** DOUBLE HELICES WERE INDIVIDUAL FLARES.

FORM DOES NOT HAVE TO EXIST ALONE. COMPOSITES OF THE CIRCULAR FLARES MAKE UP THE **BUTTONS** IN **OCARINA** -- PART OF ITS **INTERFACE** FOR HUMAN-OCARINA INTERACTION. HOW THESE FORMS **BEHAVE** AND **RESPOND** AS PART OF THE FUNCTION GIVE THE THING ITS **PERSONALITY** AND **NUANCE.**

SEVERAL FLARES MAKE UP EACH **VIRTUAL ANIMATED BUTTON.**

THEY **BLEND** TOGETHER TO CREATE A SINGLE VISUAL UNIT.

EACH FLARE CAN MOVE **INDEPENDENTLY** OF THE OTHERS...

...UPON TOUCH, THE OUTER FLARES SMOOTHLY **EXPAND** WHILE THE INNER LAYERS **CONTRACT.**

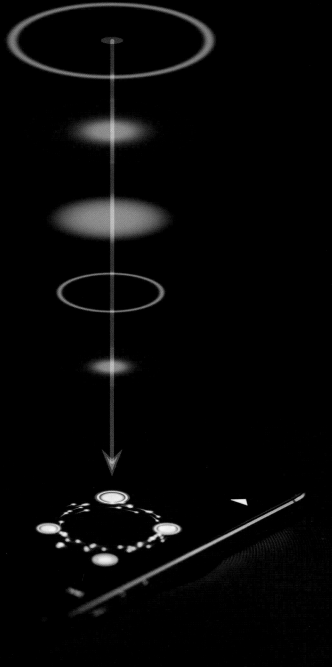

NOW THESE **FORMS** HAVE A **PURPOSE.**

FUNCTIONALLY, THE BUTTON CONTROLS PITCH. VISUALLY, IT **EXPANDS** TO PROVIDE **VISUAL FEEDBACK** AND A SENSE OF **SATISFACTION.**

PRESS!

WE'RE JUST USING OUR BASIC CIRCLES!

THIS USE OF VISUALS TO CRAFT INTERACTION BRINGS US TO OUR FIRST PRINCIPLE OF **MULTI-MODAL DESIGN!**

PRINCIPLE 3.1

DESIGN SOUND, GRAPHICS, AND INTERACTION **TOGETHER**

DESIGN SHOULD AIM TO SIMULTANEOUSLY ADDRESS THE **VISUAL**, THE **AURAL**, AND THE **INTERACTIVE.** CHANGES IN ONE DOMAIN SHOULD BE REFLECTED IN THE OTHERS. FOR EXAMPLE, BLOWING INTO THE INSTRUMENT SHOULD RESULT IN SOUND **AND** A CORRESPONDING VISUAL CHANGE (E.G., THE FLARES GLOW AND INCREASE IN SIZE). SIMILARLY, PRESSING THE BUTTON HAS A SPECIFIC INTERACTIVE AND SONIC FUNCTION, BUT IT ALSO ELICITS A SATISFYING VISUAL RESPONSE.

THIS VISUAL REACTION IS DESIGNED TO FEEL **ORGANIC.** IT'S ALSO AN ATTEMPT TO COMPENSATE FOR THE LACK OF **TACTILE FEEDBACK** ON THE FLAT TOUCHSCREEN.

OUTWARD EXPANSION!

FLARES ARE USED THROUGHOUT **OCARINA**... I **REALLY** LIKE CIRCLES.

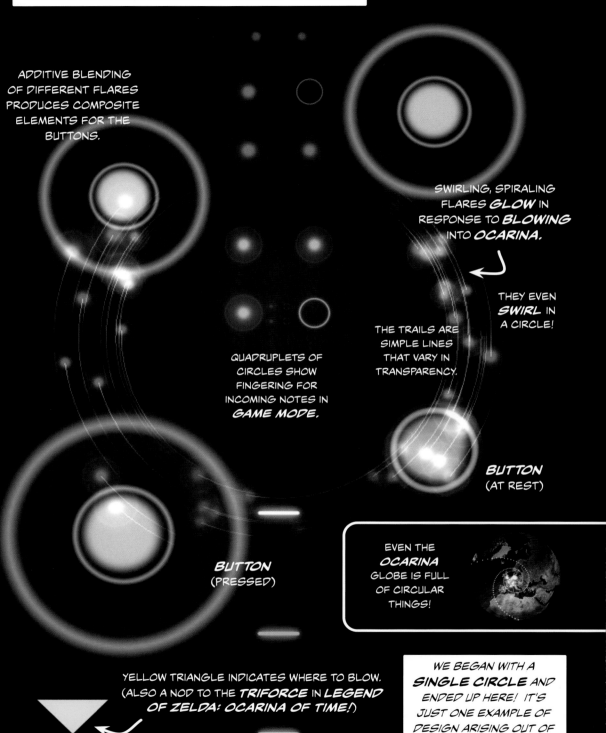

ADDITIVE BLENDING OF DIFFERENT FLARES PRODUCES COMPOSITE ELEMENTS FOR THE BUTTONS.

SWIRLING, SPIRALING FLARES *GLOW* IN RESPONSE TO *BLOWING* INTO *OCARINA.*

THEY EVEN *SWIRL* IN A CIRCLE!

THE TRAILS ARE SIMPLE LINES THAT VARY IN TRANSPARENCY.

QUADRUPLETS OF CIRCLES SHOW FINGERING FOR INCOMING NOTES IN *GAME MODE.*

BUTTON (AT REST)

BUTTON (PRESSED)

EVEN THE *OCARINA* GLOBE IS FULL OF CIRCULAR THINGS!

YELLOW TRIANGLE INDICATES WHERE TO BLOW. (ALSO A NOD TO THE *TRIFORCE* IN *LEGEND OF ZELDA: OCARINA OF TIME!*)

WE BEGAN WITH A *SINGLE CIRCLE* AND ENDED UP HERE! IT'S JUST ONE EXAMPLE OF DESIGN ARISING OUT OF SIMPLE ELEMENTS.

WE MIGHT EXTRACT ADDITIONAL ARTFUL DESIGN TAKEAWAYS FROM THE DESIGN OF *OCARINA.* SOME WERE CONSCIOUS DECISIONS MADE *DURING* THE DESIGN, OTHERS WERE UNDERSTOOD IN *RETROSPECT.*

(人) PRINCIPLE 3.2   *ANIMATE*

VISUAL DESIGN IS MORE THAN THE WAY THINGS LOOK, BUT ALSO THE WAY THEY *MOVE* OVER TIME, AND HOW -- THROUGH THEIR MOTION -- THEY *TELL THEIR STORY* AND BUILD AN EXPRESSIVE *CONNECTION* TO THE USER.

(人) PRINCIPLE 3.3   *IMBUE PERSONALITY*

GLOW, FLOW, SPIRAL, BREATHE, PULSATE, SPIN, EXPLODE... IN THE DIGITAL DOMAIN, WHERE THE DEFAULT IS *RIGID* AND *MECHANICAL*, NUANCE MUST BE *IMBUED.* THE FINISHED DESIGN SHOULD FEEL *ORGANIC* AND *LIVING.* DESIGN IN *EXPRESSIVE VERBS.*

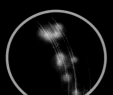

(人) PRINCIPLE 3.4   *SIMPLIFY* FORM

IDENTIFY CORE ELEMENTS.
TRIM THE REST.

(人) PRINCIPLE 3.5   *BUILD COMPLEXITY FROM SIMPLICITY*

ENDLESS VARIATIONS CAN BE CREATED FROM COMBINING BASIC ELEMENTS: LINES, SHAPES, COLORS, TRANSFORMATIONS. ARTFUL MANIPULATION OF MANY *RELATED* YET *INDEPENDENT* INSTANCES CAN GIVE RISE TO EMERGENT FORMS.

(人) PRINCIPLE 3.6   USE *VISUALS* TO REINFORCE *PHYSICAL* INTERACTION

VISUALS CAN *ENHANCE* PHYSICAL INTERACTIONS BY PROVIDING MEANINGFUL *FEEDBACK* FOR USERS.

PROMPT USERS TO EXPERIENCE *SUBSTANCE*
(人) PRINCIPLE 3.7   *(NOT TECHNOLOGY)*

GOOD DESIGN USES THE *MEDIUM* TO HIGHLIGHT A *NARRATIVE*, WHILE *HIDING* THE MEDIUM ITSELF.

THESE PRINCIPLES DERIVED FROM **OCARINA** ARE PART OF A BROADER SET OF PRINCIPLES FOR THE **VISUAL DESIGN** OF **INTERACTIVE THINGS**.

I HAVE COLLECTED THESE GENERAL PRINCIPLES OVER THE PAST 20 YEARS WHILE CRAFTING AUDIOVISUAL COMPUTER MUSIC SOFTWARE. THEY ARE NOT MEANT TO BE UNIVERSAL (OR NECESSARILY ORIGINAL). BUT THEY ARE **LENSES**, WAYS OF **THINKING**, I HAVE DERIVED THROUGH A SUSTAINED, ITERATIVE DESIGN PROCESS AND THROUGH HONING MY OWN STYLE AND AESTHETICS.

INDEED, DESIGN IS ALWAYS THE RESULT OF INTENTIONALITY **EXCEEDING** METHODOLOGY, AND IT IS ALWAYS DEPENDENT ON **CONTEXT**. THERE ARE NO EXPLICIT RECIPES FOR DESIGN. THERE ARE, HOWEVER, IDEAS, PATTERNS, AND RECURRING THEMES TO WORK WITH. AND THESE PRINCIPLES, WHICH RESIDE IN THAT TACIT DIMENSION BETWEEN THEORY AND PRACTICE, SHOULD BE VIEWED THROUGH THIS IMPLICIT LENS.

WITH THAT BRIEF INTERJECTION, HERE ARE THE REST OF THE VISUAL DESIGN **PRINCIPLES**, AND **CASE STUDIES** THAT EMBODY THEM.

PRINCIPLE 3.8 *INVITE THE SENSES*

MAKE THINGS POP! LEAD THE GAZE, MAKE USE OF **SURPRISE**, PROVIDE **NARRATIVE**.

PRINCIPLE 3.9 *MAKE IT **REAL-TIME**, WHENEVER POSSIBLE*

WE EXPERIENCE THE WORLD IN **REAL TIME**. TAKE ADVANTAGE OF THIS DYNAMISM IN DESIGN.

PRINCIPLE 3.10 *VISUALIZATION YIELDS UNDERSTANDING*

DESIGN THAT **VISUALIZES** A PROCESS, AN ALGORITHM, OR A SYSTEM HELPS ONE TO **COMPREHEND** ITS INNER WORKINGS. (IT FORCES THE DESIGNER TO HAVE UNDERSTOOD IT IN THE FIRST PLACE!)

PRINCIPLE 3.11 *PRAGMATICS: "IT'S GOTTA **READ**!"*

ELEMENTS SHOULD BE ARRANGED TO ALLOW US TO **MAKE SENSE** OF THEIR **PURPOSE** AND **RELATIONSHIPS** IN THE DESIGN. IT'S KINDA LIKE TELLING A JOKE -- YOU DON'T GET TO **EXPLAIN** IT AFTERWARD!

PRINCIPLE 3.12 *AESTHETICS: "MAKE ME **FEEL** SOMETHIN'!"*

POLISH AND **TECHNICAL PROWESS** ARE NOT AS IMPORTANT AS MAKING THE USER **FEEL SOMETHING** AS A RESULT OF THE **ENCOUNTER**. DESIGN WITH **AUTHENTICITY** AND **INTENTIONALITY**.

PRINCIPLE 3.13 *INVENT **ARTIFICIAL CONSTRAINTS***

CONSTRAINTS ARE THE UNDERLYING **RULES** THAT DEFINE HOW A SYSTEM WORKS, GIVE IT SHAPE, AND ULTIMATELY **SPECIFY** HOW A USER ENGAGES WITH IT. LIKE RULES IN A **GAME**, CONSTRAINTS MAKE A SYSTEM USEFUL, SAFE, FUN, INTERESTING. THEY PROVIDE THE BASIS FOR **CREATIVE AGENCY**.

PRINCIPLE 3.14 *SAVOR **STRANGE DESIGN LOOPS***

CONSTRUCT **FEEDBACK** AND **RECURSIVE CONNECTIONS** BETWEEN ELEMENTS. ADDRESS MULTIPLE DIMENSIONS WITH THE SAME ELEMENTS (E.G., THE PENCIL BAG/ZIPPER). BLUR THE DISTINCTION BETWEEN **MEDIUM** AND **MESSAGE**, USING SOME INTRINSIC PROPERTY OF THE DESIGN.

PRINCIPLE 3.15 *ITERATE!*

THERE IS NO **SUBSTITUTE** FOR **RELENTLESS ITERATION**. AS WE CONTINUALLY CREATE, BALANCE, ADAPT, EVALUATE, REJECT, RETRACE, WE **REFINE** OUR DESIGN'S **ALIGNMENT** WITH OUR NOTIONS OF **USEFULNESS** AND WITH HOW WE'D WANT IT TO MAKE US **FEEL**.

PRINCIPLE 3.16 *ORIGINALITY IS RECOMBINATION*

WE ARE **FASHIONED** BY WHAT WE **LOVE**, AND WE ARE **INSPIRED** FROM THE THINGS AROUND US (INCLUDING VIDEO GAMES, MOVIES, THE WAY A BRANCH SWAYS IN THE WIND). EMULATE ("STEAL LIKE AN ARTIST!") BUT ALSO ADAPT AND SPEAK THROUGH A VOICE OF YOUR OWN. **ORIGINALITY** IS THE **RECOMBINATION** OF THE THINGS WE FIND **BEAUTIFUL** AND OUR AUTHENTIC PREFERENCES.

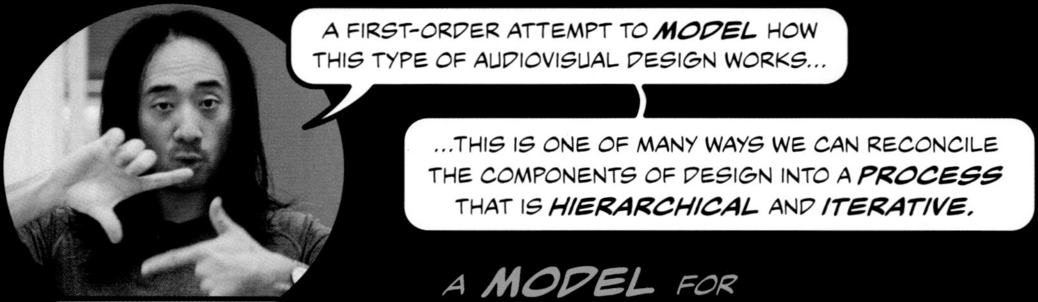

A FIRST-ORDER ATTEMPT TO **MODEL** HOW THIS TYPE OF AUDIOVISUAL DESIGN WORKS...

...THIS IS ONE OF MANY WAYS WE CAN RECONCILE THE COMPONENTS OF DESIGN INTO A **PROCESS** THAT IS **HIERARCHICAL** AND **ITERATIVE.**

MODEL 3.17

A MODEL FOR
MULTI-MODAL DESIGN

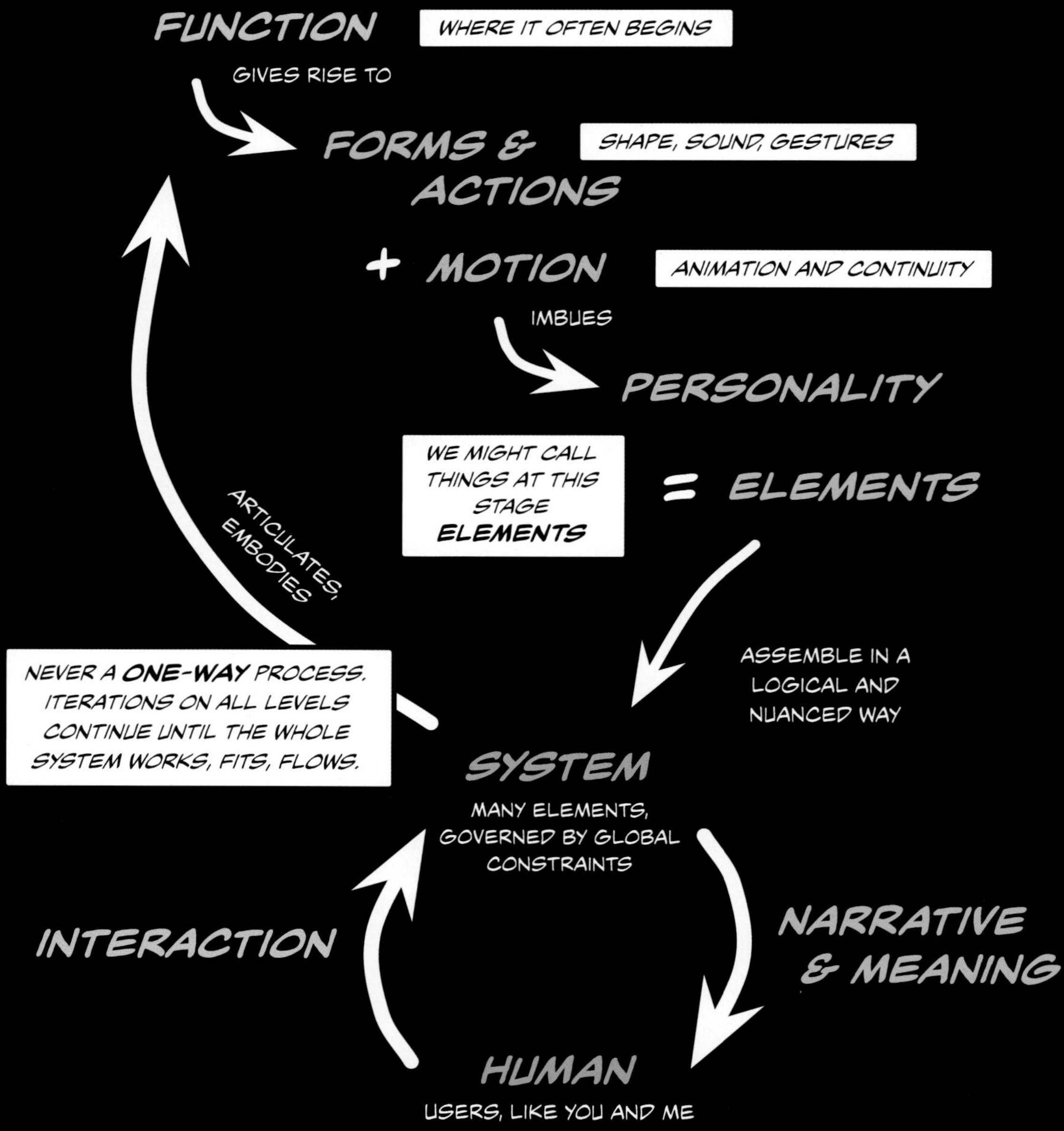

FUNCTION

WHERE IT OFTEN BEGINS

GIVES RISE TO

FORMS &
ACTIONS

SHAPE, SOUND, GESTURES

+ MOTION

ANIMATION AND CONTINUITY

IMBUES

PERSONALITY

WE MIGHT CALL THINGS AT THIS STAGE **ELEMENTS**

= ELEMENTS

ARTICULATES, EMBODIES

ASSEMBLE IN A LOGICAL AND NUANCED WAY

NEVER A **ONE-WAY** PROCESS. ITERATIONS ON ALL LEVELS CONTINUE UNTIL THE WHOLE SYSTEM WORKS, FITS, FLOWS.

SYSTEM
MANY ELEMENTS, GOVERNED BY GLOBAL CONSTRAINTS

INTERACTION

NARRATIVE
& MEANING

HUMAN
USERS, LIKE YOU AND ME

THIS TACIT **MODEL** IS A RECURRING PATTERN IN MY DESIGN. WE CAN USE THIS MODEL TO BREAK DOWN THE DESIGN OF **OCARINA**, INTO A TYPE OF CONCEPTUAL DESIGN PLAN (ON THE LEFT) AND ITS INTERPRETATION AND **INSTANTIATION** (ON THE RIGHT).

(PRE-FUNCTION) **?** *WHAT* TO DESIGN

FUNCTION AN **OCARINA**

FORMS **CIRCULAR** FORMS; RINGS, SOLIDS, TRANSLUCENT HALOS

+

MOTION

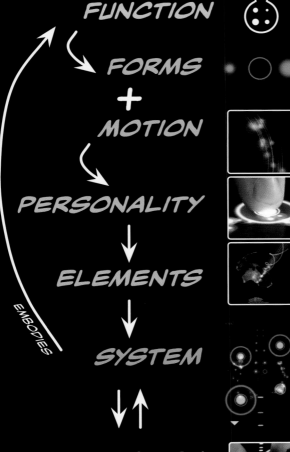

PARTICLES **SWARM** AND **GLOW** AS A RESPONSE TO **PHYSICAL INTERACTION** (BLOWING); BUTTONS ORGANICALLY EXPAND AND CONTRACT WHEN PRESSED. A DOUBLE HELIX OF MUSICAL FRAGMENTS EMANATE FROM THE **OCARINA** GLOBE.

PERSONALITY

ELEMENTS

EMBODIES

SYSTEM

VISUAL, SONIC, AND INTERACTIVE ELEMENTS MAKE UP OCARINA AS A **SINGLE ARTIFACT.** THIS IS ITS FUNCTIONAL-AESTHETIC UNITY.

INTERACTION

NARRATIVE

THE PLAYER-LISTENER **NARRATIVE** CONNECTS THE **FUNCTIONAL** AND THE **AESTHETIC** TOGETHER, INVITING THE USER TO ENGAGE **OCARINA PHYSICALLY** -- AS A TOY AND INSTRUMENT -- AND **EMOTIONALLY** AS PART OF A **SOCIAL** EXPERIENCE.

HUMAN

*NOT AN EXPLICIT RECIPE TO FOLLOW, BUT AN **IMPLICIT** PATTERN TO THINK WITH...*

SNDPEEK: A SIMPLE AUDIO VISUALIZER!

LOOKING AT SOUND IN REAL TIME!

THIS **TIME-DOMAIN WAVEFORM** PLOTS OUT THE **FLUCTUATIONS** IN AIR PRESSURE, WHICH RESULTS IN THE PHENOMENON WE CALL **SOUND.** IT IS **ANIMATED** BY REPEATEDLY DRAWING A LINE THROUGH THE MOST RECENT 1024 DIGITAL AUDIO SAMPLES (ABOUT **23MS** OF SOUND HERE), CAPTURED IN REAL TIME BY THE MICROPHONE.

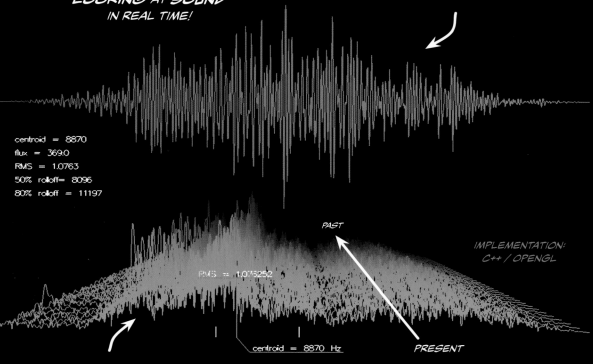

centroid = 8870
flux = 369.0
RMS = 1.0763
50% rolloff= 8096
80% rolloff = 11197

PAST

IMPLEMENTATION:
C++ / OPENGL

RMS = 1.076252

centroid = 8870 Hz

PRESENT

THE GREEN **SPECTRUM VISUALIZER** SHOWS THE **DISTRIBUTION** OF **FREQUENCIES** IN A WINDOW OF SOUND. LOWER FREQUENCIES ARE TO THE LEFT, HIGHER ONES TO THE RIGHT.

SNDPEEK RELIES ON **SMOOTH ANIMATION** AND **MOTION** TO CONVEY INFORMATION.

OLDER SPECTRA **CASCADE** INTO THE BACKGROUND IN A 3D **WATERFALL PLOT** ANIMATING THE RECENT HISTORY OF THE SPECTRUM.

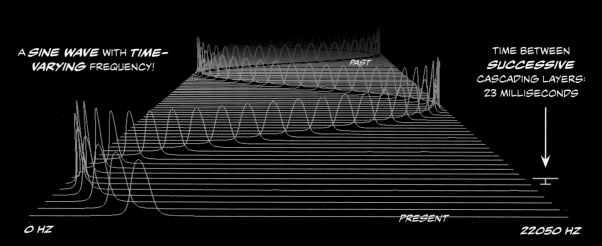

A **SINE WAVE** WITH **TIME-VARYING** FREQUENCY!

PAST

TIME BETWEEN **SUCCESSIVE** CASCADING LAYERS: 23 MILLISECONDS

PRESENT

0 HZ

22050 HZ

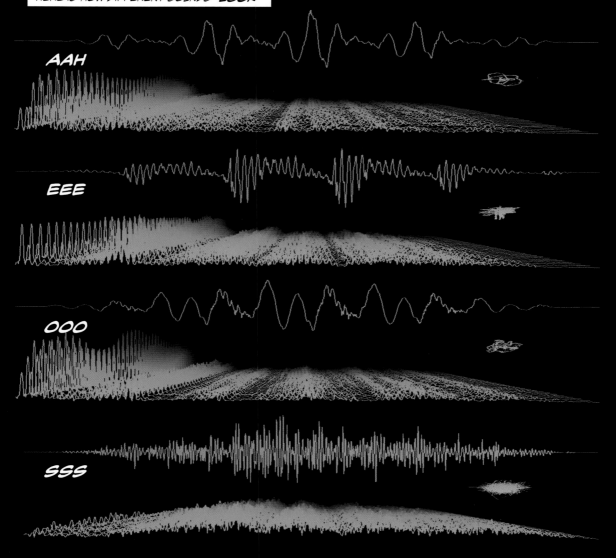

AAH

EEE

OOO

SSS

SNDPEEK BEGAN AS A PERSONAL PROGRAMMING PROJECT TO MAKE A SIMPLE **TEACHING TOOL** TO **VISUALIZE** AUDIO WAVEFORMS AND SPECTRA USING **SHORT-TIME FOURIER TRANSFORMS** (STFT'S) IN REAL TIME.

WHILE WORKING ON **SNDPEEK** AT A FAMILY GATHERING, I NOTICED HOW SMALL CHILDREN WERE DRAWN TO THE VISUALS. ONCE THE KIDS REALIZED **SNDPEEK RESPONDED** TO THEIR **VOICES** IN **REAL TIME**, THEY BROKE INTO EVERY MANNER OF LAUGHING, SHRIEKING, SQUAWKING... JUST TO SEE WHAT SOUND LOOKED LIKE.

WITHOUT MY PROMPTING, THE CHILDREN INTUITED THAT **HIGHER FREQUENCIES** APPEARED TO THE RIGHT. THERE EVEN EMERGED, BRIEFLY, A **COMPETITION** OF WHO COULD **SCREAM HIGHER** (IN BOTH LOUDNESS AND PITCH), BEFORE THE ADULTS CAME IN AND PUT AN END TO THE ENTERPRISE.

I QUIETLY (AND GLEEFULLY) OBSERVED THE PROCEEDINGS, HAVING WITNESSED THE INNATE FASCINATION HUMANS FEEL TOWARD AUDIOVISUAL PROCESSES THAT REACT TO US.

EVEN WITH SIMPLE TOOLS, THERE ARE **AESTHETICS** TO BE CONSIDERED.

THE **MINIMAL** LINE VISUALS ARE AN HOMAGE TO THE EARLY DAYS OF **VECTOR GRAPHICS** AND **MONOCHROME** DISPLAYS (GREEN ON BLACK).

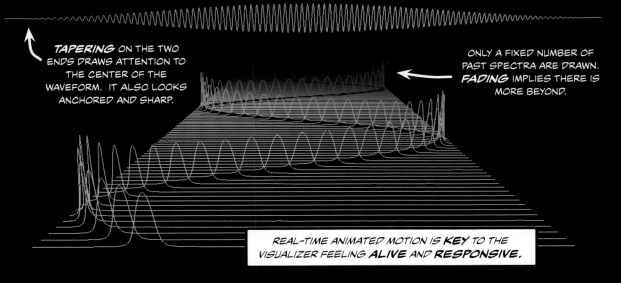

TAPERING ON THE TWO ENDS DRAWS ATTENTION TO THE CENTER OF THE WAVEFORM. IT ALSO LOOKS ANCHORED AND SHARP.

ONLY A FIXED NUMBER OF PAST SPECTRA ARE DRAWN. **FADING** IMPLIES THERE IS MORE BEYOND.

REAL-TIME ANIMATED MOTION IS **KEY** TO THE VISUALIZER FEELING **ALIVE** AND **RESPONSIVE.**

THE **WATERFALL PLOT** STORES AND DRAWS THE **N MOST RECENT** SPECTRAL **FRAMES.**

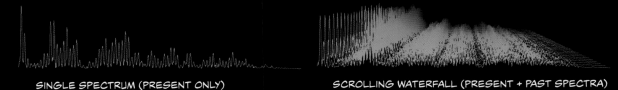

SINGLE SPECTRUM (PRESENT ONLY)

SCROLLING WATERFALL (PRESENT + PAST SPECTRA)

WE CAN MOVE AND ROTATE TO OBSERVE FROM DIFFERENT VANTAGE POINTS. FOR EXAMPLE, WE CAN ROTATE AROUND TO **BEHIND** AND **UNDER** THE WATERFALL!

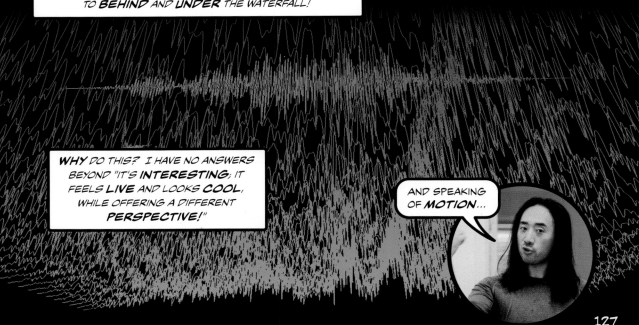

WHY DO THIS? I HAVE NO ANSWERS BEYOND "IT'S **INTERESTING**; IT FEELS **LIVE** AND LOOKS **COOL,** WHILE OFFERING A DIFFERENT **PERSPECTIVE!**"

AND SPEAKING OF **MOTION**...

THE ILLUSION OF MOTION

生命在于运动。

LIFE IS MOVEMENT.

-- CHINESE SAYING

MOTION IS THE PROCESS OF **CHANGE** THAT GIVES RISE TO EXPERIENCE. IT IS THE **TEMPORAL** CORE THAT PERVADES ANIMATION, SOUND AND MUSIC, AND PHYSICAL GESTURE.

WE'VE PONDERED THE **NATURE** OF **CHANGE** SINCE ANCIENT TIMES. GREEK PHILOSOPHER **HERACLITUS** OF **EPHESUS** (CIRCA 535-475 B.C.E.) BELIEVED THAT LIFE IS **FLUX** AND EVERYTHING **FLOWS**, WHILE **ZENO** OF **ELEA** (CIRCA 490-430 B.C.E.) DOUBLED DOWN ON AN OPPOSITE IDEA: ALL MOTION IS BUT AN **ILLUSION**. ZENO EMBODIED HIS VIEWS IN THE FORMULATION OF HIS FAMOUS **PARADOXES**:

CONSIDER AN **ARROW** MOVING **IN FLIGHT**. FOR ANY **INSTANT** IN TIME (DURATION OF 0), THE ARROW OCCUPIES A SINGLE SPACE (FOR IT CANNOT BE IN TWO PLACES AT ONCE) AND IS STATIONARY (BECAUSE THERE IS NO TIME FOR IT TO MOVE). IF IT IS SIMILARLY MOTIONLESS AT EVERY INSTANT, AND TIME IS ENTIRELY MADE OF INSTANTS, THEN IT'S **IMPOSSIBLE** FOR THE ARROW TO MOVE.

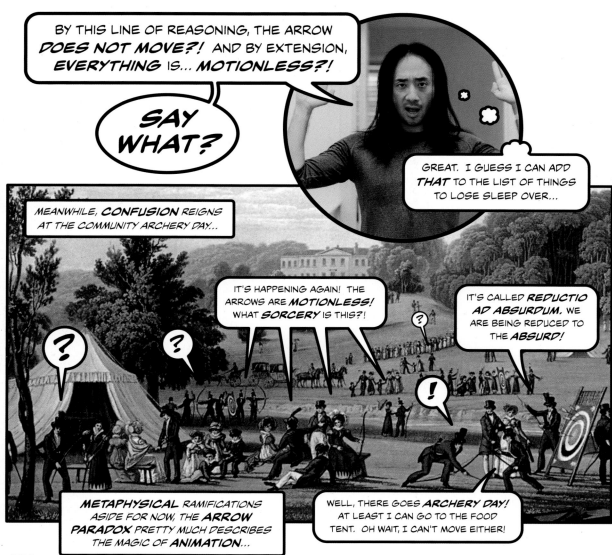

ANIMATION IS THE TIME-BASED ART OF CREATING THE **PERCEPTION** OF MOTION FROM **SEQUENCING** STATIC IMAGES. IN FILM, TELEVISION, COMPUTER GRAPHICS, MOTION ACTUALLY **IS** AN **ILLUSION.**

THERE EXISTS A **PERCEPTUAL THRESHOLD** AROUND 20 TO 30 TIMES PER SECOND (HERTZ), BELOW WHICH WE PERCEIVE **INDIVIDUAL EVENTS** AND BEYOND WHICH WE PERCEIVE **CONTINUUM** OF MOTION. MOVIES (24-48 FRAMES PER SECOND), TELEVISION (~30 FPS), AND COMPUTER GRAPHICS ALL WORK ON THIS PRINCIPLE.

COMPUTER-BASED ANIMATION, LIKE HAND-DRAWN ANIMATION, HAPPENS IN **DISCRETE TIMESTEPS.** THE **FRAMERATE** RANGES FROM 20HZ (THINGS START LOOKING **CHOPPY** BELOW THIS RATE) UP TO 90HZ (GAMES REQUIRE HIGH FRAMERATE TO LOOK FLUID AND FEEL RESPONSIVENESS; VIRTUAL REALITY HEAD-MOUNTED DISPLAYS OPERATE AT 70HZ OR HIGHER).

KA--

BOOM!

WHOA.

COMICS ARE A KIND OF ANIMATION, HAPPENING **SEQUENTIALLY** BUT AT **LESS UNIFORM** TIMESTEPS. OUR **MIND** FILLS IN THE GAPS BETWEEN FRAMES.

EVERYTHING **FLOWS!**

HERACLITUS

INTERACTIVE AUDIOVISUAL DESIGN IS VERY MUCH THE CRAFTING OF MOTION IN **REAL TIME.**

IN MANY CASES, WE CAN DIRECTLY COMPUTE INFORMATION FOR EACH TIMESTEP FROM **SIMULATION** (FOR EXAMPLE, A PLANET ROTATING AROUND A SUN).

IN OTHER SITUATIONS, OBJECTS NEED TO BE SEEN AS MOVING **BETWEEN** DISCRETE STATES, IN WHICH CASE IT'S NECESSARY TO **INTERPOLATE** TO CREATE SMOOTHNESS OF CHANGE IN COLOR, SHAPES, OPACITY -- **ANY** CONTROLLABLE PARAMETER.

THIS BRINGS US TO **ANOTHER** OF ZENO'S PARADOXES!

129

1/2   **1/4**   **1/8**   ...

GOAL

ET'S SAY WE WANT TO GET FROM OUR CURRENT POSITION (CALL IT X) TO A **GOAL.** WE MUST FIRST RAVEL A **PERCENTAGE** OF THE DISTANCE (E.G., HALFWAY). TO TRAVERSE THE REMAINING HALF, WE UST FIRST COMPLETE HALF OF THAT (OR 1/4 OF THE TOTAL DISTANCE). UPON REACHING THAT POINT, E MUST THEN COVER HALF OF THE REMAINING DISTANCE, AND SO ON... THERE ARE **INFINITE** SUCH TEPS -- AND, BY THIS REASONING, WE WOULD NEVER REACH THE GOAL!

S IS ANALOGOUS TO THE **ACHILLES** VS. E **TORTOISE** RACE PARADOX, ALSO ED BY ZENO: HAVING GIVEN THE TORTOISE EAD START, THE FASTER ACHILLES MUST ST REACH THE TORTOISE'S PRESENT ITION. BUT BY THE TIME HE GETS THERE TORTOISE (ALBEIT SLOWER) HAS **MOVED** RWARD. BY THIS REASONING, ACHILLES NEVER CATCH UP TO THE TORTOISE.

SHEESH! AND PEOPLE THINK **I'M** INTENSE! THAT ZENO IS POSITIVELY OFF HIS ROCKER!

ZENO'S PARADOX IS A CLEVER CONSTRUCTION THAT PRESENTS A **MISMATCH** BETWEEN OUR INTUITION OF MOTION VS. A WAY TO THINK ABOUT IT. THERE IS **NO** PARADOX HERE. ZENO'S CONSTRUCTION FORCES US TO THINK ABOUT **DISTANCE** IN INCREASINGLY SMALLER **TIMESTEPS** (BY A FACTOR OF 2 IN THE DICHOTOMY PARADOX)! **TRICKSTER!** THE DISTANCE **AND** TIME BETWEEN EACH ITERATION IS HALF THE PREVIOUS, RESULTING IN THE SERIES 1/2 + 1/4 + 1/8...

THIS IS A **CONVERGING SERIES**, WHOSE SUM APPROACHES A **FINITE NUMBER.** IN THE **LIMIT**, THIS SERIES ADDS UP TO EXACTLY **ONE!** WE ARE GEOMETRICALLY "**SLOWING DOWN**" TIME UP TO THE VERY MOMENT WHEN WE ACTUALLY REACH THE GOAL (OR WHEN THE FASTER OBJECT CATCHES UP WITH THE SLOWER ONE). BUT OVERALL WE ARE NONETHELESS CONSIDERING A **FINITE** DISTANCE AND AMOUNT OF TIME.

IT IS NOT HARD TO REFUTE **ZENO'S PARADOX.** YET DOING SO MIGHT BE MISSING A LARGER POINT ABOUT THE **MALLEABILITY** OF LANGUAGE, LOGIC, AND OUR COMPREHENSION OF THE WORLD. THIS TOUCHES ON NOTIONS OF INFINITY, CONVERGENT SERIES -- MUCH OF THE FOUNDATION OF CALCULUS!

FOR OUR PURPOSES OF CRAFTING MOTION, OUR P INSPIRATION FOR A PARTICULARLY USEFUL INTERPOLA I'LL CALL **ZENO'S INTERPOLATOR!** HERE

ZENO'S INTERPOLATOR

AN **INTERPOLATOR** COMPUTES **INTERMEDIATE** VALUES BETWEEN ESTABLISHED DATA POINTS. THIS SIMPLE IDEA GIVES RISE TO THE ILLUSION OF **CONTINUITY** AND **SMOOTHNESS** OF MOTION (BROADLY DEFINED) OVER TIME, ESPECIALLY WHEN THE KNOWN DATA ARE DISCRETE AND **SPARSE.**

TIME = 0

★
x

● Y

LET'S SAY WE HAVE A **VALUE** AT X THAT NEEDS TO TRAVEL, SMOOTHLY, TO **TARGET** AT Y.

SINCE WE ANIMATE MOTION FRAME BY FRAME, WE CAN USE AN **INTERPOLATOR** TO CALCULATE THE **INTERMEDIATE** POSITION VALUE AT EACH TIMESTEP ON ITS WAY TO Y. OUR APPROACH HERE ADVANCES THE VALUE TOWARD THE GOAL BY A **PERCENTAGE** OF THE **REMAINING DISTANCE** TO THE TARGET:

% OF REMAINING
DISTANCE TO Y

TIME = 1

★ - - - - - - - - - - - - - - - → ★
x
PREVIOUS X
LOCATION

● Y

% OF REMAINING
DISTANCE

TIME = 2

★ - - - - - - - - → ★
x
PREVIOUS X
LOCATION

● Y

THE VALUE X WILL ALWAYS **APPROACH** THE TARGET Y AT THE **FRAMERATE.** NOTE THAT THE VALUE X CAN BE MAPPED TO ANY **PARAMETER**, INCLUDING POSITION, FREQUENCY, TRANSPARENCY, COLOR COMPONENT, SIZE, ROTATION -- REALLY **ANYTHING** THAT CAN TAKE ON A **CONTINUOUS** RANGE OF VALUES!

WE USE **THREE NUMBERS** TO IMPLEMENT THIS INTERPOLATOR: (1) THE **CURRENT VALUE**, (2) THE **TARGET,** AND (3) THE **PERCENTAGE PROGRESS** (SLEW). AT EACH TIMESTEP, WE UPDATE THE CURRENT VALUE BY A PERCENTAGE OF THE CURRENT **DIFFERENCE** BETWEEN IT AND THE GOAL. THAT'S ALL! BY THE WAY, WHEN THE SLEW IS 0.5, THEN WE ESSENTIALLY HAVE THE DICHOTOMY PARADOX! BELOW IS THE **EQUATION** TO UPDATE THE VALUE AT EACH FRAME:

NEW VALUE = (TARGET-VALUE) * SLEW + VALUE

IT'S IMPORTANT TO NOTE THAT THE VALUE OF THE PARAMETER WILL **ALWAYS APPROACH** THE GOAL, BUT NEVER QUITE REACH IT (HENCE **ZENO'S** INTERPOLATOR)! BECAUSE IT'S A PERCENTAGE, THE CURRENT VALUE APPROACHES THE TARGET **FASTER** WHEN IT'S **FARTHER** AWAY AND MORE **SLOWLY** AS IT **NEARS** THE TARGET. THIS IS DECIDEDLY **NON-LINEAR**, SO IT DOESN'T WORK WELL FOR EVERYTHING. WHAT IS IT GOOD FOR? IN PRACTICE, IT'S A **SIMPLE, RESPONSIVE,** AND **ROBUST** (NO SPECIAL CASES) WAY TO CRAFT MOTION (VISUAL AND AUDIO PARAMETERS), IN **DISCRETE** FRAME-BY-FRAME SYSTEMS LIKE COMPUTER GRAPHICS.

WITHOUT INTERPOLATION: WE *DIRECTLY* SET THE VALUE FROM CHANGES IN THE SYSTEM STATE. CHANGES AT A DIFFERENT RATE THAN THE ANIMATION FRAMERATE MAY RESULT IN *UNEVENNESS.*

VS.

STATE CHANGE
(FROM USER INPUT OR INTERNAL SYSTEM COMPUTATION)

→ DIRECTLY SETS →

VALUE OF INTEREST
(E.G., COLOR, LOCATION, FREQUENCY)

WITH INTERPOLATION: STATE CHANGES *INDIRECTLY* INFLUENCE THE ACTUAL FRAME-TO-FRAME VALUE, BY UPDATING AN INTERPOLATOR *TARGET* AND SOMETIMES ALSO THE *SLEW* (TO CONTROL HOW AGGRESSIVELY THE INTERPOLATION HAPPENS). THE INTERPOLATOR IS *ALWAYS* RUNNING (AT A *CONSTANT* FRAMERATE) TO COMPUTE THE ONGOING VALUE OF INTEREST. *FRAMERATE* IS BROADLY DEFINED HERE. IT CAN BE THE VISUAL REFRESH RATE OR THE RATE OF COMPUTING A WINDOW OF AUDIO.

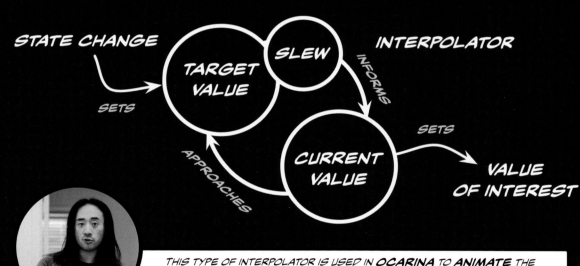

STATE CHANGE — SETS → TARGET VALUE — SLEW — INTERPOLATOR

INFORMS

CURRENT VALUE — SETS → VALUE OF INTEREST

APPROACHES

THIS TYPE OF INTERPOLATOR IS USED IN **OCARINA** TO **ANIMATE** THE EXPANSION AND CONTRACTION OF THE BUTTONS BETWEEN THE BINARY STATES OF "PRESSED" AND "RELEASED" -- AND PRETTY MUCH ANYTHING THAT MOVES.

TARGET SET TO **1.0** **SLEW** SET TO **.2**   FASTER, SMOOTH EXPANSION!

RING SIZE TARGET SET TO **0.4** **SLEW** SET TO **.1**   SLOWER RETURN TO UNPRESSED STATE

VS.

PRESSED

RELEASED

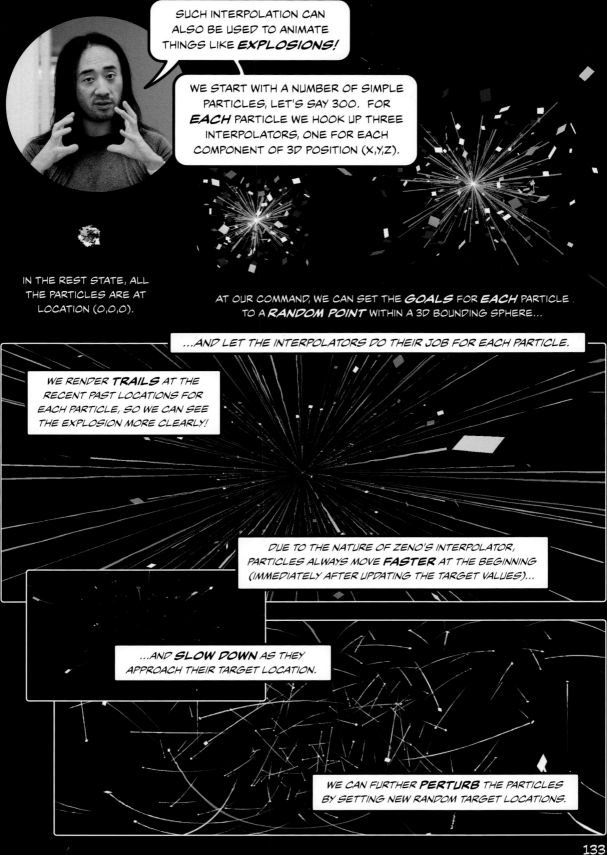

SUCH INTERPOLATION CAN ALSO BE USED TO ANIMATE THINGS LIKE **EXPLOSIONS!**

WE START WITH A NUMBER OF SIMPLE PARTICLES, LET'S SAY 300. FOR **EACH** PARTICLE WE HOOK UP THREE INTERPOLATORS, ONE FOR EACH COMPONENT OF 3D POSITION (X,Y,Z).

IN THE REST STATE, ALL THE PARTICLES ARE AT LOCATION (0,0,0).

AT OUR COMMAND, WE CAN SET THE **GOALS** FOR **EACH** PARTICLE TO A **RANDOM POINT** WITHIN A 3D BOUNDING SPHERE...

...AND LET THE INTERPOLATORS DO THEIR JOB FOR EACH PARTICLE.

WE RENDER **TRAILS** AT THE RECENT PAST LOCATIONS FOR EACH PARTICLE, SO WE CAN SEE THE EXPLOSION MORE CLEARLY!

DUE TO THE NATURE OF ZENO'S INTERPOLATOR, PARTICLES ALWAYS MOVE **FASTER** AT THE BEGINNING (IMMEDIATELY AFTER UPDATING THE TARGET VALUES)...

...AND **SLOW DOWN** AS THEY APPROACH THEIR TARGET LOCATION.

WE CAN FURTHER **PERTURB** THE PARTICLES BY SETTING NEW RANDOM TARGET LOCATIONS.

133

"NEVER MISTAKE **MOTION** FOR **ACTION**."
-- ERNEST HEMINGWAY

IMBUING

PERSONALITY

ANIMATION ALONE IS NOT ENOUGH; VISUAL DESIGN IS ALSO ABOUT **PERSONALITY**, ARISING FROM **NUANCED** MOTION IN TIME-BASED MEDIUMS OR, IN **STILL** MEDIUMS, EXPRESSIVE NON-MOTION... LIKE THE GLOW OF THE NIGHT SKY ABOVE A BUSY STREET.

IN THE DOMAIN OF THE COMPUTER, THE BUILDING BLOCKS ARE, BY DEFAULT, **SIMPLISTIC**, **RIGID**, AND **UNINTERESTING**. IT'S INCUMBENT UPON THE **DESIGNER** TO **IMBUE** PERSONALITY USING THESE VERY BUILDING BLOCKS, TO BREATHE **LIFE** INTO THE ARTIFACT.

VIBRANT **VISUAL PERSONALITIES** ARE PRESENT ALL AROUND US, IN EVERYDAY THINGS. WE HAVE BUT TO **PAUSE** AND **OBSERVE**...

LOOK AT THE WAY BRIGHT LIGHTS CREATE GLOWING **HALOS** AROUND THE EDGES OF THE BUILDINGS! THEY INSPIRE OUR EYES AND MIND TO FIND NARRATIVE AND VISUAL MEANING. LIKE THE CITY OF **NIGHT** AND **NEON** IN THE MOVIE **BLADE RUNNER**, THERE IS A SENSE OF THE ETHEREAL AND OF POSSIBILITY HERE.

LOOK AT ALL THIS **GLOW!**

HI THERE!

你好!

FUNCTIONALLY, VISUAL DESIGN DIRECTS OUR EYES AND ALLOWS US TO **MAKE SENSE** OF **SPACE** AND THE **RELATIONSHIP** OF THINGS IN IT. IN OTHER WORDS, "IT'S GOTTA **READ!**"

AESTHETICALLY, VISUAL DESIGN INSTILLS MOOD, A VIVID MENTAL VIBE OF JUST **HOW** SOMETHING IS, ALL FROM A GLANCE...

MOTION BLUR
FROM MOVING!

WHY SIMPLY "MOVE" WHEN YOU CAN...

FLOW!

IT'S HELPFUL TO DESIGN IN **EXPRESSIVE VERBS**, EACH WITH ITS OWN CHARACTER AND NUANCE. POLISH IS SECONDARY. MOST IMPORTANTLY, "MAKE ME **FEEL** SOMETHING!"

LEAP

SCURRY   WOBBLE   FLUTTER   SCUTTLE

BUSTLE

BOUNCE   WADDLE   SHUFFLE

PERCOLATE   DRIP

MELT

HOVER   WISK   EXTRUDE

ROLL   TROT   SPLASH

FLOAT   STREAK   GUSH

LOOM   SLIDE   TIPTOE   SECRETE

TUMBLE   TRICKLE   OOZE

GALLOP   LEAK   SWELL

LUMBER   GLIDE   FLOOD

SLINK   SLIDE   VENT

SLITHER   GLARE   DRENCH

WRITHE   EBB

BILLOW

DON'T JUST SIT THERE...   COIL

DOESN'T **OOZE** EVEN **SOUND** EXPRESSIVE? SLOW SLIMY FLOW...

GLOW!

FLASH   SHINE   TWINKLE

SHIMMER   SMOKE   BLAZE   FLARE

GLISTEN   GLEAM

BLOOM   BREATHE

RADIATE   SMOLDER   GLITTER

ERUPT   RIPPLE   SPARKLE

FLICKER

TWIRL

DON'T MERELY "HAPPEN" BUT...

EXPLODE!

SHATTER

MOVING IN PLACE?

OSCILLATE!

SCATTER   QUIVER

VOMIT   PIVOT   SPIRAL

BURST   PALPITATE

FLAP   SPIN   THROB

CRUMPLE   TREMBLE   FLUCTUATE

IMPLODE   UNDULATE   TWIST

SWIVEL   ROTATE

PULSATE

CASCADE

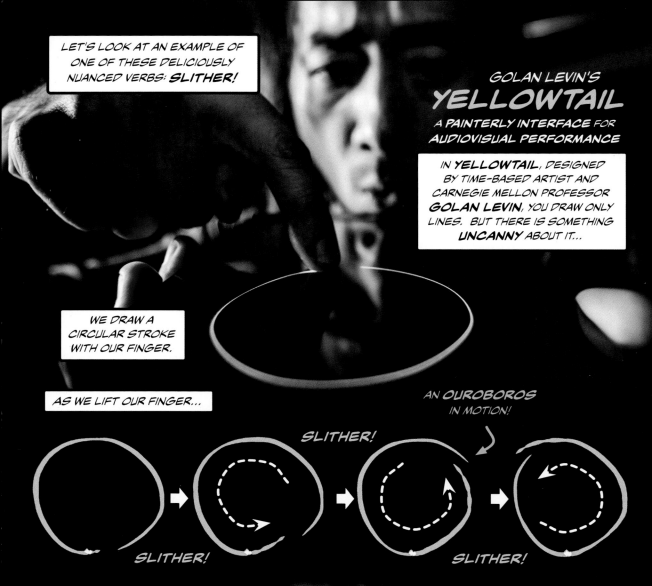

LET'S LOOK AT AN EXAMPLE OF ONE OF THESE DELICIOUSLY NUANCED VERBS: **SLITHER!**

GOLAN LEVIN'S
YELLOWTAIL
A PAINTERLY INTERFACE FOR AUDIOVISUAL PERFORMANCE

IN **YELLOWTAIL**, DESIGNED BY TIME-BASED ARTIST AND CARNEGIE MELLON PROFESSOR **GOLAN LEVIN**, YOU DRAW ONLY LINES. BUT THERE IS SOMETHING **UNCANNY** ABOUT IT...

WE DRAW A CIRCULAR STROKE WITH OUR FINGER.

AS WE LIFT OUR FINGER...

AN **OUROBOROS** IN MOTION!

SLITHER!

SLITHER!

SLITHER!

...THE STROKE COMES TO **LIFE!** WHERE WE BEGAN BECOMES THE HEAD, AND WHERE WE LIFTED OUR FINGER IS THE TAIL.

IT'S AS IF THE STROKE **REMEMBERED** HOW WE DREW IT AND USED THAT INFORMATION TO GIVE ITSELF PERSONALITY!

THIS IS YOUR DRAWING PROGRAM ON **DRUGS!**

WHOA.

THAT'S PERSONALITY!

THE STROKE IS NOT JUST MOVING, IT'S **SLITHERING**, WITH THE **SPEED** AND **MANNERISM** WITH WHICH WE DREW IT!

SLITHER!

THE MOVEMENT IS MUCH MORE **NUANCED** THAN SIMPLE TRANSLATION AND ROTATION; IT SLITHERS ALONG THE **PATH** OF **HOW** I DREW THE STROKE, THE NUANCES OF WHICH ARE REFLECTED AND EXTRAPOLATED IN ITS **CONTINUING** MOVEMENT!

SLITHER...

SLITHER...

SLITHER...

THIS MOVEMENT IS ACHIEVED BY TRACKING BOTH **TEMPORAL** AND **SPATIAL** INFORMATION ACROSS THE FULL DRAWING GESTURE.

SLITHER...

THE **NUANCES** AND **IMPERFECTIONS** OF OUR DRAWING ARE TRANSFORMED INTO A MOVING, **WRITHING**, **SLITHERING** ORGANIC THING.

SLITHER...

THE MAGIC OF YELLOWTAIL BEGINS WITH ITS INITIAL **DELIGHT** AND **SURPRISE**: IT FEELS LIKE A NORMAL PAINT PROGRAM UNTIL WE LIFT OUR FINGER AND THE STROKE COMES TO LIFE -- AND (THIS IS CRUCIAL) IT DOES SO IN A WAY THAT **MIMICS HOW** WE DREW IT IN THE FIRST PLACE! IT FEELS SO ORGANIC THAT IT'S **UNSETTLING!**

SLITHER...

WRAPS AROUND

HOW **FAST** OR **SLOW** WE DRAW EACH STROKE TRANSLATES TO ITS **THICKNESS** AND **SPEED.**

THERE'S SUCH A **SATISFYING**, CORRESPONDENCE BETWEEN **HOW** I DRAW AND THE RESULT!

A **MASTERFUL** MULTI-MODAL DESIGN OF GRAPHICS AND GESTURE.

SCURRY!

137

NOW WE COME TO A PARTICULARLY MAGICAL THING ABOUT DESIGN: THE POSSIBILITY FOR *STRANGE DESIGN LOOPS!*

THERE IS SOMETHING *ARTFUL* ABOUT DESIGN THAT INTENTIONALLY CONNECTS THE UNDERLYING *MEDIUM* TO THE DESIGN ITSELF. LIKE A PENCIL BAG MADE ENTIRELY OF ITS OWN ZIPPER, IT'S AN *AESTHETIC LEAP* THAT "CLOSES THE LOOP" BETWEEN MEDIUM AND MESSAGE.

IT'S LIKE STEPPING BACK AND DESIGNING ONE MORE LEVEL AROUND THE THING, USING SOME ASPECT *INHERENT* TO THE THING ITSELF. IT CAN BE MIND-BOGGLINGLY *META.*

THE SEARCH FOR
STRANGE DESIGN LOOPS
WHERE THE **MEDIUM** MEETS THE **MESSAGE**

STRANGE DESIGN LOOPS ENCODE AND ENACT NOTIONS OF SELF-REFERENCE, SELF-REFLEXIVITY, FEEDBACK, RECURSION, PARADOX. THEY'RE OFTEN PLAYFUL, EXPLOITING SOME *INHERENT* PROPERTY OF THE *MEDIUM.* *STRANGE LOOPS* ARISE AS *UNCANNY CONNECTIONS* BETWEEN FORM AND FUNCTION, WHERE ELEMENTS IN *CONCEPT* AND ELEMENTS IN *ACTUALITY* ARE IN CONVERSATION.

WHOA.

M. C. ESCHER'S *DRAWING HANDS* IS AN EXQUISITE EXAMPLE OF DESIGN IN *RECURSIVE NARRATIVE:* WHICH HAND STARTED DRAWING THE OTHER? DID THEY START FROM A BLANK PAGE? WHO DREW THE FIRST STROKE? HOW WAS THAT POSSIBLE WHEN THERE WEREN'T HANDS YET?!

THE APPARENT CONTRADICTION IS *RESOLVED* BY THE PRESENCE OF THE ARTIST AS A *THIRD* PARTY WHO DREW BOTH HANDS. THE STRANGE LOOP, NONETHELESS, IS SET IN MOTION. IT'S A DRAWING (MEDIUM) ABOUT DRAWING (MESSAGE)!

STRANGE LOOPS IN DESIGN GIVE YOUR *MIND* SOMEWHERE TO GO, A DESIGN CONSTRAINT AND NARRATIVE, A HIGHER STRUCTURE TO APPRECIATE AND PONDER.

THE LOGICALLY INCONCEIVABLE IS RENDERED TANGIBLE. DESIGNED TO EXIST *IN MEDIAS RES*, THIS STRANGE LOOP PROMPTS US TO PONDER TIME BEFORE AND AFTER THIS DEPICTED MOMENT.

ESCHER'S *ASCENDING AND DESCENDING:* AN EXQUISITE ARTIFACT OF VISUAL DESIGN

DOUGLAS HOFSTADTER'S BOOK: *GÖDEL, ESCHER, BACH* IS THE ORIGINAL TREATISE ON *STRANGE LOOPS,* AS AN EXPLORATION OF MIND AND THE NOTION OF SELF.

PENROSE TRIANGLE AND STAIRS -- CONSTRUCTIONS THAT MAKE USE OF *FORCED PERSPECTIVE*

DIFFERENT ANGLES

"IN THE END, WE ARE SELF-PERCEIVING, SELF-INVENTING, LOCKED-IN MIRAGES THAT ARE LITTLE MIRACLES OF SELF-REFERENCE."

-- DOUGLAS HOFSTADTER
I AM A STRANGE LOOP

THE INFINITE STAIRWAY HAS A SIBLING IN SOUND: **SHEPARD TONES.** FIRST DEVELOPED BY COGNITIVE SCIENTIST **ROGER SHEPARD,** THIS **AUDITORY ILLUSION,** LIKE THE ROTATING STRIPES ON A **BARBER POLE,** SEEMS TO ASCEND OR DESCEND INDEFINITELY!

INTERESTING ASIDE: IN **MEDIEVAL TIMES,** BARBERS ALSO PERFORMED MEDICAL PROCEDURES TO HEAL THE SICK, INCLUDING LEECHING, TOOTH EXTRACTION, AND **BLOODLETTING** (USED TO TREAT AN ALARMINGLY LARGE RANGE OF SICKNESSES, FROM SORE THROAT TO THE PLAGUE). THE **STRIPES** ON A ROTATING BARBER POLE SYMBOLIZE BLOOD; THE "CORRECT" SPIN DIRECTION IS THE ONE WHERE THE "BLOOD" APPEARS TO FLOW DOWNWARD.

UH... I JUST WANT A **HAIRCUT?**

YOU SURE? SPECIAL PROMOTION TODAY: **FREE BLOODLETTING** WITH EVERY HAIRCUT!

SHEPARD TONES ARE CONSTRUCTED WITH A SET OF **PARALLEL TONES,** TUNED **OCTAVES** APART. THE **INTENSITY** OF EACH TONE IS DETERMINED BY A **NORMAL** (GAUSSIAN) **CURVE.** ALL PARALLEL TONES MOVE TOGETHER, WHETHER ASCENDING OR DESCENDING. IN A DESCENDING SHEPARD TONE, LOWER TONES SUBTLY FADE OUT WHILE HIGHER TONES **SNEAKILY** FADE IN, ALL CONSTRUCTING THE ILLUSION OF A SINGLE TONE THAT DESCENDS FOREVER.

THIS IS A 3D PLOT OF THE FREQUENCY CONTENTS OF 30 SECONDS OF **CONTINUOUSLY DESCENDING** SHEPARD TONE (ALSO CALLED A SHEPARD-RISSET GLISSANDO).

EACH RED LINE REPRESENTS A SINGLE PARALLEL TONE. MANY TONES (AT VARYING INTENSITIES) ARE PRESENT AT ANY MOMENT IN TIME!

OUR **AUDITORY SYSTEM** INTERPRETS THESE COMPONENTS AS PARTIALS OF A **SINGLE** TONE. IT'S A CONVINCING ILLUSION!

WARMER COLORS DENOTE HIGHER INTENSITY

LOG-SCALE INTENSITY ENVELOPE

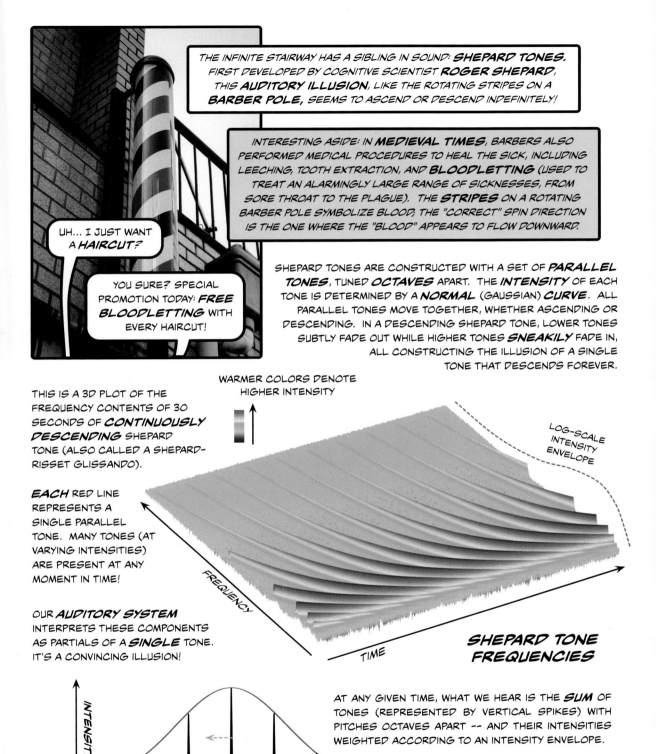

FREQUENCY

TIME

SHEPARD TONE FREQUENCIES

INTENSITY

PITCH

AT ANY GIVEN TIME, WHAT WE HEAR IS THE **SUM** OF TONES (REPRESENTED BY VERTICAL SPIKES) WITH PITCHES OCTAVES APART -- AND THEIR INTENSITIES WEIGHTED ACCORDING TO AN INTENSITY ENVELOPE.

OVER TIME, ALL PARALLEL TONES MOVE TOGETHER IN THE SAME DIRECTION (E.G., LEFT IF DESCENDING, AS SHOWN ABOVE). NEW TONES ARE INTRODUCED ON THE RIGHT, WHILE TONES BECOME GRADUALLY INAUDIBLE ON THE LEFT.

HERE IS ANOTHER STRANGE DESIGN
LOOP: A SONG *DESIGNED* TO BE
ENDLESSLY ITERATIVE...

START HERE

THIS IS THE SONG THAT NEVER ENDS, YES, IT GOES ON AND ON MY FRIEND! SOME PEOPLE STARTED SINGING IT NOT KNOWING WHAT IT WAS, AND THEY'LL CONTINUE SINGING IT FOREVER JUST BECAUSE:

SONG THAT NEVER ENDS
BY NORMAN MARTIN

THIS IS THE SONG THAT NEVER ENDS,
YES, IT GOES ON AND ON MY FRIEND!
SOME PEOPLE STARTED SINGING IT NOT KNOWING WHAT IT WAS,
AND THEY'LL CONTINUE SINGING IT FOREVER JUST BECAUSE...

———————————————————————

THERE IS A LOT
OF DESIGN TO
THIS SEEMINGLY
SIMPLE SONG!

DESIGN CONSTRAINTS (WHAT MUST IT DO?)
• MUST REPEAT WITHOUT LOGICAL ENDPOINT
• MELODY MUST MAKE MUSICAL SENSE
• LYRICS MUST MAKE SEMANTIC SENSE

DESIGN DECISIONS
• MUSICALLY CONNECT END TO BEGINNING
WITH AND WITHOUT A SENSE OF FINALITY
• LYRICALLY, THIS SONG IS ABOUT A SONG
THAT DOESN'T END (THAT'S SO META!)

FOR YOUR ENDLESS READING/SINGING PLEASURE,
ROTATE THE BOOK AS YOU READ AROUND THE RING
(YOU CAN THEREBY COMPLETE YET ANOTHER LAYER
OF THE STRANGE DESIGN LOOP).

DESIGN LOOPS DON'T HAVE TO BE **OVERT**, BUT THEY DO NEED TO BE **PLANNED**, LIKE A PUN, A PALINDROME, OR EVEN AN APPARENT CONTRADICTION. THEY ARE ELEMENTS OF DESIGN THAT HOLD MEANING SIMULTANEOUSLY ON MORE THAN ONE LEVEL.

Ceci n'est pas une pipe.

I AM NOT GE! I'M A PICTURE OF GE!

MAGRITTE'S **THE TREACHERY OF IMAGES** FAMOUSLY OBSERVES "THIS IS NOT A PIPE" -- INVITING THE VIEWER TO CONSIDER BOTH A PIPE AND AN IMAGE OF IT. THE ART BREAKS THE **FOURTH WALL** -- BETWEEN THE WORLD OF IMAGES AND THE WORLD IN WHICH THEY ARE VIEWED.

AS DESIGNERS, WE **SAVOR** SUCH UNCANNY CONNECTIONS. THERE IS SOMETHING **ELEGANT** AND **BEAUTIFUL** ABOUT **FEEDBACK** AND **ECONOMY** OF DESIGN. IT'S LIKE LOOKING INTO A HALL OF MIRRORS, ARRANGED WITH FINITE COMPONENTS, TO CONCEPTUALIZE **INFINITY**!

POETICALLY, STRANGE DESIGN LOOPS CAN COMMENT AND REFLECT ON **INTERRELATIONSHIPS** THAT CONNECT OUR PHYSICAL UNIVERSE TO ABSTRACT CONCEPTS, OR AN ARTIFACT TO ITS MEDIUM.

AN ART **INSTALLATION**... OF **FIRE**, CAPTURED IN **BIRDCAGES.**

ONLY THE **PEOPLE** ARE **IMMORTAL!!!**

FLAMES ACT AS OMNIDIRECTIONAL **LOUDSPEAKERS** OF SURPRISING LOUDNESS AND CLARITY.

EACH FLAME IS ELECTRICALLY **MODULATED** TO THE AUDIO OF IMPASSIONED SPEECH OF DICTATORS.

STALIN. HITLER. MUSSOLINI.

Жить стало лучше, товарищи*!!!*

THE FLAMES **BURN** AND **SPEAK** WITH PHYSICAL AND METAPHORICAL INTENSITY. THE **BIRDCAGES** REFERENCE THE AFTERLIFE, AS IF THE CAGES CAPTURED THE **VOICES** AFTER THEIR OWNERS HAVE GONE.

PAUL DEMARINIS'S
FIREBIRDS

THE UNCANNY CONNECTIONS BETWEEN FIRE AND LANGUAGE FEED INTO EACH OTHER IN STRANGE DESIGN LOOPS, **SUBTLE** AND **POIGNANT.**

I SUPPOSE **THIS BOOK** IS A KIND OF STRANGE DESIGN LOOP. IT'S A BOOK ON **DESIGN**, AND IT'S AS MUCH **WRITTEN** AS IT IS **DESIGNED**. IF THE **MEDIUM** IS THE **MESSAGE**, THEN DESIGNING A "STRANGE" COMIC BOOK IS BOTH AN **ILLOGICAL** EXTREME BUT ALSO A STRANGELY **LOGICAL** THING TO DO.

BY WRITING **THIS PAGE**, WE MAY HAVE CREATED YET ANOTHER STRANGE LOOP.

STRANGE DESIGN LOOPS ARE **SUCCINCT DESCRIPTIONS** OF INFINITY (WITHOUT HAVING TO BE INFINITE), DESIGN **NARRATIVES** THAT OFFER THE MIND "ROOM TO ROAM"! YOU REALLY ONLY NEED TWO INHERENT ELEMENTS WITHIN FORM OR FUNCTION TO CREATE A LOOP (SOMETIMES YOU ONLY NEED ONE!). THE IDEA IS TO SOMEHOW **EVOKE** A SENSE AND NOTION OF INFINITY, THROUGH THE ESTABLISHMENT OF CO-RELATIONSHIPS.

WHILE TRAVELING IN **INNER MONGOLIA**, I ONCE SAW A RED TRUCK IN THE DISTANCE...

ARE WE THERE YET?

CHILD, WE ARE **TRUCKS**, WE ARE **ALWAYS** ON THE ROAD.

THE **TRUCK** WAS CARRYING A **BABY TRUCK** -- AN OFFSPRING CARRIED BY ITS PARENT. THIS RECURSIVE SETUP PROVIDES A VISUAL AND ANTHROPOMORPHIC **CO-RELATIONSHIP**. THE TRUCK-RECURSION NARRATIVE MAKES YOU PAUSE, THINK, FEEL.

AUDIOVISUAL DESIGN IS AN ART OF COMPOSITION AND NARRATIVE. IT SHOULD STRIVE TO GIVE THE SENSES AND THE MIND "A PLACE TO GO."

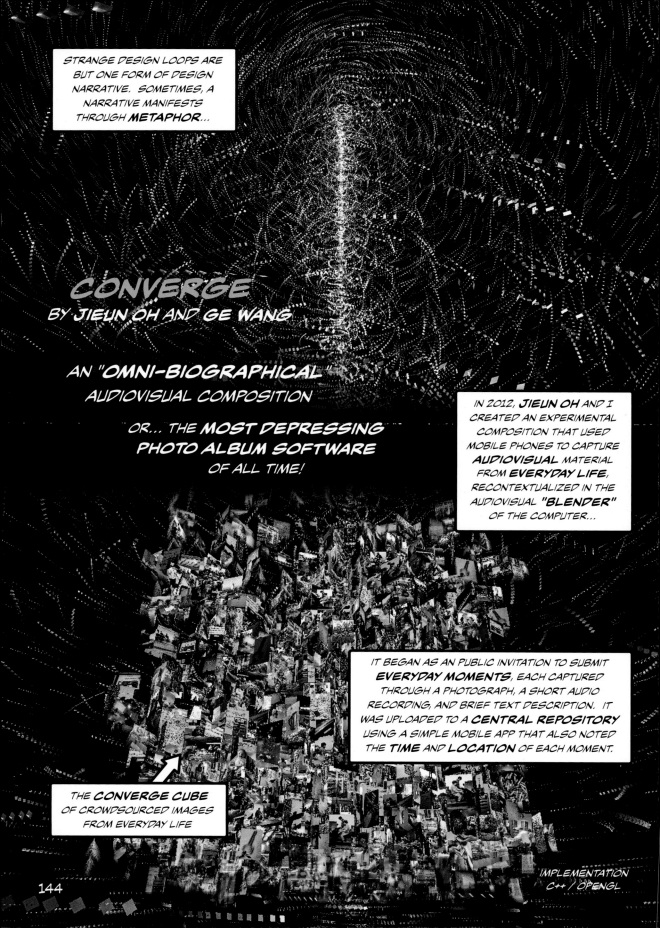

STRANGE DESIGN LOOPS ARE BUT ONE FORM OF DESIGN NARRATIVE. SOMETIMES, A NARRATIVE MANIFESTS THROUGH **METAPHOR**...

CONVERGE
BY *JIEUN OH AND GE WANG*

AN *"OMNI-BIOGRAPHICAL"* AUDIOVISUAL COMPOSITION

OR... THE **MOST DEPRESSING PHOTO ALBUM SOFTWARE** OF ALL TIME!

IN 2012, **JIEUN OH** AND I CREATED AN EXPERIMENTAL COMPOSITION THAT USED MOBILE PHONES TO CAPTURE **AUDIOVISUAL** MATERIAL FROM **EVERYDAY LIFE**, RECONTEXTUALIZED IN THE AUDIOVISUAL **"BLENDER"** OF THE COMPUTER...

IT BEGAN AS AN PUBLIC INVITATION TO SUBMIT **EVERYDAY MOMENTS**, EACH CAPTURED THROUGH A PHOTOGRAPH, A SHORT AUDIO RECORDING, AND BRIEF TEXT DESCRIPTION. IT WAS UPLOADED TO A **CENTRAL REPOSITORY** USING A SIMPLE MOBILE APP THAT ALSO NOTED THE **TIME** AND **LOCATION** OF EACH MOMENT.

THE **CONVERGE CUBE** OF CROWDSOURCED IMAGES FROM EVERYDAY LIFE

IMPLEMENTATION C++ / OPENGL

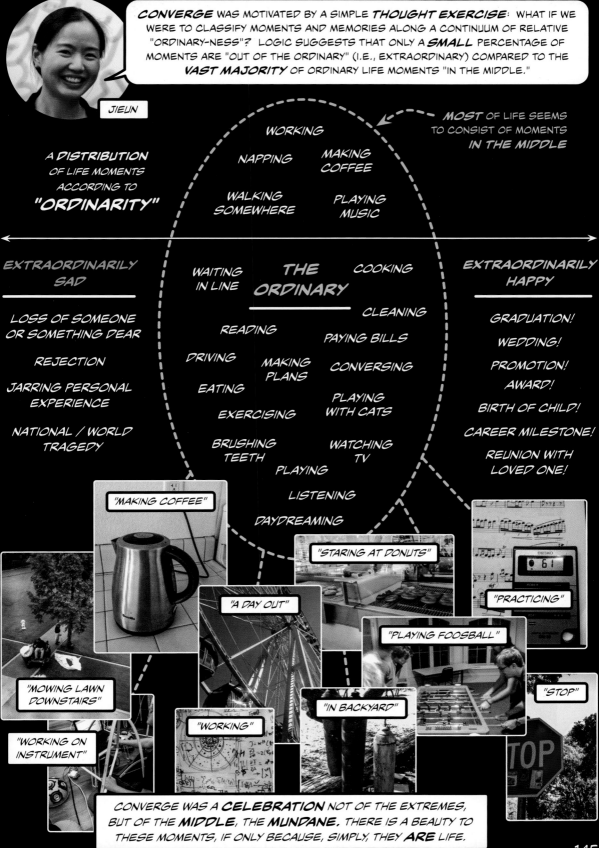

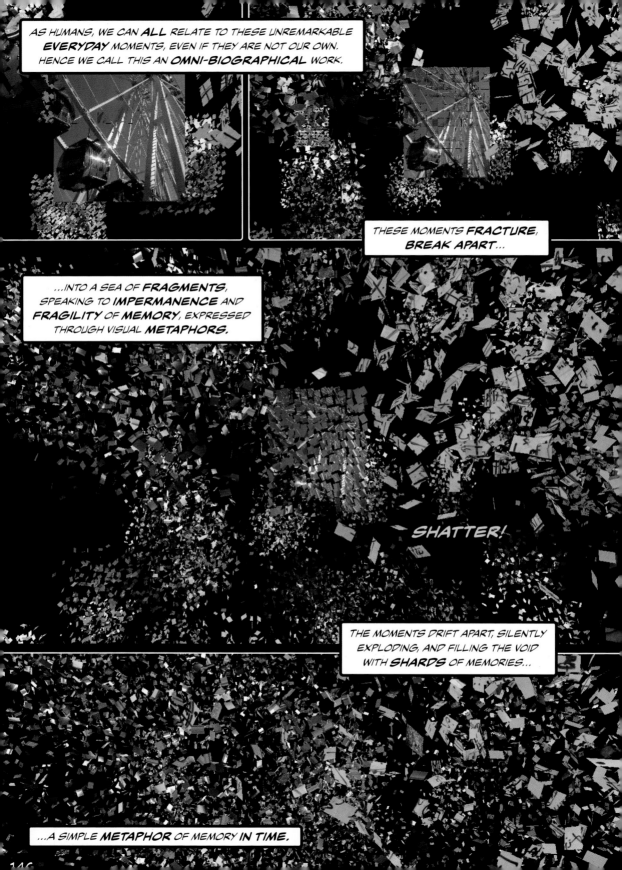

AS HUMANS, WE CAN **ALL** RELATE TO THESE UNREMARKABLE **EVERYDAY** MOMENTS, EVEN IF THEY ARE NOT OUR OWN. HENCE WE CALL THIS AN **OMNI-BIOGRAPHICAL** WORK.

THESE MOMENTS **FRACTURE, BREAK APART**...

...INTO A SEA OF **FRAGMENTS,** SPEAKING TO **IMPERMANENCE** AND **FRAGILITY** OF **MEMORY,** EXPRESSED THROUGH VISUAL **METAPHORS.**

SHATTER!

THE MOMENTS DRIFT APART, SILENTLY EXPLODING, AND FILLING THE VOID WITH **SHARDS** OF MEMORIES...

...A SIMPLE **METAPHOR** OF MEMORY **IN TIME.**

BUT THERE IS TACIT **ORDER** IN THE **CHAOS**...

...AS "CURRENTS" BEGIN TO FORM...

...AND GRADUALLY PULL FRAGMENTS INTO A GENTLE **SPIRAL DANCE**.

THE SUM OF MANY SIMPLE ELEMENTS GIVES RISE TO AN **EMERGENT** VISUAL SYSTEM.

FRAGMENTS MOVE AT A SPEED **PROPORTIONAL** TO THEIR DISTANCE TO THE CENTER COLUMN (AS AN APPROXIMATION OF **KEPLER'S 3RD LAW** OF PLANETARY MOTION). THIS CREATES A DYNAMIC **WHIRLPOOL**

SIMPLE ELEMENTS, IN **MOTION**, CAUGHT IN A COSMIC VORTEX...

147

ONE AT A TIME, FRAGMENTS **COALESCE** BACK INTO COHERENT MEMORIES.

2 km away from here

"COALESCE!"

(THEIR MOTION GOVERNED BY MANY **ZENO INTERPOLATORS**)

2 km away from here

6 years, 2 months, 25 days, 4 hours, 16 minutes, 9 seconds ago.

"MAKING COFFEE"

THE **TIME** OF EACH MEMORY IS EXPRESSED, NOT IN ABSOLUTES, BUT **RELATIVE** TO THE **PRESENT MOMENT!**

A **DISTORTED** VERSION OF THE ASSOCIATED **SOUND** PLAYS, DETUNED, WARPED YET FAMILIAR, SPEAKING TO THE INACCURACY OF MEMORY.

SIMILARLY THE GEOGRAPHIC **LOCATION** IS EXPRESSED RELATIVE TO THE VIEWER.

2 km away from here

6 years, 2 months, 25 days, 4 hours, 16 minutes, 10 seconds ago..

6 years, 2 months, 25 days, 4 hours, 16 minutes, 11 seconds ago.

THE TIMER COUNTS BACK **RELATIVE** TO THE PRESENT MOMENT, **TICKING UP** IN REAL TIME, A REMINDER THAT THESE MOMENTS ARE CONSTANTLY **MOVING AWAY** FROM US... AND THAT WE'LL **NEVER** BE AS CLOSE TO THEM AS WE ARE **NOW.**

IT REALLY **IS A DEPRESSING** PHOTO ALBUM...

THIS IS A DESIGN LOOP THAT BREAKS THE FOURTH WALL AND INVITES YOU TO THINK ABOUT THESE MOMENTS RELATIVE TO **YOUR** PRESENT MOMENT.

BUT ALSO KIND OF **BEAUTIFUL** -- THESE SMALL MOMENTS THAT WE MIGHT MISS ONCE THEY ARE GONE, FOR THEY REMIND US OF LIFE AND TIMES THAT CONTAIN THEM.

2 km away from here

5Y 1M 26D 2H 5M 30S AGO

"AUDIO EDITING IN MORNING"

2 KM AWAY

7Y 6M 4D 12H 5M 9S AGO

"DRIVING DOWN PAGE MILL"

4 KM AWAY

4Y 7M 3D 8H 19M 5S AGO

"DINNER TIME!"

8 KM AWAY

6Y 1M 26D 2H 3M 5S AGO

"MOWING LAWN DOWNSTAIRS"

2 KM AWAY

5Y 10M 1D 31H 58M 24S AGO

"DARLING HARBOUR, SYDNEY"

11942 KM

6Y 1M 16D 19H 20M 23S AGO

"DONUTS"

12 KM AWAY

5Y 7M 21D 1H 45M 25S AGO

"CCRMA 3RD FLOOR BOARD"

RIGHT HERE

3Y 7M 23D 15H 58M 0S AGO

"A RIVER OF THE MIND"

WITHIN 1 KM

6Y 7M 21D 1H 53M 14S AGO

"PLAYING FOOSBALL"

RIGHT HERE

5Y 3M 21D 15H 45M 41S AGO

"STOP"

WITHIN 1 KM

6Y 1M 18D 15H 19M 56S AGO

"ROB AND NICK"

NEARBY

6Y 1M 18D 15H 19M 56S AGO

"THE DISH, STANFORD"

3 KM AWAY

5Y 9M 8D 22H 36M 3S AGO

"MY KEY TO SUCCESS!"

2 KM AWAY

6Y 7M 21D 1H 53M 14S AGO

"NOON KNOLL STROLL"

RIGHT HERE

6Y 0M 26D 22H 29M 5S AGO

"NICK @ NIME"

11963 KM AWAY

6Y 0M 18D 16H 23M 11S AGO

"ISAAC AND ROB"

NEARBY

5Y 8M 8D 21H 2M 18S AGO

"IN BRAUN PRACTICE ROOM"

WITHIN 1 KM

6Y 1M 12D 16H 12M 30S AGO

"REHEARSING WITH SLORK"

WITHIN 1 KM

7Y 3M 23D 14H 19M 6S AGO

"COOING"

12 KM AWAY

3Y 11M 2D 23H 8M 37S AGO

"IN BACKYARD"

RIGHT HERE

5Y 1M 4D 10H 56M 32S AGO

"MICHAEL @ EG"

95 KM AWAY

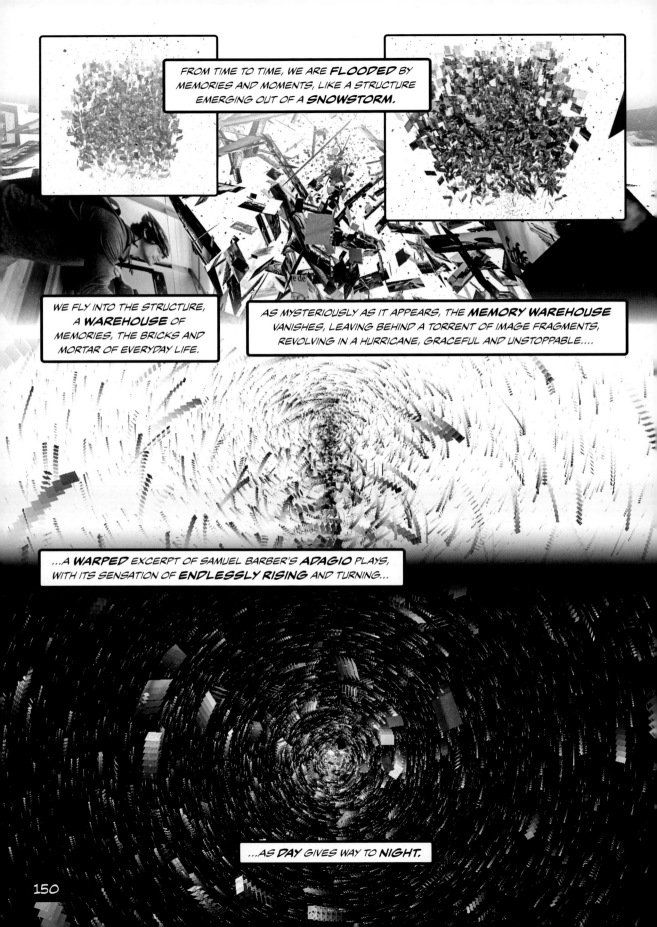

FROM TIME TO TIME, WE ARE **FLOODED** BY MEMORIES AND MOMENTS, LIKE A STRUCTURE EMERGING OUT OF A **SNOWSTORM.**

WE FLY INTO THE STRUCTURE, A **WAREHOUSE** OF MEMORIES, THE BRICKS AND MORTAR OF EVERYDAY LIFE.

AS MYSTERIOUSLY AS IT APPEARS, THE **MEMORY WAREHOUSE** VANISHES, LEAVING BEHIND A TORRENT OF IMAGE FRAGMENTS, REVOLVING IN A HURRICANE, GRACEFUL AND UNSTOPPABLE....

...A **WARPED** EXCERPT OF SAMUEL BARBER'S **ADAGIO** PLAYS, WITH ITS SENSATION OF **ENDLESSLY RISING** AND TURNING...

...AS **DAY** GIVES WAY TO **NIGHT.**

AS WE ZOOM FURTHER AWAY, IMAGE FRAGMENTS TURN INTO **STARS**, SWIRLING IN A **GALAXY** OF MEMORIES.

COMPOSED OF EVERYDAY IMAGES, THE **SPIRAL** FORMS VISUALLY FROM THE DIFFERENCE IN ROTATIONAL SPEEDS, AS A FUNCTION OF DISTANCE FROM THE CENTER.

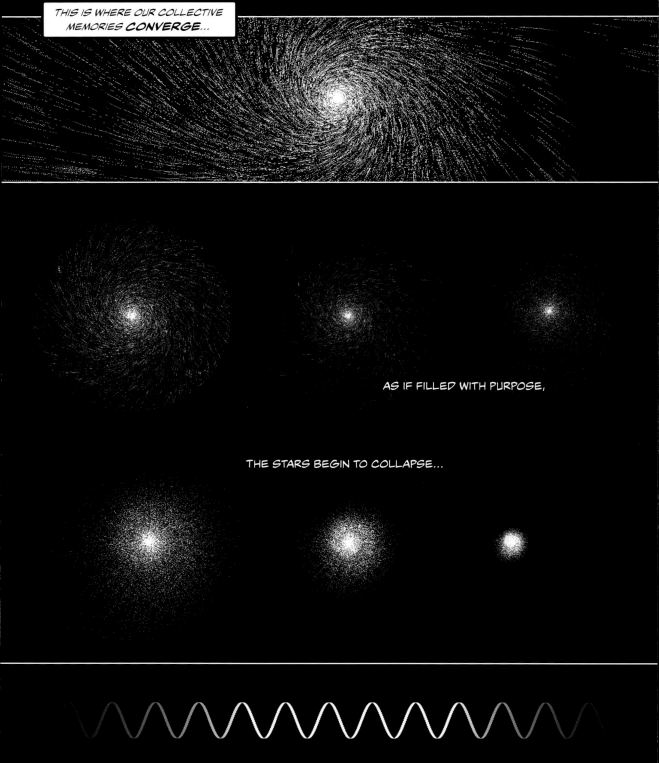

THIS IS WHERE OUR COLLECTIVE MEMORIES **CONVERGE**...

AS IF FILLED WITH PURPOSE,

THE STARS BEGIN TO COLLAPSE...

MEANWHILE, THE SOUND IS REDUCED TO THE SIMPLEST AND PUREST OF TONES... A SMALL, **FRAIL SINUSOID.** THE BUILDING BLOCK OF SOUND, IT IS THE SONIC ANALOGUE OF SIMPLICITY ITSELF.

UNTIL THE STARS *CONVERGE*
TO A *SINGLE POINT*

THEN... SILENCE.

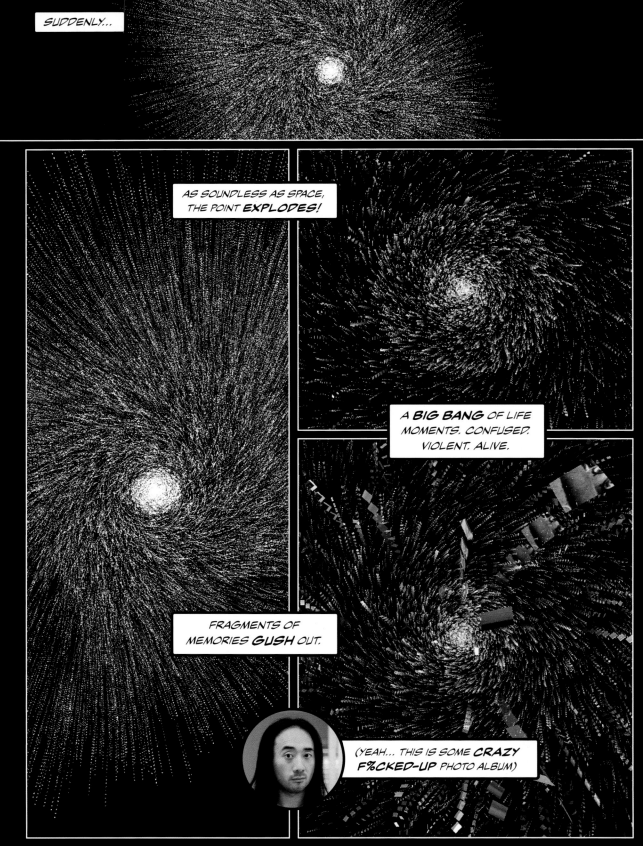

SUDDENLY...

AS SOUNDLESS AS SPACE, THE POINT **EXPLODES!**

A **BIG BANG** OF LIFE MOMENTS. CONFUSED. VIOLENT. ALIVE.

FRAGMENTS OF MEMORIES **GUSH** OUT.

(YEAH... THIS IS SOME **CRAZY F%CKED-UP** PHOTO ALBUM)

THE EXPLOSION STABILIZES, AND LIKE THE SKIES AFTER A STORM, IT CLEARS.

WHEN WE THINK BACK ON A FAVORITE TIME OR PERSON, IT'S OFTEN **NOT** THE CATACLYSMICALLY GOOD OR BAD THAT WE REMEMBER, BUT THE SIMPLE, **EVERYDAY MOMENTS**: TALKING WITH A LOVED ONE, SHARING A MEAL -- IN THE MIDST OF EVERYDAY LIFE.

A PEACEFUL CHIME PLAYS (RECORDED FROM A REHEARSAL) AS MEMORIES COME BACK INTO FOCUS.

THE IMAGES **COALESCE**, EXCEPT THIS TIME, IT'S NOT ONLY ONE IMAGE...

...BUT **ALL** OF THEM.

155

THE FRAGMENTS -- INFINITESIMAL AND YET INFINITE REPRESENTATIONS OF MEMORY OF EVERY SMALL NOTHING -- **COALESCE**, BUT ONLY BRIEFLY...

BEFORE THEY **CRUMPLE**...

...IN THE EPHEMERALITY OF TIME.

ALL THE PIECES COLLAPSE AND GATHER...

COUNTLESS FRAGMENTS **FUSE** TOGETHER, AN AMALGAM OF EVERYDAY MOMENTS.

THE BALL RECEDES IN SPACE AND TIME.

LIKE A GIANT **KATAMARI BALL**, ROLLED UP ACROSS THE FRAGMENTS OF OUR COLLECTIVE MEMORY.

WHERE IT GOES, WE DO NOT KNOW...

WELL, THERE YOU HAVE IT: **CONVERGE**, AN OMNI-BIOGRAPHICAL AUDIOVISUAL DESIGN, A SOFTWARE, AND A **BITTERSWEET** PHOTO ALBUM. I'M NOT SAYING EVERYTHING SHOULD BE SAD OR DEPRESSING, BUT I GOTTA WONDER...

WHY DOESN'T SOFTWARE **MAKE US FEEL** MORE OFTEN?

AS I SAID IN THE BEGINNING OF THIS CHAPTER, DESIGN ONLY SEEMS INEVITABLE AFTER IT'S DONE. ALONG THE WAY, MANY DECISIONS HAVE TO BE MADE: ELEMENTS, MOTION, PERSONALITY, NARRATIVE, AND MANY MORE. EACH OF THESE HAS A HAND IN THE FINAL RESULT.

WE HAVE ALSO STARTED ADDRESSING THE **FUNCTIONAL-AESTHETIC UNITY** OF GRAPHICS, SOUND, AND INTERACTION: **OCARINA** AS A COMPACT SYNTHESIS OF THE VISUAL, AUDIO, AND INTERACTIVE TO CREATE A COHESIVE TOY AND EXPERIENCE; THE VISUALIZATION OF SOUND IN **SNDPEEK**; ANIMATION AND GESTURE IN GOLAN LEVIN'S **YELLOWTAIL**; THE NARRATIVE OF IMAGE AND SOUND IN **CONVERGE**.

OUTSIDE INSPIRATIONS INFLUENCE FUNCTIONAL, AESTHETIC, AND TECHNICAL SENSIBILITIES. OUR DESIGN SENSE IS **HONED** BY WHAT WE FIND BEAUTIFUL, USEFUL, AND GOOD, AND INSPIRATION IS TAKEN FROM **WHEREVER** ONE HAPPENS TO FIND IT. MORE OBVIOUS SOURCES COME FROM VIDEO GAMES, CARTOONS, MOVIES, OR SIMPLY NATURE. LESS OBVIOUS, WE TAKE INSPIRATION FROM ALGORITHMS, PHILOSOPHY, OR SOME UNIQUE ASPECTS OF A TECHNOLOGY.

GRATEFULLY (AND MORE OFTEN THAN I'D CARE TO ADMIT), I DRAW INSPIRATION FROM **POPULAR CULTURE** (WE ARE ALL CHILDREN BORN OF OUR EXPERIENCE). I'VE LOVED CHANNELING WHIMSY FROM CARTOONS LIKE **SORCERER'S APPRENTICE**, WHICH I'VE ALWAYS FELT WAS AN **ALLEGORY** ON **TECHNOLOGY**. IN IT, MICKEY MOUSE (THE APPRENTICE!) WIELDS HIS ABSENT MASTER'S MAGIC WAND IN AN ATTEMPT TO **AUTOMATE** HIS CHORES, ONLY LEADING TO WACKINESS AND MAYHEM. THE VISUALS AND ANIMATION DESIGN FOR THIS SEGMENT ARE METICULOUSLY AND ARTFULLY WOVEN IN TANDEM WITH THE MUSICAL SCORE (BY PAUL DUKAS, 1896), A MASTERFUL EXAMPLE OF **WHIMSY** AND **MAGIC** THROUGH THE SONIC AND THE VISUAL (AS BROOMS AND OTHER HOUSEHOLD OBJECTS TAKE ON **PERSONALITIES**). AT THE SAME TIME, THE APPRENTICE'S ROLE SEEMS **SYMBOLIC** OF OUR OWN AS RESEARCHERS AND DESIGNERS OF **TECHNOLOGY** BEYOND OUR FULL UNDERSTANDING AND CONTROL, WHICH COMMANDS OUR FASCINATION AND AT TIMES CAN FEEL LIKE **MAGIC**... BUT CAN END UP IN A BIG **MESS**!

CHAPTER 3 DESIGN ETUDE

VIEWING: WATCH "SORCERER'S APPRENTICE" -- LOOK FOR THE MELDING OF VISUALS AND MUSIC IN ITS DESIGN.

BEFORE WE EXPLORE PROGRAMMABILITY AND SOUND IN CHAPTER 4...

SEEING SOUND!

DESIGN AND PROTOTYPE A REAL-TIME **SOUND VISUALIZATION SOFTWARE.** SOUND GOES IN, GRAPHICS RESPOND IN REAL TIME, CONVEYING SOMETHING ABOUT THE AUDIO. IN PARTICULAR, DESIGN SOMETHING TO VISUALIZE BOTH THE TIME-VARYING SOUND (WAVEFORM) AS WELL AS THE INSTANTANEOUS FREQUENCY CONTENT (SPECTRUM). IF YOU'D LIKE, USE SNDPEEK AS A STARTING POINT (HTTPS://ARTFUL.DESIGN/).

• PART 1: WAVEFORM

TO START, THINK ABOUT HOW TO VISUALIZE THE **TIME DOMAIN WAVEFORM.** CONCEIVE, SKETCH, PLAN.

IN A WAY, THE WAVEFORM IS NOTHING MORE THAN A SET OF POINTS THAT, WHEN CONNECTED, SKETCH OUT A SIGNAL. THIS SET OF POINTS CHANGES RAPIDLY (DOZENS OF TIMES PER SECOND), GIVING RISE TO THE FLUIDITY AND FLOW OF ANIMATION.

WAVEFORM
(TIME-VARYING AIR PRESSURE)

SPECTRUM
(FREQUENCY CONTENT)

HOW MIGHT IT **LOOK?** HOW TO **EXPRESS** THE SIGNAL? IN SNDPEEK, THE WAVEFORM IS SIMPLY CONNECTED BY LINE SEGMENTS. WHAT ARE OTHER WAYS? PERHAPS THESE POINTS CAN TRACE OUT THE RIDGE ON A STRANGE AUDIO-MODULATED 3D MOUNTAIN? ALSO DOES IT HAVE TO BE LINEAR? CAN A WAVEFORM BE REPRESENTED **CIRCULARLY**, OR AS A **SPIRAL** FORM?

OR HOW ABOUT SOMETHING LIKE **THIS?**

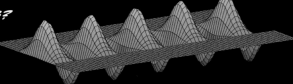

• PART 2: SPECTRUM

FIND A WAY TO VISUALLY REPRESENT EACH SPECTRUM THAT CORRESPONDS WITH EACH WAVEFORM IN TIME. FOR EXAMPLE, SNDPEEK KEEPS TRACK OF THE MOST RECENT SPECTRA AND DRAWS ALL OF THEM IN A WATERFALL PLOT TO VISUALIZE THE CHANGING OF THE SPECTRUM OVER TIME. HOW MIGHT YOU REPRESENT THIS?

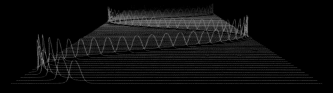

• PART 3: PROTOTYPE, PERSONALITY, NARRATIVE

PROTOTYPE IT. MAKE IT. USE WHATEVER TOOLS AND LANGUAGES MAY BE AVAILABLE TO YOU (E.G., C++, OPENGL, PROCESSING) -- OR RECRUIT A FRIEND WHO PROGRAMS! WHATEVER YOU DO, DO IT WITH **PERSONALITY** AND AESTHETICS. **TRY IT** ON SOUND. YELL OR PLAY MUSIC INTO IT, RUN A SHEPARD TONE GENERATOR THROUGH IT! HOW DOES IT LOOK? IS IT USEFUL? IS IT FUN? DOES IT CAPTURE SOMETHING BEAUTIFUL? HOW CAN YOU MAKE IT MORE NUANCED? FUNCTIONALLY, **DOES IT READ?** AESTHETICALLY, DOES IT MAKE YOU **FEEL SOMETHING?** DELIGHTED? MESMERIZED? PLEASED?

"We can forgive a man for making a useful thing as long as he does not admire it. The only excuse for making a useless thing is that one admires it intensely.

All art is quite useless."

— **Oscar Wilde** (1890)
The Picture of Dorian Gray

SUPERFLUITY AND FREEDOM

All art is superfluous: it serves no practical use, no purpose beyond itself. Yet therein lies an aspect of beauty and freedom, for not being bound to an external purpose is precisely the necessary condition of being **free**. It is an auto-recursive strange design loop of doing something for its own sake, where the pursuit itself is its reward. Indeed, art may be sold—thereby gaining a "use"—but that has nothing to do with the essence of the art itself.

Art belongs in that category of ends-in-themselves, along with hobbies, the idea of play, aesthetics. All are either superfluous outside themselves or useless by definition (like play). Yet they are among the most cherished things— to protect them we would give up life's necessities. Being purposeless ought not be confused with being frivolous.

And if human beings have no preordained purpose (at least none we seem to agree on), then might we not belong— for all intents and purposes (or lack thereof)—in the same category? Are we ends-in-ourselves? And are we...free?

CHAPTER 4
PROGRAMMABILITY
& SOUND DESIGN

LINES OF **CODE** IN A SOFTWARE **PROGRAM** ARE LIKE **SENTENCES** IN A **BOOK**: THEY ARE NOT ONLY **BUILDING BLOCKS** THAT, PIECE BY PIECE, MAKE UP A LARGER SYSTEM, BUT ALSO CELLS THAT ENCODE **MEANING** AND **NUANCE**.   JUST AS A BOOK CAN BE **MORE** THAN A VEHICLE FOR CONVEYING INFORMATION, A SOFTWARE SYSTEM IS **MORE** THAN LOGIC AND SHEER FUNCTIONALITY...

...IT'S **IDEAS** IN MOTION.

ALL DESIGN NEEDS A *MEDIUM.* THE ARTFUL DESIGNER IN THE AGE OF *COMPUTABLE TECHNOLOGY* ALSO CONTENDS WITH *PROGRAMMING,* WHICH THE DESIGNER WIELDS AS A *TOOL* AND *CANVAS.*

PROGRAMMABILITY CAN FEEL DOWNRIGHT *MAGICAL.* IT GIVES RISE TO FORMS, PROCESSES, AND EXPERIENCES THAT OTHERWISE WOULD NOT EXIST OUTSIDE OUR IMAGINATION. AS A DESIGN MEDIUM, CODE IS BOTH THE *BLUEPRINT* AND THE *END PRODUCT.* IT IS AN *INSTRUMENT* OF *ARTICULATION: HOW* SOMETHING LOOKS, SOUNDS, FEELS, INTERACTS -- IN SHORT, HOW SOMETHING IS TO *EXIST.*

TO MAKE A MUSICAL ANALOGY, CODE IS LIKE BOTH THE MUSICAL *SCORE* AND THE *ENACTMENT* OF THAT SCORE. THIS CHAPTER EXPLORES PROGRAMMABILITY THROUGH THE LENS OF *COMPUTERIZED SOUND* AND *MUSIC DESIGN* (PROGRAMMING EXPERIENCE *NOT* REQUIRED).

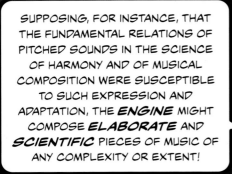

SUPPOSING, FOR INSTANCE, THAT THE FUNDAMENTAL RELATIONS OF PITCHED SOUNDS IN THE SCIENCE OF HARMONY AND OF MUSICAL COMPOSITION WERE SUSCEPTIBLE TO SUCH EXPRESSION AND ADAPTATION, THE **ENGINE** MIGHT COMPOSE **ELABORATE** AND **SCIENTIFIC** PIECES OF MUSIC OF ANY COMPLEXITY OR EXTENT!

I **TOTALLY** JUST PREDICTED COMPUTER MUSIC HERE IN 1843! THIS MIGHT JUST PAVE THE WAY FOR SOME **EPIC** RAVES AND **DANK** DANCE PARTAYZ! SOME DAY...

THE CONCEPT OF **COMPUTER-BASED MUSIC** CAN BE TRACED BACK TO **ADA LOVELACE** -- COUNTESS AND DAUGHTER OF THE POET LORD BYRON. ADA DESCRIBED HER APPROACH AS A KIND OF **POETICAL SCIENCE.**

COUNTESS
ADA LOVELACE
POETICAL SCIENTIST AND WORLD'S FIRST PROGRAMMER
(1815-1852)

LADY **ADA LOVELACE** WAS LIKELY THE WORLD'S **FIRST TRUE COMPUTER PROGRAMMER**... LIKE, **EVER.** SHE WROTE AN ALGORITHM IN COMPLETE DETAIL TO COMPUTE BERNOULLI NUMBERS FOR THE THEORETICAL **ANALYTICAL ENGINE**, WHICH SHE WORKED ON WITH **CHARLES BABBAGE.** WHILE IT WAS NEVER BUILT, THE ANALYTICAL ENGINE HAD PROVISIONS FOR **DECISION** AND **LOOPING**, FUNDAMENTAL CONSTRUCTS FOR COMPUTERS AS WE KNOW THEM.

OVER 100 YEARS LATER, MUSICIANS **WERE INDEED** USING COMPUTERS TO MAKE SOUND!

WHY USE A **COMPUTER?**

...TO DO THINGS THAT YOU COULDN'T DO **WITHOUT** IT. THE COMPUTER IS A KIND OF **WORKSHOP!**

JEAN-CLAUDE RISSET
COMPOSER AND COMPUTER MUSIC PIONEER
(1938-2016)

OH, AND BY THE WAY...

IN FRENCH, THE WORD FOR COMPUTER IS **ORDINATEUR**, MEANING "ONE WHO PUTS THINGS **IN THE RIGHT ORDER.**" IT HAS ROOTS IN THEOLOGY AND REFERS TO SOMEONE WHO PRESIDES OVER AN **ORDINATION** AND CONFERS THE SACRAMENT. THE SAME WORD EVEN CONNOTES **GOD**, AS IN "GOD WHO PUTS ORDER IN THE WORLD."

THAT "ORDINATEUR" MEANS BOTH **GOD** AND **COMPUTER** CONJURES MEGALOMANIACAL NOTIONS OF THE PROGRAMMER AS A CREATOR OF THEIR SOFTWARE **UNIVERSE**, OR PERHAPS A MORE HUMBLING CONJECTURE THAT OUR UNIVERSE IS ONE GIANT COMPUTER SIMULATION -- THE WORK OF SOME **GREAT PROGRAMMER.** MORE SIMPLY, THE ETYMOLOGY OF THE WORD SPEAKS TO THE CORE ACT OF **PROGRAMMING** AS "PUTTING THINGS IN THE RIGHT PLACE" -- NOT UNLIKE **DESIGN ITSELF.**

COMPUTER MUSIC RESEARCHERS ARE LIKE **EXPERIMENTAL COOKS.** THEY USE THEIR HANDS AND IMAGINATION TO CONCOCT NEW EXPERIENCES.

COMPOSING MUSIC WITH PROGRAMMING ISN'T MUCH DIFFERENT THAN COOKING WITH RAW INGREDIENTS. THE COMPUTER GIVES YOU THE ABILITY TO SYNTHESIZE SOUND **FROM SCRATCH.** WITH PROGRAMMING, YOU CAN **COOK UP** YOUR OWN SOUNDS, INSTRUMENTS, AND **MUSICAL SPACES** THAT WOULDN'T EXIST ELSEWHERE.

JOHN CHOWNING
INVENTOR OF **FM SYNTHESIS** (TECHNOLOGY BEHIND "THAT 1980s SYNTHESIZER SOUND"), FOUNDER OF STANFORD UNIVERSITY'S CENTER FOR COMPUTER RESEARCH IN MUSIC AND ACOUSTICS (CCRMA), COMPOSER, COMPUTER MUSIC PIONEER, COOK.

FM SYNTHESIS, INVENTED BY JOHN IN 1967, IS AT THE HEART OF THE **YAMAHA DX7**, THE WORLD'S FIRST COMMERCIALLY VIABLE **DIGITAL SYNTHESIZER**, USHERING IN A NEW ERA OF PROGRAMMABLE SOUND SYNTHESIS AND FOREVER CHANGING HOW ELECTRONIC MUSIC IS MADE.

WHEN JOHN CHOWNING WAS A **GRADUATE STUDENT** AT STANFORD UNIVERSITY IN THE 1960s, HE TOOK A COURSE (IN THE THEN-NASCENT **COMPUTER SCIENCE** DEPARTMENT) THAT WAS ESSENTIALLY "COMPUTER SCIENCE FOR NON-ENGINEERS." THE SECTION WAS **CLOSED** TO ENGINEERING STUDENTS AND WAS OFFERED SPECIFICALLY TO THOSE IN THE **HUMANITIES.** THE COURSE WAS TAUGHT BY **GEORGE FORSYTHE**, THE FOUNDER OF STANFORD'S COMPUTER SCIENCE DEPARTMENT.

STANFORD UNIVERSITY COURSE BULLETIN 1963-1964

COMPUTER SCIENCE
COURSES FOR UNDERGRADUATE AND GRADUATE STUDENTS

CS5. Computer Programming for Engineers—This course is an introduction to a problem-oriented language for describing computational processes. There will be practice in solving elementary problems on Stanford's automatic digital computers. The course is limited to freshman students. Prerequisites: Mathematics A and C, or equivalents.

2 units, autumn, (Van Zoeren), MW 3
or spring, (———), MW 3

CS136. Use of Automatic Digital Computers—Methods of utilizing automatic digital computers in the solution of problems. Study of problem-oriented languages for description of algorithms. Programming of elementary problems from mathematics and other fields, and testing the programs on a computer. Freshmen and sophomores with strong backgrounds in high school mathematics may enroll with consent of instructor. In spring quarter, the section MWF at 11 is directed to students of social science and the humanities, and is closed to students of mathematics, engineering, or the physical sciences, and to students with prior experience in computing.

3 units, autumn, (Herriot), MWF 11, or (———), MWF 1
or (———), TTh 9:35–10:50
or winter, (———), MWF 10, or (———), MWF 1
or spring, (Forsythe), MWF 11, or (———), MWF 1

GEORGE FORSYTHE

CS136. Use of Automatic Digital Computers—Methods of utilizing automatic digital computers in the solution of problems. Study of problem-oriented languages for description of algorithms. Programming of elementary problems from mathematics and other fields, and testing the programs on a computer. Freshmen and sophomores with strong backgrounds in high school mathematics may enroll with consent of instructor. In spring quarter, the section MWF at 11 is directed to students of social science and the humanities, and is closed to students of mathematics, engineering, or the physical sciences, and to students with prior experience in computing.

MAX MATHEWS

JOHN WAS PROMPTED TO SIGN UP FOR FORSYTHE'S COURSE AFTER READING AN ARTICLE IN **SCIENCE** MAGAZINE BY **MAX MATHEWS**, "THE DIGITAL COMPUTER AS MUSICAL INSTRUMENT." JOHN WROTE TO MAX, AND MAX SENT BACK A BOX OF PUNCH CARDS, WHICH COMPRISED A PROGRAM TO SYNTHESIZE SOUNDS. AND THUS BEGAN COMPUTER MUSIC RESEARCH AT STANFORD.

MAX MATHEWS IS REMEMBERED AS THE **FIRST COMPUTER MUSICIAN**, THE GRANDFATHER OF COMPUTER MUSIC, AND THE INVENTOR OF THE **RADIO BATON** AND **MUSIC I** (THE FIRST MUSIC PROGRAMMING LANGUAGE). MAX WAS A RESEARCHER AT **AT&T BELL LABS** BEFORE JOINING THE FACULTY AT STANFORD.

WHEN **ARTHUR C. CLARKE** VISITED BELL LABS, HE HEARD A COMPUTERIZED RENDITION OF "DAISY BELL (BICYCLE BUILT FOR TWO)," FOR WHICH MAX SYNTHESIZED THE ACCOMPANIMENT. CLARKE THOUGHT IT WOULD BE FITTING THAT ALL COMPUTERS IN THE FUTURE LEARN TO SING "DAISY" AS THEIR **FIRST SONG** -- THAT IS WHY **HAL 9000** SINGS IT IN **2001: A SPACE ODYSSEY!**

AS **MAX** LIKED TO SAY:

WITH THE DIGITAL COMPUTER, WE CAN, IN THEORY, GENERATE ABSOLUTELY **ANY** SOUND, SIMPLY BY COMPUTING THE **SEQUENCE** OF **NUMBERS** REPRESENTING IT. HOWEVER, THE VAST **MAJORITY** OF SOUNDS ARE HARSH, ABRASIVE, OR DOWNRIGHT ANNOYING.

THE **CHALLENGE**, BOTH SCIENTIFICALLY **AND** ARTISTICALLY, IS TO **FIND** THE **GOOD** ONES!

IN COMPUTER MUSIC, PROGRAMMABILITY CAN GIVE RISE TO **FANTASTICAL AUTOMATIONS** AND **SOUNDS** WE CAN'T ACCESS OTHERWISE. BUT IT'S UP TO **US** TO DEVELOP OUR OWN AESTHETIC COMPASS IN WORKING WITH THIS UNIQUE MEDIUM.

MORE IMPORTANT THAN SIMPLY KNOWING HOW TO PROGRAM IS KNOWING **WHAT IT MEANS** TO PROGRAM. THERE IS AN **ART** TO IT, AND TO WHAT WE **DO WITH IT.** THE MORE **NUANCED** WE MAKE THE INSTRUCTIONS, THE MORE POTENTIAL THERE IS FOR **RICHNESS** IN THE RESULT!

WE SHOULD NOT BE WARY OF CODE, NOR SHOULD WE DO (OR LEARN) PROGRAMMING FOR ITS OWN SAKE; INSTEAD, DO IT TO **EXPRESS** SOMETHING!

Ⓧ PRINCIPLE 4.1

PROGRAMMING IS A **CREATIVE** ENDEAVOR

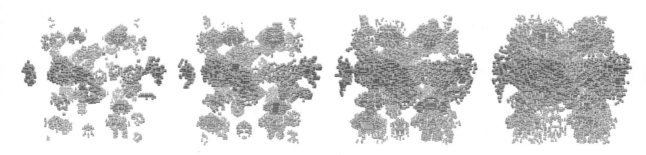

A FEW GOOD
PROGRAMMABLE SONIC PARAMETERS

→ *TO EVOLVE OVER TIME!*

• FREQUENCY *RELATED: PITCH*

MEASURED IN **HERTZ** (HZ), FREQUENCY HAS TO DO WITH THE NUMBER OF **OSCILLATIONS** PER UNIT TIME, WHICH WE HEAR AS PITCH; FUNDAMENTAL FREQUENCIES OF MUSICAL PITCHES VARY, TYPICALLY, BETWEEN 20 HZ TO 5000 HZ.

• DYNAMICS *RELATED: VOLUME, INTENSITY, LOUDNESS*

A SINGLE **MULTIPLIER** ON THE **AMPLITUDE** OF THE SOUND, **DYNAMICS** IS **ALL TOO EASY** TO OVERLOOK AND UNDER-UTILIZE IN COMPUTER-BASED SOUND DESIGN. IT ALSO HAS TO DO WITH THE RELATIVE LOUDNESS OF ELEMENTS, AS WELL AS **TRAJECTORIES** OF CHANGE IN LOUDNESS (E.G., CRESCENDO, DECRESCENDO, TREMOLO). SUBTLE VARIATIONS IN AMPLITUDE CAN HAVE PRONOUNCED EFFECT OVER TIME; AMPLITUDE ENVELOPES CAN BE APPLIED TO CREATE EXPRESSIVE **CONTOURS** AS A FUNCTION OF TIME.

• TIMBRE *RELATED: TONE, FREQUENCY SPECTRUM OF SOUND*

NOTORIOUSLY DIFFICULT TO **DEFINE**, TIMBRE IS SOMETIMES DESCRIBED AS... AN **ESSENTIAL QUALITY** OF SOUND THAT IS **DISTINCT** FROM ITS PITCH AND INTENSITY. (NOT THAT HELPFUL.) FORTUNATELY, IT IS EASIER TO EMPIRICALLY **CREATE** AND **MODULATE** TIMBRE THAN TO DEFINE IT. WE CAN COMPUTATIONALLY CRAFT TIMBRE BY ADDING COMPONENTS, SUBTRACTING (THROUGH FILTERING), MODELING, AND MORE.

• TEXTURE *RELATED: INTERACTION OF "VOICES" IN A MIXTURE*

IN MUSIC, TEXTURE REFERS TO HOW INDIVIDUAL **VOICES** STACK UP (E.G., MONOPHONY, POLYPHONY, COUNTERPOINT) AND **INTERACT** WITH ONE ANOTHER TO FORM A COHERENT OVERALL SOUND. MORE GENERALLY, TEXTURE IS HOW DISPARATE SOUNDS **LAYER** TOGETHER IN A **MIXTURE.** DESCRIPTIONS OF TEXTURE INCLUDE "GRITTY," "SMOOTH," "POINTILLISTIC," "THIN," "FULL," ...

• RHYTHM *RELATED: PULSE, BEAT, METER, TEMPO, GROOVE, ARHYTHM*

ANY DISCERNIBLE **REPETITION** OF ARRIVAL OR CHANGE OF SOUND. CAN BE AS IN-YOUR-FACE AS A REPEATING KICK DRUM, OR AS SUBTLE AS A GENTLE FLUCTUATION, PULSE, OR FLUTTER.

• HARMONY *RELATED: CONSONANCE, DISSONANCE, CHORDS, TONALITY*

THE **SIMULTANEITY** OF PITCHES, AND HOW THEY EVOLVE, PROGRESS, MOVE FROM ONE STATE TO ANOTHER. THE **PROGRESSION** OF HARMONIES OVER TIME IS A PROMINENT MUSICAL QUALITY TO MANIPULATE AND DESIGN.

• SPATIALIZATION *RELATED: PANNING, STEREO, MULTI-CHANNEL, 3D*

THE PLACEMENT, SIMULATION, AND MOVEMENT OF **SOUND** IN **SPACE**, TAKING ADVANTAGE OF THE FACT WE HAVE **TWO** EARS, CAPABLE OF SONICALLY POSITIONING OBJECTS IN A 3D ENVIRONMENT.

SOUND IS A **TIME-BASED** MEDIUM, AND MUSICAL SOUND IS ABOUT **CHANGE** OVER TIME. **STATIC** SOUNDS, NO MATTER HOW COMPLEX AT A GIVEN MOMENT, GET INVARIABLY AND PROFOUNDLY **BORING.** **INTERESTING** SOUNDS ARE THE **EVOLUTION** OF SOMETHING **IN** THE SOUND, HOWEVER PROMINENT OR **SUBTLE**, SUDDEN OR GRADUAL, DISCRETE OR CONTINUOUS. **SOUND DESIGN**, LIKE ANIMATION, IS AN **ART** OF **CHANGE.**

(INTERESTING)

☯ PRINCIPLE 4.2　*SOUND IS MOTION*

(OVER TIME)

AND THE WAY WE MAKE **MUSICAL** USE OF THE PROGRAMMABLE **PARAMETERS** OF SOUND IS BY FINDING INTERESTING, **NUANCED** WAYS TO **CONTROL** THEM **PRECISELY OVER TIME.**

AT SOME LEVEL, PROGRAMMING **SOUND**...

...IS ABOUT **PROGRAMMING** *TIME!*

SO, WHAT DOES IT **MEAN** TO CONTROL SOUND PARAMETERS OVER TIME? LET'S LOOK AT A SIMPLE **EXAMPLE**...

UP AHEAD ARE A FEW MORE **BLUEPRINT PAGES**, ON WHICH YOU'LL FIND CODE AND OTHER TECHNICAL CONTENT. I'VE ENDEAVORED TO MAKE THEM AS LUCID AS POSSIBLE, BUT DON'T WORRY IF YOU DON'T UNDERSTAND THE CODE -- IT'S MORE IMPORTANT TO GET A **GENERAL SENSE** OF WHAT IT'S TRYING TO DO!

HERE IS A SIMPLE PROGRAM TO **GENERATE** A **SINE WAVE**, **RANDOMLY** CHANGING ITS FREQUENCY **10 TIMES A SECOND**:

BLEEP

```
// sound scheme
SinOsc ada => dac;

// loop
while( true )
{
    // random number as frequency
    Math.random2f(30,1000) => ada.freq;
    // wait 100 milliseconds
    100::ms => now;
}
```

BL°°P

BLEEP BLEEP

IT SOUNDS LIKE **BLEEPS** AND **BLOOPS** REMINISCENT OF HOW COMPUTERS SOUND IN SCIENCE FICTION MOVIES FROM THE 1950s.

BL°°°P

HOW DOES THIS WORK? LET'S **BREAK IT DOWN**, STARTING WITH THE **LANGUAGE** IT'S WRITTEN IN...

169

CHUCK
A STRONGLY-TIMED MUSIC PROGRAMMING LANGUAGE!

CHUCK IS A **PROGRAMMING LANGUAGE** FOR **SOUND GENERATION** AND **MUSIC CREATION.** IT WAS DESIGNED AS A **TOOL** FOR RESEARCHERS, COMPOSERS, AND SONIC TINKERERS TO PROGRAM MUSICAL SOUNDS BY WORKING DIRECTLY WITH A NOTION OF **TIME** ITSELF. IT IS **OPEN-SOURCE** AND FREELY AVAILABLE. (AND, AS I LIKE TO SAY, IT CRASHES EQUALLY WELL ON ALL COMMODITY OPERATING SYSTEMS!) IT HAS A **PERSONALITY**, AND IS PRETTY EASY TO LEARN.

I STARTED DESIGNING **CHUCK** BACK IN 2002 (WHEN I WAS IN GRAD SCHOOL). SINCE THAT TIME, **CHUCK** HAS BEEN USED TO CRAFT INSTRUMENTS FOR **LAPTOP ORCHESTRAS** AND IS THE AUDIO ENGINE IN **OCARINA**, RUNNING INSIDE MILLIONS OF PHONES...

2002, IN THE BOWELS OF THE COMPUTER SCIENCE DEPARTMENT AT PRINCETON...

CHUCK REPRESENTS AN EXTREME EXPRESSION OF **IMPERATIVE PROGRAMMING**, ASKING THE PROGRAMMER TO EXPLICITLY **SPECIFY** EVEN THE PASSAGE OF **TIME** TO CONTROL AUDIO SYNTHESIS. WHAT **POSSESSED** YOU TO **DESIGN** IT LIKE THAT?

GEORG ESSL
FELLOW COMPUTER MUSIC RESEARCHER, BIG BROTHER

PERRY R. COOK
ADVISOR, LIFE MENTOR, **CHUCK** CO-CONSPIRATOR, ZEN MASTER

CHUCK'S DESIGN CHOICES PRESENT A DIFFERENT **WAY OF THINKING**, A DIFFERENT **AESTHETIC** OF PROGRAMMING SOUND. I WANTED TO CREATE A TOOL THAT COULD SPECIFY PRECISELY **HOW** AND **WHEN** THINGS HAPPEN. THE WAY **CHUCK** HANDLES **TIME** AND **PARALLELISM** IS DESIGNED AS A WAY TO **THINK** ABOUT MUSIC ITSELF...

WE CAN *DISSECT* OUR BLEEP/BLOOP CODE EXAMPLE, WRITTEN IN *CHUCK*, AND POINT OUT SOME OF ITS *FUNCTIONALITIES*...

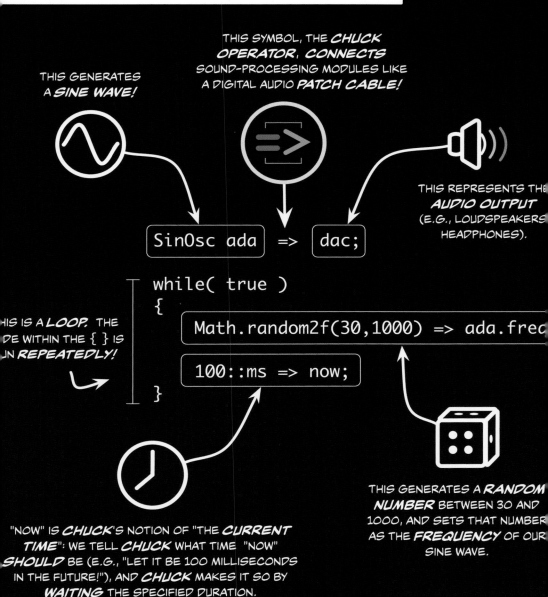

THIS GENERATES A *SINE WAVE!*

THIS SYMBOL, THE *CHUCK OPERATOR*, *CONNECTS* SOUND-PROCESSING MODULES LIKE A DIGITAL AUDIO *PATCH CABLE!*

THIS REPRESENTS THE *AUDIO OUTPUT* (E.G., LOUDSPEAKERS HEADPHONES).

```
SinOsc ada  =>  dac;

while( true )
{
    Math.random2f(30,1000) => ada.freq

    100::ms => now;
}
```

HIS IS A *LOOP*. THE DE WITHIN THE { } IS N *REPEATEDLY!*

THIS GENERATES A *RANDOM NUMBER* BETWEEN 30 AND 1000, AND SETS THAT NUMBER AS THE *FREQUENCY* OF OUR SINE WAVE.

"NOW" IS *CHUCK*'S NOTION OF "THE *CURRENT TIME*": WE TELL *CHUCK* WHAT TIME "NOW" *SHOULD* BE (E.G., "LET IT BE 100 MILLISECONDS IN THE FUTURE!"), AND *CHUCK* MAKES IT SO BY *WAITING* THE SPECIFIED DURATION.

MOST COMPUTER LANGUAGES HAVE WAYS TO DEAL WITH TIME (E.G., A "*WAIT*" DIRECTIVE), BUT THESE APPROACHES ARE OFTEN *COARSE* AND *UNPREDICTABLE*. IN CHUCK, *TIME* IS *ULTRA-PRECISE* BECAUSE IT IS INFERRED FROM THE *DIGITAL AUDIO STREAM* ITSELF. SOUND IN CHUCK IS BOTH THE *OUTPUT* AND THE MEANS BY WHICH CHUCK KEEPS TRACK OF *TIME*.

THE *FUNCTIONALITIES* OF A PROGRAMMING LANGUAGE DETERMINE WHAT YOU CAN *DO* WITH IT. THE *WAYS* IN WHICH A LANGUAGE *PRESENTS* ITS FUNCTIONALITIES TO YOU CONSTITUTE ITS *AESTHETICS* -- THEY SHAPE HOW YOU *THINK* ABOUT WHAT YOU WANT TO DO.

```
create a SinOsc
 called ada
nOsc ada => dac;

 execute loop?
hile( true )

    // generate random number as frequency
    Math.random2f(30,1000) => ada.freq;
    // advance time
    100::ms => now;
```

CODE

MENTAL FLOW CHART

MAY BE AVAILABLE IN DIFFERENT PROGRAMMING LANGUAGES. BUT THE *SPECIFIC* WAY A PARTICULAR LANGUAGE EXPRESSES THAT FUNCTIONALITY HAS TO DO WITH THE *AESTHETICS* OF THAT LANGUAGE -- AND CHANGES *HOW* YOU THINK ABOUT THE TASK AT HAND!

I THINK THIS MAY BE WHY WE HAVE SO *MANY* PROGRAMMING LANGUAGES. IT'S NOT SO MUCH THAT THEY ALL *DO* DIFFERENT THINGS, BUT THAT EACH ONE MAKES YOU *THINK* DIFFERENTLY...

FOR EXAMPLE, THE WAY YOU WORK WITH *TIME* IN *CHUCK* LENDS ITSELF TO *REASONING CLEARLY* ABOUT *WHEN* THINGS HAPPEN -- IT'S STRAIGHTFORWARD TO BUILD A *MENTAL FLOW CHART* OF WHAT HAPPENS AND WHEN...

START **HERE**

CREATE A SINE WAVE GENERATOR CALLED "ADA"; **CONNECT** IT TO "DAC"

EXECUTE LOOP BODY? — NO — END **HERE***

YES

*BY THE WAY, THIS PARTICULAR CODE I PROGRAMMED TO **NEVER** REACH THI STATE: IT ALWAYS TAKES THE *YES* PAT

GENERATE A RANDOM NUMBER -- AND **SET** IT AS ADA'S FREQUENCY

THIS IS AN *INFINITE LOOP* (THAT MAKES SOUND)!

WAIT 100 MILLISECONDS, MOVING *CHUCK'S INTERNAL CLOCK* FORWARD AND *SYNTHESIZING* AUDIO

BUILDING ON THIS WAY OF THINKING ABOUT **TIME**, **CHUCK** ADDS ONE MORE DIMENSION: THE ABILITY TO RUN MULTIPLE PROGRAMS **IN PARALLEL**, EACH MANAGING TIME IN ITS OWN WAY. CHUCK USES THIS TIME INFORMATION TO **AUTOMATICALLY** AND PRECISELY **SYNCHRONIZE** THESE PROGRAMS.

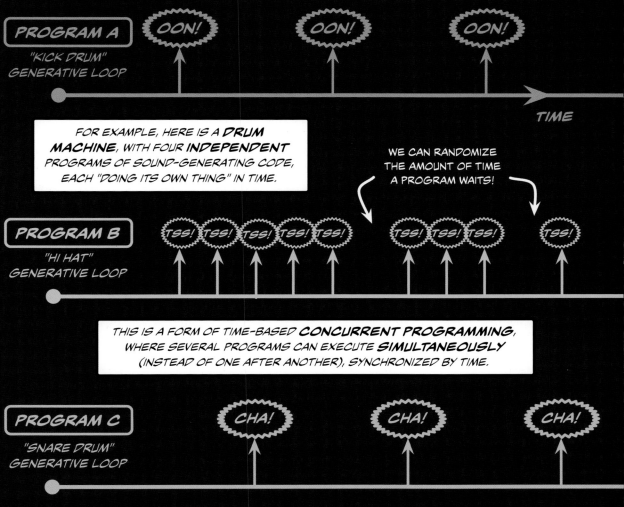

PROGRAM A

"KICK DRUM" GENERATIVE LOOP

OON! OON! OON!

TIME

FOR EXAMPLE, HERE IS A **DRUM MACHINE**, WITH FOUR **INDEPENDENT** PROGRAMS OF SOUND-GENERATING CODE, EACH "DOING ITS OWN THING" IN TIME.

WE CAN RANDOMIZE THE AMOUNT OF TIME A PROGRAM WAITS!

PROGRAM B

"HI HAT" GENERATIVE LOOP

TSS! TSS! TSS! TSS! TSS! TSS! TSS! TSS! TSS!

THIS IS A FORM OF TIME-BASED **CONCURRENT PROGRAMMING**, WHERE SEVERAL PROGRAMS CAN EXECUTE **SIMULTANEOUSLY** (INSTEAD OF ONE AFTER ANOTHER), SYNCHRONIZED BY TIME.

PROGRAM C

"SNARE DRUM" GENERATIVE LOOP

CHA! CHA! CHA!

PROGRAM D

"SINE BLOOPS" GENERATIVE LOOP

BLOOOP BLEEP BLOOOP

IN ESSENCE, **TIME** AND **CONCURRENCY** IN **CHUCK** ARE TWO SEPARATE DIMENSIONS THAT CAN **WORK TOGETHER** -- AS A GENERALIZED WAY TO MODEL **SOUND** BOTH AS A **PHENOMENON OVER TIME** AND AS A **MIXTURE** OF MANY ELEMENTS HAPPENING AT THE **SAME TIME**.

CHUCK CODE, BY DESIGN, IS A COMPLETE SPECIFICATION OF NOT ONLY **WHAT**, BUT ALSO **WHEN** THINGS HAPPEN. IT **COMPELS** THE PROGRAMMER TO BE EVER **AWARE** OF TIME WHEN **WRITING** CODE, AND TO BE ABLE TO **REASON** ABOUT TIME PRECISELY WHEN **READING** IT. THIS **WAY OF WORKING** AIMS TO FOCUS THE PROGRAMMER'S MENTAL EFFORTS ON SOUND AND HOW IT CHANGES OVER TIME, LEAVING THE LOW-LEVEL DETAILS FOR **CHUCK** TO HANDLE.

WHEN YOU RUN THE CODE, **CHUCK** GOES TO WORK! HERE'S A VISUALIZATION OF HOW OUR TIME-BASED CODE BECOMES SOUND...

OUR **CHUCK** CODE RUNS AT SPECIFIC **POINTS IN TIME** (E.G., EVERY 100::MS), CONTROLLING SOUND **PARAMETERS** (E.G., SINE WAVE FREQUENCY); CHUCK INTERNALLY KEEPS ITS OWN REPRESENTATION OF **TIME** (KNOWN TO THE PROGRAMMER AS "NOW") AND AUTOMATICALLY USES THIS INFORMATION TO DETERMINE **WHEN** THINGS HAPPEN AS SPECIFIED BY THE CODE.

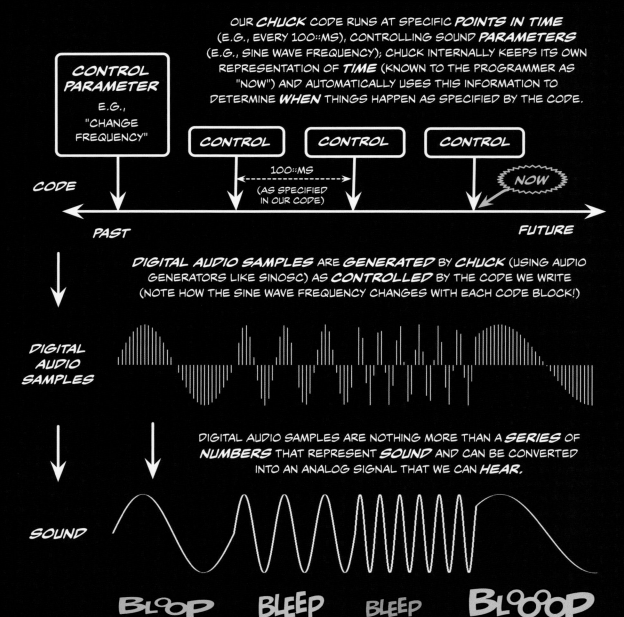

CONTROL PARAMETER

E.G., "CHANGE FREQUENCY"

CONTROL   **CONTROL**   **CONTROL**

100::MS
(AS SPECIFIED IN OUR CODE)

NOW

CODE

PAST   FUTURE

DIGITAL AUDIO SAMPLES ARE **GENERATED** BY CHUCK (USING AUDIO GENERATORS LIKE SINOSC) AS **CONTROLLED** BY THE CODE WE WRITE (NOTE HOW THE SINE WAVE FREQUENCY CHANGES WITH EACH CODE BLOCK!)

DIGITAL AUDIO SAMPLES

DIGITAL AUDIO SAMPLES ARE NOTHING MORE THAN A **SERIES** OF **NUMBERS** THAT REPRESENT **SOUND** AND CAN BE CONVERTED INTO AN ANALOG SIGNAL THAT WE CAN **HEAR.**

SOUND

BLOOP   BLEEP   BLEEP   BLOOOP

TIME IN **CHUCK** ILLUSTRATES HOW DESIGN IS CRUCIAL WHEN MAKING A NEW **TOOL.** TOOLS DO **MORE** THAN SERVE A PURPOSE -- THEY SHAPE OUR **THINKING.** A USEFUL TOOL SUGGESTS PARTICULAR **WAYS OF WORKING** (E.G., "IF ALL YOU HAVE IS A HAMMER, THEN EVERYTHING LOOKS LIKE A NAIL").

THAT'S WHY **DESIGN** IS **RELEVANT** IN CREATING ANY TOOL. THE **CHOICES** THAT GO INTO ITS DESIGN **IMPACT** HOW USERS **THINK** ABOUT WHAT THEY DO WITH THE TOOL AND HOW THEY **FEEL** WHEN THEY ARE USING IT, ULTIMATELY **SHAPING** THE KINDS OF THINGS THEY **CREATE** WITH THE TOOL.

MY **INTENTION** IN DESIGNING **CHUCK** WAS TO DEVELOP A **SIMPLE** YET **FLEXIBLE** WAY TO PROGRAM **TIME** AS A MUSICAL CONSTRUCT. IT WAS AN EFFORT TO **HIDE** THE LOW-LEVEL COMPLEXITIES OF DIGITAL AUDIO SYNTHESIS WHILE **EXPOSING** AN ULTRA-PRECISE HIGH-LEVEL WAY TO **CONTROL** IT. THE RESULT IS BOTH A **TOOL** AND A **WAY OF THINKING** TO WORK WITH MUSICAL SOUND, CENTERED AROUND THE FLOW OF TIME.

...I CALL IT **STRONGLY-TIMED** PROGRAMMING!

AS YOU CAN SEE, IT IS **ABOUT TIME!**

WITH THIS IN MIND, LET'S EXAMINE A MORE COMPLEX CASE STUDY OF CONTROLLING SOUND OVER TIME...

THE THX DEEP NOTE!

TO ILLUSTRATE SOUND SYNTHESIS BY WAY OF PARAMETRIC EVOLUTION, WE ARE GOING TO **RECREATE** ONE OF THE MOST RECOGNIZABLE PIECES OF COMPUTER-GENERATED SOUND EVER DESIGNED: THE **THX DEEP NOTE!**

DESIGNED AND PROGRAMMED IN 1982 BY **JAMES ANDY MOORER** (ALSO A FOUNDING MEMBER OF CCRMA), THE **DEEP NOTE** WAS FIRST INTRODUCED WITH THE 1983 PREMIER OF **RETURN OF THE JEDI** AND HAS BEEN HEARD IN COUNTLESS THX TRAILERS FOR MOVIES AND VIDEO GAMES!

WHEN ANDY CREATED THE **DEEP NOTE**, HE WAS AN EMPLOYEE OF LUCASFILM'S COMPUTER DIVISION (WHICH NOT ONLY LED TO THX BUT EVENTUALLY PIXAR). THX CREATOR **TOM HOLMAN** ASKED ANDY TO CREATE A **SOUND LOGO** THAT "COMES OUT OF NOWHERE AND GETS **REALLY, REALLY BIG,**"

IN 1982, IT TOOK ANDY MOORER 325 LINES OF C CODE RUNNING ON A SPECIALIZED HARDWARE AND SOFTWARE **AUDIO SIGNAL PROCESSOR.** HERE WE ARE GOING TO RECREATE IT IN **CHUCK!** IT WON'T BE EXACTLY THE SAME, BUT WE WILL TRY TO CAPTURE THE ESSENCE OF THE SOUND DESIGN!

THE **DEEP NOTE** WAS **SYNTHESIZED** USING **30** VOICES WITH RANDOMIZED STARTING FREQUENCIES BETWEEN 40HZ TO 350HZ. THESE VOICES SMOOTHLY **GLIDE** TOWARD A PREDETERMINED **CHORD** SPANNING 6 OCTAVES, OVER A DURATION OF 30 SECONDS.

IT'S A WONDERFUL DEMONSTRATION OF THE POWER OF **PRECISELY** CONTROLLING TIME-VARYING AUDIO -- AND USING **SIMPLE** BUILDING BLOCKS TO CREATE A **COMPLEX** SOUND!

A PLAN...

0 **SETUP STAGE:** CREATE PROVISIONS FOR 30 VOICES. IN OUR CASE, WE WILL INSTANTIATE 30 SAWTOOTH WAVE GENERATORS, **RANDOMIZING** THEIR RESPECTIVE **STARTING** FREQUENCIES (OUR EMULATION WILL USE 160-360HZ AS THE STARTING RANGE). EACH VOICE WILL EVENTUALLY REACH ONE OF 9 PREDETERMINED **TARGET** FREQUENCIES.

1 **INITIAL STAGE:** BEGIN THE SOUND BY **RAMPING** UP THE VOICES IN **AMPLITUDE** (WHILE **HOLDING** THE STARTING FREQUENCIES CONSTANT). THE ORIGINAL DEEP NOTE DOES SOMETHING MORE SOPHISTICATED -- WE'LL ONLY APPROXIMATE IT HERE. THE GOAL IS TO CREATE THE PART OF THE SOUND THAT "COMES OUT OF NOWHERE."

2 **CONVERGING STAGE:** GRADUALLY **CHANGE** THE FREQUENCIES OF ALL THE VOICES **TOWARD** THEIR RESPECTIVE **TARGET FREQUENCIES,** ACCOMPLISHED BY UPDATING EACH VOICE'S FREQUENCY EVERY SO OFTEN (EVERY 10::MS), SO THAT IT SMOOTHLY APPROACHES THE TARGET (MUCH LIKE OUR ZENO'S INTERPOLATOR IN CHAPTER 3, EXCEPT THIS INTERPOLATION IS **LINEAR**). HERE, THE SOUND GETS "REALLY BIG"!

3 **TARGET STAGE:** ALL VOICES **REACH** THEIR TARGET FREQUENCIES AT PRECISELY THE **SAME TIME,** SOUNDING OUR PREDETERMINED **CHORD** AND CREATING AN **EPIC** AND UNMISTAKABLE SENSE OF **ARRIVAL** AND **RESOLUTION!** WE WILL HOLD THIS CHORD BRIEFLY BEFORE FADING OUT.

WE CAN ILLUSTRATE THE PROGRAM *GRAPHICALLY* -- 30 *LINES* REPRESENT THE *FREQUENCIES* OF THE 30 VOICES OVER TIME. OBSERVE THE *THREE STAGES* THE SOUND GOES THROUGH!

③ *"ORDER + RESOLUTION"* (AND A *BIG CHORD!*)

THE TARGET FREQUENCIES STACK UP TO A *BIG CHORD* SPANNING MULTIPLE OCTAVES AND GIVING A SENSE OF EPIC *RESOLUTION* AND *ARRIVAL* -- WHOA.

② *"CONVERGENCE"*

THE VOICES SMOOTHLY *GLIDE* TOWARD THEIR RESPECTIVE TARGET FREQUENCIES. OVERALL, 30 VOICES CONVERGE ON 9 TARGET FREQUENCIES IN A GIANT 30-WAY *GLISSANDO*, BUILDING A SENSE OF INTENSE MOTION -- "IT'S HAPPENING!"

① *"CHAOS"*

30 FREQUENCIES RANDOMIZED BETWEEN 160-350HZ GIVE AN UNSETTLING, *BROODING* FEELING -- "SOMETHING IS ABOUT TO HAPPEN..."

1800 HZ

1500 HZ

1200 HZ

900 HZ

600 HZ

360 HZ

300 HZ

150 HZ

75 HZ
37.5 HZ

160 HZ

⓿

FREQUENCY

TIME

THE 9 *TARGET FREQUENCIES* ARE *JUST-INTONED*: THE INTERVALS BETWEEN THEM ARE *TUNED* AS RATIOS OF SMALL INTEGERS. MATHEMATICALLY, THIS "LINES UP" HARMONICS IN THE NOTES AND, SONICALLY, RESULTS IN A BIG, STABLE, AND PURE SOUND.

APPROXIMATE PIANO KEYS FOR THE TARGET FREQUENCIES; MIDDLE C FOR REFERENCE

177

9 TARGET FREQUENCIES
WE ASSOCIATE EACH OF 30
VOICES WITH ONE OF THESE

SETUP

```chuck
// D1,  D2, D3,  D4,  D5,  A5,  D6,   F#6,  A6
[ 37.5, 75, 150, 300, 600, 900, 1200, 1500, 1800,
  37.5, 75, 150, 300, 600, 900, 1200, 1500, 1800,
  37.5, 75, 150, 300, 600, 900, 1200,       1800,
            150, 300,      900, 1200
] @=> float targets[];

float initials[30];
3.0::second => dur CHAOS_HOLD_TIME;
5.5::second => dur CONVERGENCE_TIME;
3.5::second => dur TARGET_HOLD_TIME;
2.0::second => dur DECAY_TIME;

SawOsc saw[30];
Gain gainL[30];
Gain gainR[30];
NRev reverbL => dac.left;
NRev reverbR => dac.right;
0.075 => reverbL.mix => reverbR.mix;

for( 0 => int i; i < 30; i++ )
{
    saw[i] => gainL[i] => reverbL;
    saw[i] => gainR[i] => reverbR;
    1.0 - gainL[i].gain() => gainR[i].gain;
    0.1 => saw[i].gain;
    Math.random2f( 160, 360 ) => initials[i] => saw[i].freq;
    Math.random2f( 0.0, 1.0 ) => gainL[i].gain;
}
```

 DURATION FOR
VARIOUS STAGES

MAKE **30** SAWTOOTH
GENERATORS AS OUR VOICES,
ALONG WITH ADDITIONAL SOUND
OBJECTS FOR SIGNAL ROUTING

RANDOMIZE INITIAL
FREQUENCIES AND **PAN** EACH
SAWTOOTH (ALSO RANDOMLY)
IN THE STEREO FIELD

FOR **REFERENCE**, THIS IS OUR DEEP
NOTE EMULATION **ALGORITHM** AS A
CHUCK PROGRAM, IN FOUR SECTIONS
CORRESPONDING TO OUR INITIAL PLAN.

DON'T WORRY IF YOUR EYES START **WATERING** FROM LOOKING AT THIS CODE -- THIS IS JUST TO GIVE A GENERAL IDEA OF HOW WE CAN USE CODE TO CONTROL SOUND OVER TIME.

"CHAOS"

(1)

```
now + CHAOS_HOLD_TIME => time end;
while( now < end )
{
    1 - (end-now) / CHAOS_HOLD_TIME => float progress;
    for( 0 => int i; i < 30; i++ ) {
        0.1 * Math.pow(progress,3) => saw[i].gain;
    }
    10::ms => now;
}
```

RAMP UP VOLUME FOR EACH VOICE WHILE HOLDING ITS INITIAL FREQUENCY.

"CONVERGENCE"

(2)

```
now + CONVERGENCE_TIME => end;
while( now < end )
{
    1 - (end-now)/CONVERGENCE_TIME => float progress;
    for( 0 => int i; i < 30; i++ ) {
        initials[i] + (targets[i]-initials[i])*progress
            => saw[i].freq;
    }
    10::ms => now;
}
```

IN SMALL TIME INCREMENTS (10::MS) **UPDATE** FREQUENCIES TO APPROACH TARGETS SMOOTHLY!

VOICES **ARRIVE** AT THE **TARGET** FREQUENCIES SIMULTANEOUSLY; **HOLD** THE RESULTING CHORD.

"RESOLUTION"

(3)

```
TARGET_HOLD_TIME => now; // hold the chord!

now + DECAY_TIME => end;
while( now < end )
{
    (end-now) / DECAY_TIME => float progress;
    for( 0 => int i; i < 30; i++ ) {
        0.1 * progress => saw[i].gain; // fade
    }
    10::ms => now;
}
```

FADE TO SILENCE.

THERE ARE SEVERAL PROGRAMMABILITY AND DESIGN IDEAS *IN MOTION* HERE, INCLUDING *PRECISION* OF CONTROL, SONIC *NARRATIVE*, AND *STRENGTH* IN *NUMBERS* OF SIMPLE ELEMENTS ACTING TOGETHER TO CULMINATE IN A SINGLE PRONOUNCED EFFECT.

PRINCIPLE 4.3

BUILD COMPLEXITY AS THE *SUM* OF *SIMPLE* ELEMENTS

AN AUDIO-SPECIFIC VERSION OF VISUAL DESIGN PRINCIPLE 3.5: BUILD COMPLEXITY FROM SIMPLICITY

COMPUTERS ARE REALLY GOOD AT *MAKING COPIES.* ONCE WE CAN PROGRAM *ONE* THING, IT'S TRIVIAL TO INSTANTIATE *MORE* OF IT. THE AIM IS NOT MERELY TO HAVE MORE, BUT TO CREATE SOMETHING *NEW* IN THE *AMALGAM.*

FOR EXAMPLE, OUR *DEEP NOTE* EMULATION IS ACHIEVED THROUGH THE ADDITION OF 30 BASIC SAWTOOTH VOICES, MODULATING THEIR FREQUENCIES IN A SPECIFIC AND SYNCHRONIZED WAY. THIS PRODUCES THE SENSE OF A SINGLE, COHERENT SOUND! WE MIGHT STILL HEAR INDIVIDUAL VOICES IN THE MIX, BUT WE ALSO HEAR THE SUM TOTAL OF THE VOICES AS A *CULMINATING,* COHESIVE SOUND.

THE KEY HERE IS NOT ONLY THAT WE HAVE *MANY* VOICES, BUT THAT EACH ONE IS BOTH *INDEPENDENTLY* CHANGING IN FREQUENCY AND *GLOBALLY COORDINATED* WITH THE OTHER VOICES.

> THIS CASE STUDY ALSO REINFORCES PERHAPS THE MOST IMPORTANT *ETHOS* IN THIS CHAPTER...

PRINCIPLE 4.5

DESIGN THINGS WITH A COMPUTER THAT WOULD NOT BE POSSIBLE WITHOUT!

DO NOT SIMPLY COPY, PORT, DIGITIZE, OR EMULATE. RATHER, CREATE SOMETHING NOVEL AND UNIQUE TO THE *MEDIUM* -- SOMETHING THAT *COULD NOT EXIST* WITHOUT IT.

IT'S *TEMPTING* TO *REMAKE* WHAT ALREADY EXISTS. WHILE THAT REMAINS A USEFUL EXERCISE, MANY PEOPLE DO THAT BECAUSE IT'S *OBVIOUS.* BUT WITH NEW TECHNOLOGICAL MEDIUMS ALSO COME THE OPPORTUNITY AND *RESPONSIBILITY* TO DISCOVER WHAT THE MEDIUM IS *INNATELY* GOOD AT. *DESIGN TO THE MEDIUM!*

THIS IS AN ESSENTIAL *GUIDING PRINCIPLE* OF ARTFUL DESIGN (WITH *ANY* MEDIUM OR TECHNOLOGY). LET'S APPLY THIS LENS AND DECONSTRUCT A COMPUTER MUSIC COMPOSITION -- ONE THAT USES THE COMPUTER AS A KIND OF *PERSONAL MUSICAL FILTER* TO THE WORLD.

WHILE VIOLINS AND PIANOS ARE *SUBLIME VEHICLES* OF *MUSICAL THOUGHT*, PEOPLE HAVE OFTEN LISTENED WITH MUSICAL EARS TO THE SOUNDS OF WINE GLASSES, CRICKETS ON A SUMMER NIGHT, THE WIND IN THE TREES, STEPS ON THE PAVEMENT, BIRD SONG, SPEECH, CONCH SHELLS, CHURCH BELLS, ETC., AND COMPOSERS FROM *MONTEVERDI* TO *MESSIAEN* HAVE TINKERED WITH *WORLDNOISE* IN THEIR MUSIC.

UNTIL RECENTLY, HOWEVER, IT HAS BEEN DIFFICULT TO CAPTURE SOUNDS OF THE NATURAL WORLD AND TAKE THEM INTO OUR COMPOSITION WORKSHOPS. BUT *NOW*, WITH THE CONVERGENCE OF RECORDING AND COMPUTER TECHNOLOGIES, WE HAVE THE ABILITY TO PLAY THESE *INSTRUMENTS* OF THE WORLD AS NEVER BEFORE.

PAUL LANSKY
Homebrew

THE FIVE PIECES ON THIS RECORDING ARE ATTEMPTS TO VIEW THE MUNDANE, EVERYDAY NOISES OF DAILY LIFE THROUGH A *PERSONAL MUSICAL FILTER*. THERE ARE NO OTHER-WORLDLY SOUNDS USED HERE -- JUST THE COMINGS AND GOINGS WHICH GREET OUR EARS AS WE MAKE IT THROUGH THE DAY. WITH THE ASSISTANCE AND INTERVENTION OF COMPUTER TECHNOLOGY, THESE PIECES MODESTLY TRY TO MAKE THE ORDINARY SEEM EXTRAORDINARY, THE UNMUSICAL, MUSICAL. THEY TRY TO FIND *IMPLICIT MUSIC* IN THE WORLDNOISE AROUND US!

PAUL LANSKY'S 1992 ALBUM *HOMEBREW* USED THE SOUNDS OF KITCHENWARE, TRAFFIC, HANDS CLAPPING, AND A *MALL* IN PRINCETON, NEW JERSEY. IT REIMAGINED THEM WITH THE COMPUTER TO CREATE SOUNDS THAT WERE AT ONCE *FAMILIAR* AND *FANTASTICAL.* WHEN I FIRST HEARD THIS ALBUM YEARS AGO, I WAS *MESMERIZED.* IT *SHATTERED* MY EARLIER NOTION THAT COMPUTERIZED SOUNDS HAVE TO SOUND COLD AND MECHANICAL. IN PAUL'S MUSIC, I HEARD A *WORLD* THAT WAS CLOSER TO THE EVERYDAY, AND A *MUSIC* THAT WOULDN'T BE POSSIBLE WITHOUT A COMPUTER.

182

THE FIRST PIECE ON THE ALBUM WAS CALLED "TABLE'S CLEAR." (THE IDEA OF A **SOUND KITCHEN** MADE **LITERAL**!)

MUSICIANS HAVE ALWAYS LOOKED AT THE DINNER TABLE WITH **GREEDY EARS** (PARDON THE METAPHORICAL CONFUSION), BUT IT'S HARD NOT TO TREAT BOTTLES AND GLASSES AS IF THEY WERE **PERCUSSION INSTRUMENTS.** "TABLE'S CLEAR" IS A DIGITAL EXPLORATION OF THIS DOMAIN -- HERE NOTHING IS BREAKABLE, AND WE CAN PLAY AS FAST AND HARD AS WE LIKE.

THE PIECE HAD ITS ORIGIN ONE EVENING AFTER DINNER IN OCTOBER 1990, WHEN MY TWO SONS, **JONAH** AND **CALEB** (AGES 14 AND 9 AT THE TIME) TOOK OUR KITCHEN APART, RECORDING THE SOUNDS OF EVERYTHING THEY COULD FIND THAT WOULD MAKE NOISE (INCLUDING THEMSELVES). I RAN THE TAPE MACHINE AND **HANNAH** RAN FOR COVER. I THEN TRANSFERRED ALL THE SOUNDS TO MY COMPUTER, SPENT A FEW MONTHS WORKING, AND CAME UP WITH THIS PIECE.

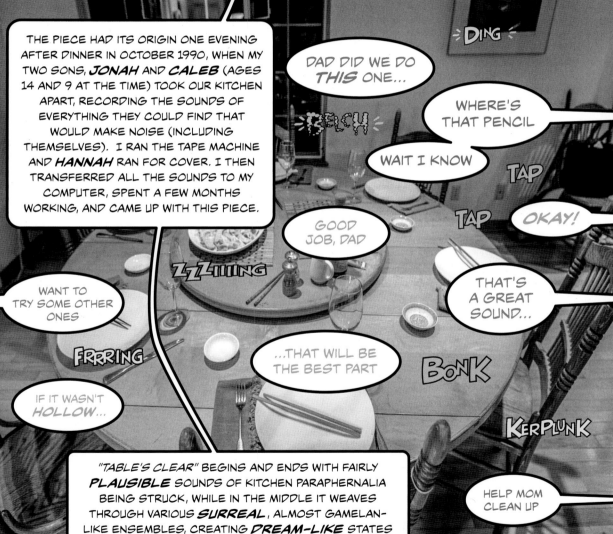

DAD DID WE DO **THIS** ONE...

WHERE'S THAT PENCIL

WAIT I KNOW

GOOD JOB, DAD

OKAY!

THAT'S A GREAT SOUND...

WANT TO TRY SOME OTHER ONES

...THAT WILL BE THE BEST PART

IF IT WASN'T **HOLLOW**...

HELP MOM CLEAN UP

"TABLE'S CLEAR" BEGINS AND ENDS WITH FAIRLY **PLAUSIBLE** SOUNDS OF KITCHEN PARAPHERNALIA BEING STRUCK, WHILE IN THE MIDDLE IT WEAVES THROUGH VARIOUS **SURREAL**, ALMOST GAMELAN-LIKE ENSEMBLES, CREATING **DREAM-LIKE** STATES FROM WHICH WE FINALLY AWAKE, ONLY TO BE REMINDED OF OUR OWN AWKWARD PHYSICAL LIMITATIONS.

```
/* 22c    soft pop*/770956,
/* 23c    softpop 2*/778982,
/* 24c    soft pop 3*/788816,
/* 25c    soft pop 4*/793270
/* 26c
/* 27c
/* 28c    water gurgle*/809146,
/* 2                         349,
/* 3                         401,
/* 3                    0.0
/* 3                    21
/* 33a    clear glass 10.071 */102
/* 34a    clear glass2 10.071*/104
/* 35a    ringingtrangle 10.032 */
/* 36a    clear lid 10.002*/112609
/* 37a    high clear lid 10.071*/1
/* 38b    woodblock*/1183996,
/* 39b    woodblock2*/1197
/* 40b    high block*/12
/* 41c    one notetube
/* 42c    two note tu
/* 43c    mouth pop
/* 44c    mouth pop
/*
/*
/*
/*
/*
/*
/*
/*
/*
/*
/*
/* 54c    softpop*
/* 55d    soft fry
/* 56d    harder fry
/* 57c     clap1 */
/* 58c    clap2 */ 43.
/* 59     end */44.944*

input("s
output("
punch(1,

oldpch =
clearglass=29
inskip = get_array(0,clearglass)/SR
endnote = get_array(0,clearglass+1)/
newpitches = load_array(1,7.10,8.07)
time=0
olddur = endnote-inskip
for(i=0; i<newpitches; i=i+1) {
    newpch = get_array(1,i)
    transposition = (octpch(newpch)
octpch(oldpch)) * 12
```

```
float SR   SR=44100
phrases = load_array(0,
/* 0    big belch 1 */          0
/* 1    big belch 2 */          6
/* 2    okay */                 111572
/* 3    where that pencil */    1
/* 4    wait I know */          2
/* 5    put all these things down */2
/* 6    if it wasnt hollow */   4
/* 7    want to try some other ones *
/* 8    dad wait dad */         6
```

```
    newdur = olddur * ratio
    xrans(inskip=0,outskip=time,sampleend=newdur*(-1)
        transposition,amp=1,outdur=999)
    time = time + newdur
}
old_pitches=load_array(0,10.03)
newpitches = load_array(1,9.03,8.10,8.08,8.03)
input("sf/woodblock.snd")
olddur = dur(0)
for(i=0; i<newpitches; i=i+1) {
    newpch = get_array(1,i)
    oldpch = get_array(0,0)
```

PAUL'S ORIGINAL CODE WAS WRITTEN IN **CMIX**, A PROGRAMMING LANGUAGE HE CREATED.

INDIVIDUAL SOUNDS ARE **ISOLATED** FROM THE RECORDING, EDITED, AND **TRANSFORMED** INTO USABLE **MUSICAL ATOMS!**

MANY SOUNDS FROM THE KITCHEN ARE **PITCHED.** POTS, PANS, LIDS, PLATES, GLASSES ALL HAVE DISCERNIBLE **RESONANT** FREQUENCIES, DUE TO THEIR **MATERIALS** AND **SHAPES.** THEY ARE GENERALLY NOT, HOWEVER, **MUSICALLY** TUNED. THE COMPUTER ALTERS THIS, CAPABLE OF STRETCHING, TUNING SOUNDS, AND TURNING THEM INTO BUILDING BLOCKS OF PITCH, HARMONY, AND RHYTHM.

IT'S A WAY TO **TRANSFORM REALITY** WITH COMPUTERS, TAKING THE FAMILIAR EVERYDAY AND **RESHAPING** IT THROUGH A PERSONAL MUSICAL FILTER!

COMPUTERS WERE ORDERS OF MAGNITUDE **SLOWER** BACK IN 1992, AND PROGRAMS APPEARED CLOSER TO MACHINE CODE THAN HUMAN-READABLE DESCRIPTION. BUT THE **BASIC PRINCIPLES** WERE THE SAME. NO MATTER HOW ADVANCED THE TECHNOLOGY, IT TAKES HUMAN **INTENTIONALITY** TO USE THE COMPUTER AS A TOOL AND A LABORATORY FOR NEW IDEAS.

CODE IS ITSELF A TYPE OF **DYNAMIC MUSICAL SCORE.** INDIVIDUAL SOUNDS, LIKE ≳DING≴ FROM HITTING A WINE GLASS, CAN BE PRECISELY SEQUENCED THROUGH **ALGORITHMS** THAT TELL THE COMPUTER WHAT TO PLAY, WHEN, AND HOW. THIS ALGORITHMIC PROCESS CAN EXPRESS A MIXTURE OF DELIBERATE **ORDER** AND CONTROLLED **CHANCE.** PAUL, FOR EXAMPLE, USES A LOT OF **RANDOM-WITHOUT-REPLACEMENT** ALGORITHMS, WHICH DISALLOW DUPLICATES WHILE RANDOMLY CHOOSING FROM A SET OF POSSIBLE SOUNDS.

A MIXTURE OF PREDETERMINED AND **DYNAMICALLY** GENERATED MUSICAL MATERIAL CREATES THE EXPRESSIVE RHYTHMS AND GROOVES IN "TABLE'S CLEAR."

THIS **SPECTROGRAM** BELOW VISUALIZES THE **FREQUENCY CONTENT** THROUGHOUT THE 18 MINUTES OF "TABLE'S CLEAR," AND REVEALS ITS **STRUCTURAL** FORM. **TIME** PROCEEDS FROM LEFT TO RIGHT. LOWER FREQUENCIES APPEAR TOWARD THE BOTTOM, COLOR-CODED FOR INTENSITY.

 LAYER BY LAYER AND WITH INCREASING COMPLEXITY, AN **IMPROBABLE** RHYTHM BEGINS TO EMERGE... AS IF THE SOUNDS CANNOT HELP BUT MAGICALLY FALL INTO RHYTHM WITH ONE ANOTHER.

 IMPROBABLE GIVES WAY TO **FANTASTICAL**, AS THE SOUNDS OF THE KITCHEN COME TOGETHER IN A FULL-ON **SYMPHONIC DANCE.**

VERTICAL LINES VISUALIZE BROAD-SPECTRUM SIGNALS, LIKE THE SOUND OF A POT BEING STRUCK.

THE FIRST HARMONIC SHIFT / CHORD CHANGE IS A PROMINENT MOMENT. UP TO NOW, EVERYTHING HAD BEEN IN RHYTHMIC DEVELOPMENT; THE INTRODUCTION OF NEW HARMONIC LANGUAGES ADDS ANOTHER DIMENSION.

WE CAN SEE THE ASCENDING VOICES IN THE UPWARD SLOPING RED LINES.

3 MINUTES 45 SECONDS

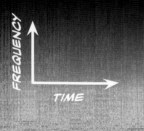

FREQUENCY

TIME

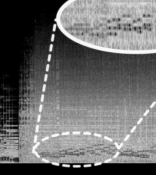

(A) **"THE ORDINARY"**

STARTS OUT ORDINARY, JUST SOUNDS OF KITCHENWARE BEING STRUCK, **HAPHAZARDLY**, **CHAOTICALLY**...

"SELF-ORGANIZATION" (B)

THE MAGIC OF THIS PIECE IS THAT THE EVERYDAY KITCHEN GRADUALLY COMES **ALIVE** AND ASSEMBLES ITSELF INTO COHERENT MUSICAL PITCHES AND SHIFTING RHYTHMIC PATTERNS. THIS SECTION **TRANSFORMS** AND **DEFAMILIARIZES** THE ORDINARY INTO THE **EXTRAORDINARY.**

(C) **"PERCUSSIVE THROWDOWN"**

SOUNDS ARE REVISITED IN AN UNPITCHED PERCUSSIVE HOOTENANNY: YET ANOTHER RECONTEXTUALIZATION OF THE MATERIAL!

(D)

"ASCENSION"

ETHEREAL ASCENDING VOICES PROVIDE A SENSATION OF **WEIGHTLESSNESS**, CONTRASTED WITH MORE URGENT ENERGY OF RAPIDLY STRUCK SOUNDS OF PRIOR SECTIONS. THE FANTASTICAL TRANSFORMATION REACHES A CLIMAX.

(E) **"PITCHED & PERCUSSIVE"**

RINGING SOUNDS AND UNPITCHED PERCUSSIVE ELEMENTS MAKE FOR A PENULTIMATE EXPLORATION OF THE SOURCE MATERIAL. IT'S OUR **RETURN TRIP** BACK TO EVERYDAY REALITY...

"BACK TO START"

ENDS WHERE WE STARTED, AS IF WAKING FROM A **DREAM**...

I FIRST HEARD "TABLE'S CLEAR" IN SCOTT LINDROTH'S ELECTRONIC MUSIC COURSE AT DUKE UNIVERSITY (WHERE I DID MY UNDERGRAD). AS THE EVERYDAY SOUNDS OF KITCHEN UTENSILS MAGICALLY ARRANGED THEMSELVES AS IF BY THE WAVE OF A WIZARD'S WAND, I WAS *ENTRANCED.* UP TO THAT POINT AND IN MY THEN *LIMITED* EXPOSURE TO COMPUTERIZED SOUND, I HAD ONLY HEARD COMPUTER MUSIC THAT WAS INTERESTING *CONCEPTUALLY*...

...AND HERE WAS *THIS* PIECE OF MUSIC -- VISCERAL, ORGANIC, *PLAYFUL* -- THAT I COULD SIMPLY *LIKE.* MOREOVER, IT WAS *UNMISTAKABLE* THAT THIS MUSIC WAS ONLY POSSIBLE THROUGH SOME *ARTFUL INTERVENTION* OF THE COMPUTER.

I REALIZED IF THIS WAY OF MAKING MUSIC IS AN *AESTHETIC*, THEN AESTHETICS *CANNOT* MERELY BE A *PASSIVE* THING, BUT AN *ACTIVE* AGENT FOR EXPRESSION, EMBRACING BOTH THE WONDERS OF TECHNOLOGY AND THE HUMAN MIND THAT WORKS WITH IT.

PAUL'S NOTION OF A *PERSONAL MUSICAL FILTER* IS A *LENS* THROUGH WHICH TO *HEAR* THE *IMPLICIT MUSIC* OF EVERYDAY LIFE, TO RE-ENGAGE WITH THE ORDINARY SOUNDS THAT WE BARELY THINK ABOUT, WITH *NEW EARS.*

VAROOOM

VAROOOM

VAROOOM

IT INVITES US TO *LISTEN* TO SOUNDS THAT NATURALLY EXIST AROUND US AND IMAGINE HOW WE MIGHT *TRANSFORM* THEM INTO SOMETHING *EXTRAORDINARY*, TO NOTICE A TYPE OF POETRY IN EVERYDAY LIFE.

THESE IDEAS INSPIRE US. WHEN THE *STANFORD LAPTOP ORCHESTRA* TRAVELED TO *BEIJING* FOR A RESIDENCY, WE EXPERIMENTED WITH *FOUND SOUNDS.* JOHN GRANZOW AND KITTY SHI WORKED WITH THE SOUND AND INTERACTION OF EVERYDAY OBJECTS LIKE CHOPSTICKS, BOWLS, AND FANS...

AS THEY EXPLORED THE *UNLIKELY* JUXTAPOSITIONS OF FAMILIAR SOUNDS, A *GYROSCOPE* SENSOR (ON A PHONE) TRACKED THE *ROTATION* OF THE TABLE, WHILE THE NEARBY LAPTOPS *SAMPLED* AND *TRANSFORMED* THE ACOUSTICALLY MADE SOUNDS!

UMM... I WONDER IF THIS *AWESOME FOG* WILL *SHORT OUT* THE ELECTRONICS OF OUR LAPTOP ORCHESTRA...

JOHN

ROMAIN

KITTY

ROTARY *FAN*

CERAMIC JARS

METAL BOWLS

ERHU BOW

LAZY SUSAN (ROTATING SERVING TABLE)

SHUTTLECOCK

CHOPSTICKS

IN A WORK SIMPLY CALLED "BEIJING," **MADELINE HUBERTH** AND I TRANSFORMED THE SOUNDS OF STREETS, PEOPLE, BICYCLES, AND SUBWAY TRAINS, INSPIRED BY PAUL'S **HOMEBREW** ETHOS AND A SYNTHESIS TECHNIQUE THAT PAUL USED MUSICALLY.

MADELINE

WE RECORDED, ISOLATED, TRANSFORMED, AND REASSEMBLED THE SOUNDS OF THE **CITY**, RECONTEXTUALIZING THE FAMILIAR THROUGH A METAPHORICAL MUSICAL FILTER!

请站稳扶好！
PLEASE STAND FIRM AND HOLD THE HANDRAILS!

PEOPLE

TRICYCLES & BICYCLES

TRAFFIC

SUBWAY

蛋夹馍是四块！

STREET FOOD VENDORS

BACK STREETS

ENGINES REVVING

SIZZLING FRIED DOUGH

THE DESIGN WEAVED THE TRANSFORMED SOUNDS INTO A **SONIC TAPESTRY** THAT WAS BOTH THE CITY AND A MORE ABSTRACT VERSION OF IT IN OUR **MIND'S EAR**...

187

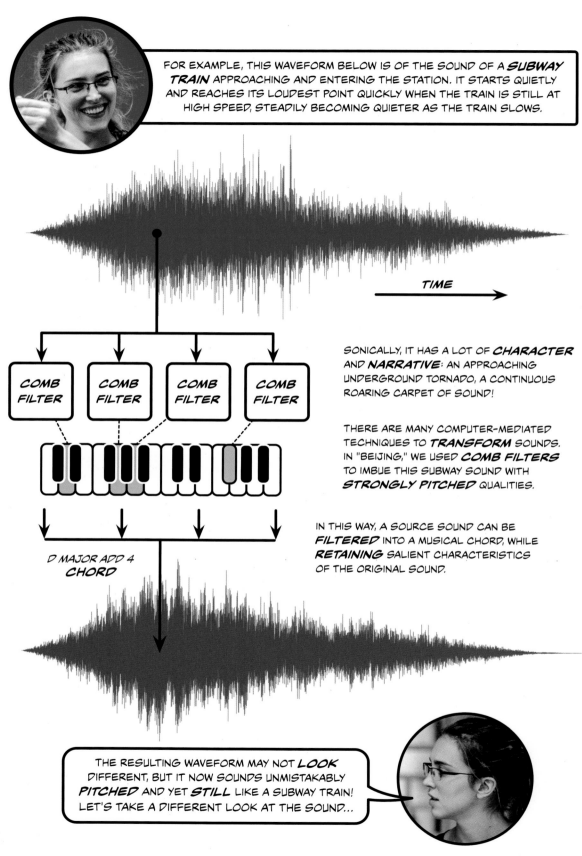

FOR EXAMPLE, THIS WAVEFORM BELOW IS OF THE SOUND OF A *SUBWAY TRAIN* APPROACHING AND ENTERING THE STATION. IT STARTS QUIETLY AND REACHES ITS LOUDEST POINT QUICKLY WHEN THE TRAIN IS STILL AT HIGH SPEED, STEADILY BECOMING QUIETER AS THE TRAIN SLOWS.

TIME

COMB FILTER

COMB FILTER

COMB FILTER

COMB FILTER

D MAJOR ADD 4 *CHORD*

SONICALLY, IT HAS A LOT OF *CHARACTER* AND *NARRATIVE*: AN APPROACHING UNDERGROUND TORNADO, A CONTINUOUS ROARING CARPET OF SOUND!

THERE ARE MANY COMPUTER-MEDIATED TECHNIQUES TO *TRANSFORM* SOUNDS. IN "BEIJING," WE USED *COMB FILTERS* TO IMBUE THIS SUBWAY SOUND WITH *STRONGLY PITCHED* QUALITIES.

IN THIS WAY, A SOURCE SOUND CAN BE *FILTERED* INTO A MUSICAL CHORD, WHILE *RETAINING* SALIENT CHARACTERISTICS OF THE ORIGINAL SOUND.

THE RESULTING WAVEFORM MAY NOT *LOOK* DIFFERENT, BUT IT NOW SOUNDS UNMISTAKABLY *PITCHED* AND YET *STILL* LIKE A SUBWAY TRAIN! LET'S TAKE A DIFFERENT LOOK AT THE SOUND...

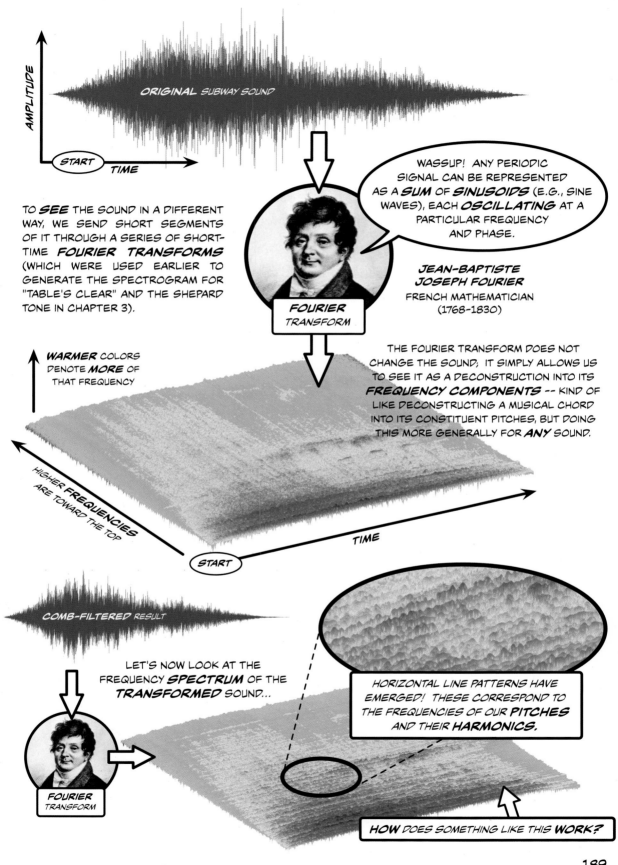

AMPLITUDE

ORIGINAL SUBWAY SOUND

START TIME

TO **SEE** THE SOUND IN A DIFFERENT WAY, WE SEND SHORT SEGMENTS OF IT THROUGH A SERIES OF SHORT-TIME **FOURIER TRANSFORMS** (WHICH WERE USED EARLIER TO GENERATE THE SPECTROGRAM FOR "TABLE'S CLEAR" AND THE SHEPARD TONE IN CHAPTER 3).

WASSUP! ANY PERIODIC SIGNAL CAN BE REPRESENTED AS A **SUM** OF **SINUSOIDS** (E.G., SINE WAVES), EACH **OSCILLATING** AT A PARTICULAR FREQUENCY AND PHASE.

JEAN-BAPTISTE JOSEPH FOURIER FRENCH MATHEMATICIAN (1768-1830)

FOURIER TRANSFORM

WARMER COLORS DENOTE **MORE** OF THAT FREQUENCY

THE FOURIER TRANSFORM DOES NOT CHANGE THE SOUND; IT SIMPLY ALLOWS US TO SEE IT AS A DECONSTRUCTION INTO ITS **FREQUENCY COMPONENTS** -- KIND OF LIKE DECONSTRUCTING A MUSICAL CHORD INTO ITS CONSTITUENT PITCHES, BUT DOING THIS MORE GENERALLY FOR **ANY** SOUND.

HIGHER **FREQUENCIES** ARE TOWARD THE TOP

TIME

START

COMB-FILTERED RESULT

LET'S NOW LOOK AT THE FREQUENCY **SPECTRUM** OF THE **TRANSFORMED** SOUND...

HORIZONTAL LINE PATTERNS HAVE EMERGED! THESE CORRESPOND TO THE FREQUENCIES OF OUR **PITCHES** AND THEIR **HARMONICS.**

FOURIER TRANSFORM

HOW DOES SOMETHING LIKE THIS **WORK?**

189

AN **INVERSE COMB FILTER REINFORCES** SPECIFIC FREQUENCIES IN AN EXISTING SOUND THROUGH A **RECURSIVE FEEDBACK LOOP.** AN ELECTRICAL ENGINEER MIGHT REPRESENT IT AS THIS **BLOCK DIAGRAM:**

INPUT

OUTPUT

(FEEDBACK)

Z^{-L}
DIGITAL DELAY

R^L

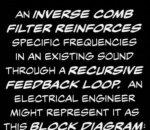

WE CAN PROGRAM A COMB FILTER WITH A FEW LINES OF CHUCK CODE!

```
// feedforward: input to output
adc => Gain node => dac;
// feedback: from output back to input
node => Delay delay => Gain attenuation => node;

// set amount of delay
500::samp => delay.delay;
// set attenuation
0.8 => attenuation.gain;
```

THIS IS THE **INTERNAL SCHEMATIC** OF A SOUND FILTER! SOUND GOES IN, IT GETS CHANGED SOMEHOW, AND PROCESSED SOUND COMES OUT -- KINDA LIKE AN **EFFECT PEDAL** FOR AN ELECTRIC GUITAR...

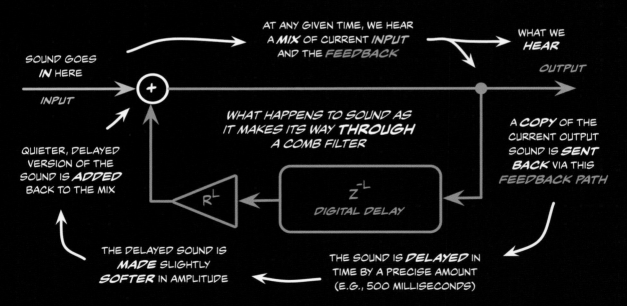

AT ANY GIVEN TIME, WE HEAR A **MIX** OF CURRENT **INPUT** AND THE *FEEDBACK*

WHAT WE **HEAR**

OUTPUT

SOUND GOES **IN** HERE

INPUT

A **COPY** OF THE CURRENT OUTPUT SOUND IS **SENT BACK** VIA THIS *FEEDBACK PATH*

WHAT HAPPENS TO SOUND AS IT MAKES ITS WAY **THROUGH** A COMB FILTER

Z^{-L}
DIGITAL DELAY

R^L

QUIETER, DELAYED VERSION OF THE SOUND IS **ADDED** BACK TO THE MIX

THE DELAYED SOUND IS **MADE** SLIGHTLY **SOFTER** IN AMPLITUDE

THE SOUND IS **DELAYED** IN TIME BY A PRECISE AMOUNT (E.G., 500 MILLISECONDS)

ESSENTIALLY, WE HAVE CREATED **DIGITAL ECHOES!** BY SETTING THE DELAY AMOUNT (L), WE CAN TUNE THE **RATE** AT WHICH ECHOES ARE HAPPENING TO THE FREQUENCY THAT WE WANT TO REINFORCE. WE CAN PROGRAM THE ECHOES TO RECUR **SO QUICKLY** (E.G., HUNDREDS OF TIMES PER SECOND) THAT WE STOP PERCEIVING THEM AS INDIVIDUAL COPIES, BUT INSTEAD AS A **PITCH.**

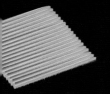

THIS KIND OF FILTER **REINFORCES** SPECIFIC FREQUENCIES IN AN EXISTING SOUND BY **REDUCING** ALL OTHER FREQUENCIES IN THE SIGNAL. IT IS A FORM OF **SUBTRACTIVE SYNTHESIS.**

THE SHAPE BELOW VISUALIZES WHAT THE COMB FILTER DOES TO DIFFERENT FREQUENCIES IN THE INPUT SOUND -- IT REINFORCES FREQUENCIES CORRESPONDING TO THE SHARP **TEETH** (OR PEAKS), WHILE TAKING OUT FREQUENCIES IN THE "VALLEYS" BETWEEN THEM.

ST PEAK HOLDS THE
AMENTAL AND IS
LY ASSOCIATED WITH
TCH WE HEAR FROM
OVERALL SOUND.

THE **PEAKS** OF A COMB FILTER'S **FREQUENCY RESPONSE** ARE EQUALLY SPACED FREQUENCIES AT INTEGER MULTIPLES, OR **HARMONICS**, OF THE FIRST FULL PEAK.

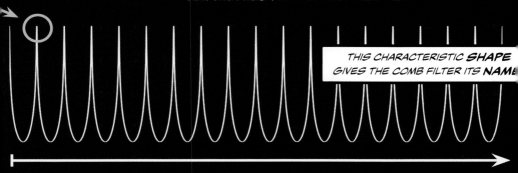

THIS CHARACTERISTIC **SHAPE** GIVES THE COMB FILTER ITS **NAME**

HERTZ (HZ)   **FREQUENCY**

HOW TO MUSICALLY **CONTROL** A COMB FILTER

EQUENCY OF THE FIRST PEAK (AND PITCH WE PERCEIVE) IS ENTIRELY DETERMINED BY TH
OF **DELAY.**   **SHORTER** DELAYS RESULT IN **HIGHER** FUNDAMENTAL FREQUENCIE
GOUS TO HOW SHORTER VIBRATING STRINGS PRODUCE HIGHER PITCHES).   THE **FORMUL**
GIVES US A WAY TO **TUNE** A COMB FILTER TO IMBUE A **SPECIFIC PITCH:**

$$\text{DELAY}_{\text{(\# OF SAMPLES)}} = \frac{\text{SAMPLE RATE}}{\text{FREQUENCY OF DESIRED PITCH}}$$

A TYPICAL AUDIO **SAMPLE RATE** OF 44100HZ (I.E., SAMPLES PER SECOND, OR **HOW**
NY VALUES ARE USED TO REPRESENT A SECOND OF AUDIO), THE FOLLOWING MUSICAL
ORD CAN BE CONSTRUCTED USING FOUR COMB FILTERS, WHOSE DELAYS ARE TUNED TO
FOLLOWING FREQUENCIES:

DELAY (# OF SAMPLES)	300.3	225.0	200.5	119.2
	COMB FILTER	COMB FILTER	COMB FILTER	COMB FILTER

OSER THE FEEDBACK ATTENUATION IS TO 1, THE **SHARPER** THE PEAKS IN THE RESPON
ORE PRONOUNCED THE AUDIBLE EFFECT).   THE CLOSER THE ATTENUATION IS TO
ORE THE FILTER LEAVES THE SIGNAL UNCHANGED.

PUTTING A **SOUND** THROUGH A **FILTER** IS LIKE
POURING A MIXTURE INTO A **SIEVE**: SOME THINGS FALL
THROUGH, OTHER THINGS REMAIN. WITH SOUND, WHAT
REMAINS IS WHAT YOU **HEAR.** FOR EXAMPLE...

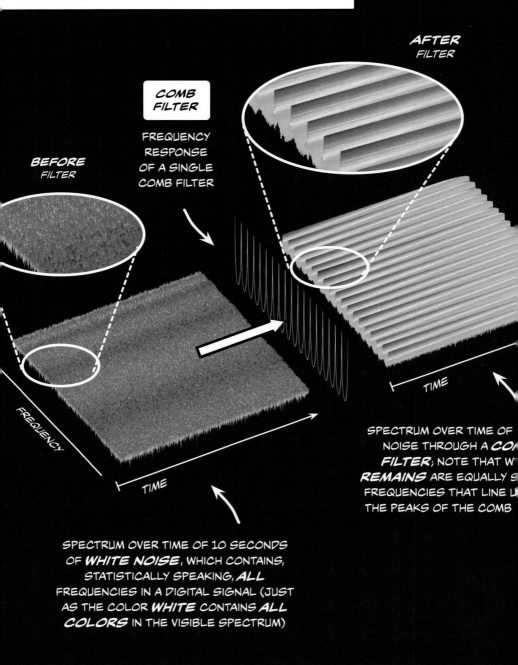

AFTER
FILTER

COMB
FILTER

FREQUENCY
RESPONSE
OF A SINGLE
COMB FILTER

BEFORE
FILTER

FREQUENCY

TIME

TIME

SPECTRUM OVER TIME OF
NOISE THROUGH A **COM**
FILTER; NOTE THAT W
REMAINS ARE EQUALLY S
FREQUENCIES THAT LINE L
THE PEAKS OF THE COMB

SPECTRUM OVER TIME OF 10 SECONDS
OF **WHITE NOISE**, WHICH CONTAINS,
STATISTICALLY SPEAKING, **ALL**
FREQUENCIES IN A DIGITAL SIGNAL (JUST
AS THE COLOR **WHITE** CONTAINS **ALL**
COLORS IN THE VISIBLE SPECTRUM)

IF OUR **THX** SOUND IS SYNTHESIS THROUGH THE **ADDITION** OF 30 VOICES,
THEN **FILTERING** IS A TYPE OF **SUBTRACTIVE SYNTHESIS**, IN WHICH WE
SCULPT A SOUND BY SELECTIVELY FILTERING OUT PARTS OF THE ORIGINAL.

THERE ARE SEVERAL **VARIANTS** OF A COMB FILTER. THE ONE USED IN "BEIJING" IS A **KARPLUS-STRONG PLUCKED STRING** FILTER, WHICH HAS AN **EXTRA LOWPASS FILTER** IN THE FEEDBACK LOOP THAT ATTENUATES HIGH FREQUENCIES OVER TIME. THIS GIVES IT A WARMER, LESS ABRASIVE SOUND -- IT ALSO IS CLOSER TO HOW SOUND PROPAGATES THROUGH AIR, WITH HIGHER FREQUENCIES DISSIPATING FIRST. ANOTHER DETAIL: AN **ALL-PASS DELAY** CAN BE USED TO PRODUCE **FRACTIONAL** (NON-INTEGER) SAMPLE DELAY FOR MORE PRECISE PITCH TUNING.

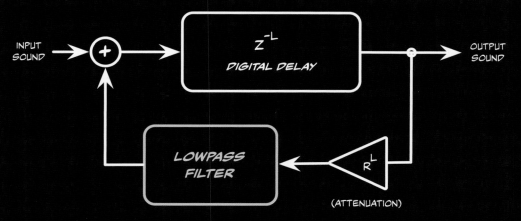

IF WE **PRELOAD** THE DELAY WITH A **WHITE NOISE** BURST AND LET IT RECIRCULATE IN THE FILTER'S FEEDBACK LOOP, WE CAN SEE THE LOWPASS FILTER'S EFFECT ON THE RESULTING SIGNAL OVER TIME...

...THE PLUCKED STRING FILTER **REDUCES** HIGHER FREQUENCIES **FASTER** THAN LOWER FREQUENCIES!

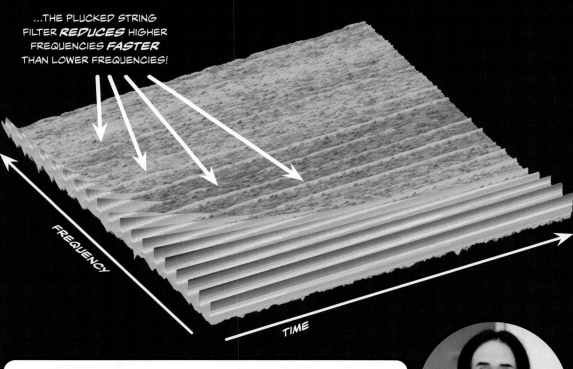

AND **THIS** IS HOW THE PITCHES ARE **IMPRINTED** ON THE SUBWAY SOUNDS OF "BEIJING"! AS YOU CAN SEE, THERE IS QUITE A BIT OF **ENGINEERING PRECISION** INVOLVED IN THIS KIND OF THING. BUT JUST AS IMPORTANTLY, IT'S ABOUT WHAT WE CAN **DO** WITH IT -- THESE TECHNIQUES ALLOW US TO TRANSFORM SOUND WITH A **COMPUTER** IN WAYS THAT WOULD **NOT BE POSSIBLE** WITHOUT!

PRINCIPLE 4.6

USE THE COMPUTER AS *AGENT* OF *TRANSFORMATION*

THE COMB FILTER IS JUST ONE OF *MANY* PROGRAMMABLE TECHNIQUES TO *TRANSFORM REALITY* WITH THE COMPUTER, TO EXPERIENCE THE WORLD THROUGH A *DIFFERENT LENS!*

IT IS ESSENTIAL TO FIGURE OUT *WHAT* TO TRANSFORM. FOR EXAMPLE, IT IS UNWIELDY TO TRANSFORM ENTIRE FIELD RECORDINGS; INSTEAD WE MIGHT FIRST *ISOLATE* INDIVIDUAL SOUNDS (E.G., A TAP OF A FRYING PAN), THEN *TRANSFORM* THEM (E.G., PITCH SHIFT OR FILTER THE SOUND), *RE-ARRANGE* THEM IN TIME (LIKE PAUL LANSKY'S KITCHENWARE IN A HIGHLY RHYTHMIC ARRANGEMENT), AND *MANIPULATE* EACH COMPONENT INDEPENDENTLY.

ONE POSSIBLE *BLUEPRINT* FOR TRANSFORMATION!

SOUND COLLECTION

LISTEN TO THE WORLD AROUND US

FOUND SOUNDS

TIME MODIFICATIONS CAN SHORTEN OR LENGTHEN A SOUND TO DESIRED MUSICAL DURATION; CAN ALSO MASSIVELY STRETCH SOUNDS TO ACHIEVE A SPECIAL EFFECT -- E.G. STRETCHING A *YELL* TO BE 20 TIMES ITS ORIGINAL LENGTH

ISOLATION

TUNE THE SOUND, TRANSPOSE ITS PITCH, OR IMBUE A SOUND WITH PITCH

INDIVIDUAL SOUNDS

TRIM INDIVIDUAL SOUNDS FROM RECORDING

TRANSFORMATION

SEQUENCING AND *JUXTAPOSITION* OF SOUND AT VARIOUS TIMESCALES; CREATE SONIC GESTURES FROM COMPOSITING INDIVIDUAL SOUNDS, OR CRAFTING UNCANNY REPETITION

MODIFIED SOUNDS

IDENTIFY STARTING POINT OF EACH SOUND (WHEN A SOUND BEGINS IN OUR PERCEPTION; ESPECIALLY USEFUL FOR SOUNDS WE INTEND TO USE PERCUSSIVELY)

ASSEMBLY

SONIC GESTURES

MUSICAL PHRASES

(ANOTHER POTENTIAL PLACE TO EMPLOY *RANDOMNESS*)

PREPARE THE SOUND FOR RE-USE: *NORMALIZE* THE VOLUME, *EQUALIZE* ITS FREQUENCY CONTENTS (SO IT *POPS* THE RIGHT WAY), *FADE* IN/OUT THE BEGINNING AND END (SO IT WON'T SOUND ABRUPT WHEN SPLICED INTO THE MIX), ETC.

ARTFULLY *ASSEMBLE* TRANSFORMED SOUNDS IN *TIME* (E.G., RHYTHMIC PATTERNS) AND *LAYERS* (HOW WE HEAR THEM TOGETHER)

STRUCTURE

THE HIGH-LEVEL STRUCTURAL PLAN AND OVERALL *FORM* OF THE SONIC / MUSICAL DESIGN

ADDITIVE SYNTHESIS

SINCE ANY PERIODIC SOUND CAN BE DESCRIBED AS A SUM OF SINUSOIDS (THANKS, FOURIER), WE CAN, IN THEORY, SYNTHESIZE ANY SOUND BY ADDING SPECIFIC SINE WAVES TOGETHER

SUBTRACTIVE SYNTHESIS

SCULPT SOUNDS THOUGH THE APPLICATION OF **FILTERS** (E.G., COMB FILTER)

FREQUENCY MODULATION SYNTHESIS

TECHNIQUE PIONEERED BY **JOHN CHOWNING**, WHEREBY OSCILLATORS **MODULATE** MORE OSCILLATORS TO PRODUCE **RICH TIMBRES** EFFICIENTLY. RESPONSIBLE FOR "THAT 80S SYNTH SOUND" IN POP MUSIC

PHYSICAL MODELING

USE **DIGITAL SIGNAL PROCESSING** (DSP) TO MODEL THE PHYSICS OF HOW SOUND MOVES IN VARIOUS ACOUSTIC MEDIUMS (E.G., VIBRATING STRINGS, AIR COLUMNS, ETC.)

BASIC SYNTHESIS

GENERATE SOUNDS FROM BASIC SYNTHESIS ELEMENTS: OSCILLATORS, AMPLITUDE ENVELOPES, NOISE, FILTERS, ETC.

GRANULAR SYNTHESIS

BREAK A SOUND DOWN INTO SHORT "**MOLECULES**" AND RECONSTITUTE THEM BACK, LIKE AN **IMPRESSIONISTIC** PAINTING OF SOUND PARTICLES

SPATIALIZATION

MODEL **SOUND IN SPACE**, ROOM ACOUSTICS, 3D SOUND, AND HOW OUR BODY (HEAD, SHOULDER, EARLOBES) AFFECT HOW WE HEAR. (DIRECT APPLICATIONS INCLUDE MULTI-CHANNEL COMPOSITION, GAMES, VIRTUAL REALITY)

VOCODERS

A FAMILY OF TECHNIQUES TO MANIPULATE SOUND THROUGH ITS **FREQUENCY BANDS**, THESE CAN BE USED TO **CROSS-SYNTHESIZE** SIGNALS (E.G., LION'S ROAR + ELECTRIC GUITAR). **PHASE VOCODERS** OPERATE ON SIGNALS IN THE **FREQUENCY DOMAIN** FOR HIGH-QUALITY TIME AND PITCH TRANSFORMATIONS

THE MULTITUDES OF TECHNIQUES AND ALGORITHMS ARE THE BUILDING BLOCKS IN A **SONIC WORKBENCH!** OUR TOOLSHED IS QUITE WELL-STOCKED, AND WE CONTINUE TO DISCOVER NEW TECHNIQUES ALL THE TIME.

SPECTRAL MODELING SYNTHESIS

EMPIRICAL APPROACH TO MODELING SOUND **BY EXAMPLE**: IT EXTRACTS INFORMATION FROM THE SOUND ITSELF (AND NOT THE **PHYSICAL MECHANICS** OF HOW IT'S GENERATED, IN CONTRAST TO **PHYSICAL MODELING**)

SINGING / VOICE SYNTHESIS

THE HUMAN **VOICE**, IN ITS LIMITLESS NUANCE, IS SPECIAL TO US. MANY TECHNIQUES HAVE BEEN DEVELOPED TO EXPRESSIVELY SYNTHESIZE SPEECH, SINGING VOICE, AND EVEN LAUGHTER. FM, FILTER BANKS, LINEAR PREDICTIVE CODING, ARTICULATORY TRACT MODELING, FORMANT WAVE FUNCTIONS -- TO NAME A FEW

TOOLS & TECHNIQUES

ALGORITHMS, PROGRAMMING LANGUAGES, APPS, SYSTEMS, ETC.

INSPIRE

"WE BECOME WHAT WE BEHOLD.
WE SHAPE OUR TOOLS, AND
THEN OUR TOOLS SHAPE US."
-- MARSHALL MCLUHAN

LEAD TO

AESTHETIC POSSIBILITIES

WAYS OF THINKING; SENSIBILITIES; MODES OF INDIVIDUAL EXPRESSION; MANNER IN WHICH WE LIVE WITH OUR TECHNOLOGY (FOR BETTER OR WORSE)

...WHICH IN TURN INSPIRE NEW **TOOLS.**

NEW WAYS OF WORKING WITH THE COMPUTER GIVE RISE TO
GENUINELY NEW **AESTHETIC POSSIBILITIES** -- FOR
EXAMPLE, PAUL LANSKY'S ARTISTIC PROCESS OF
TRANSFORMING WORLD SOUNDS THROUGH THE COMPUTER,
PROCLAIMING BY EXAMPLE A **NEW TYPE** OF MUSIC THAT
WASN'T POSSIBLE WITHOUT A **NEW MEDIUM.**

THESE NEW FORMS ARE ANYTHING
BUT **STATIC**; THEY **CO-EVOLVE** WITH
THEIR UNDERLYING **MEDIUMS!**

☯ PRINCIPLE 4.7

AESTHETICS IS NOT A *PASSIVE* THING -- BUT AN *ACTIVE AGENT* OF *DESIGN!*

AESTHETICS IS NOT SOMETHING YOU ADD TO A NEARLY FINISHED PRODUCT, BUT AN **ACTIVE**
FORCE AND **INTENTIONALITY** THAT SHAPE DESIGN FROM THE **START.** IT TAKES THE
FORM OF **CREATIVE CONSTRAINTS** AND **ARTICULATIONS OF PREFERENCE**
THAT PUSH A DESIGN FORWARD IN A SEA OF **PETRIFYING POSSIBILITIES**, OR
GUIDING PRINCIPLES THROUGH WHICH DECISIONS ARE CONSIDERED THROUGH A
HUMAN LENS. IT BINDS CREATIVE THOUGHT TO AN EXPRESSION OF THAT THOUGHT,
OFTEN TAKING ON A LIFE OF ITS OWN. ONE **FOLLOWS** AESTHETICS IN DESIGN JUST AS
A NOVELIST **FOLLOWS** A CHARACTER'S **TRUE NATURE** TO AN INEVITABLE OUTCOME.

TAKE, FOR EXAMPLE, LIVE COMPUTER-MUSIC PERFORMANCE, WHICH HAS GONE THROUGH ITS SHARE OF CO-EVOLUTION WITH TIME AND TECHNOLOGY. IT CAN BE TRACED BACK TO GROUPS LIKE THE **LEAGUE OF AUTOMATIC MUSIC COMPOSERS**, WHO CREATED MUSIC WITH **LIVE ELECTRONICS**.

JIM HORTON

JOHN BISCHOFF

TIM PERKIS

NOT A TOASTER!

1970s

THE LEAGUE GAVE RISE TO **THE HUB**, WHO WORKED WITH **NETWORKS** OF COMPUTERS, EXPLORING COMPUTER-MEDIATED MUSIC AS A KIND OF LIVE **BAND!**

1980s

TIM PERKIS

PHIL STONE

PAUL BROWN

SCOT GRESHAM-LANCASTER

MARK TRAYLE

JOHN BISCHOFF

I SEE THE **AESTHETICS** INFORMING THIS WORK AS PERHAPS COUNTER TO OTHER TRENDS IN COMPUTER MUSIC: **INSTEAD** OF ATTEMPTING TO GAIN MORE COMPLETE CONTROL OVER EVERY ASPECT OF THE MUSIC, WE SEEK MORE **SURPRISE** THROUGH THE LIVELY AND **UNPREDICTABLE** RESPONSE OF THESE SYSTEMS AND HOPE TO ENCOURAGE AN ACTIVE RESPONSE TO SURPRISE IN THE PLAYING. AND INSTEAD OF TRYING TO ELIMINATE THE **IMPERFECT** HUMAN PERFORMER, WE TRY TO USE THE ELECTRONIC TOOLS AVAILABLE TO **ENHANCE** THE **SOCIAL** ASPECT OF MUSIC-MAKING.

1990s   PERSONAL LAPTOP PERFORMANCE

2000s   RISE OF **LIVE CODING**

COMPUTERS HAVING BECOME SO **FAST** AND **UBIQUITOUS**, PEOPLE BEGAN EXPLORING **CODE** AS AN INSTRUMENT FOR MUSICAL PERFORMANCE...

ALEX MCLEAN LIVE-HACKING **PERL** IN **NIGHTCLUBS**...

...RESULTING IN NEW LANGUAGES AND PERSPECTIVES ON THE **ACT** AND **ROLE** OF **PROGRAMMING**.

LAPTOP ORCHESTRAS AND MOBILE PHONE MUSIC WERE DIRECTLY INSPIRED BY THEIR RESPECTIVE NAMESAKE *TECHNOLOGIES*, EXPLORING NOVEL AND *NUANCED* WAYS IN WHICH SUCH TECHNOLOGIES COULD BE USED.

2000s –> 2010s

"WHAT IF WE HAVE *MANY* HUMANS AND LAPTOPS IN AN ENSEMBLE?"

A NEW FORM OF TECHNOLOGY-MEDIATED *PERSON-TO-PERSON* PERFORMANCE?

PROGRAMMABILITY IS A MEDIUM OF PRECISION BUT ALSO *RAPID EXPERIMENTATION* (E.G., SIMPLY CHANGING A NUMBER IN CODE CAN LEAD TO *WILDLY DIFFERENT* RESULTS). IF *CURIOSITY* DRIVES EXPERIMENTATION, THEN *TASTE* DETERMINES HOW TO "PULL IT BACK."

(人) PRINCIPLE 4.8

EXPERIMENT TO ILLOGICAL EXTREMES! (AND PULL BACK ACCORDING TO TASTE)

PUSHING TO EXTREMES CAN TAKE MANY FORMS: PHYSICALLY IMPOSSIBLE RHYTHMS, *IMPROBABLE SOUNDS* (E.G., MODELING A PLUCKED STRING THE SIZE OF THE GOLDEN GATE BRIDGE; THE CHOICE OF SOUNDS IN "TABLE'S CLEAR" AND "BEIJING"), *SUPERHUMAN* PRECISION (THE 30-WAY GLISSANDO IN THE *THX* SOUND LOGO), *MASS INSTANTIATION* OF SIMPLE ELEMENTS, AND UNCANNY MANIPULATION OF SONIC AND MUSICAL PROPERTIES. DOING IT THROUGH CODE CAN BE QUITE A *POWER TRIP*, BUT MORE IMPORTANTLY, IT LEADS TO NEW *WAYS OF THINKING*, HEARING, SEEING.

TAKE FOR INSTANCE *LIVE CODING* AS AN EXTREME EXPRESSION OF COMPUTER-MEDIATED *PERFORMANCE*, PUTTING THE MEDIUM ITSELF ON STAGE -- IT'S PROGRAMMABILITY AS *INSIDER ART!*

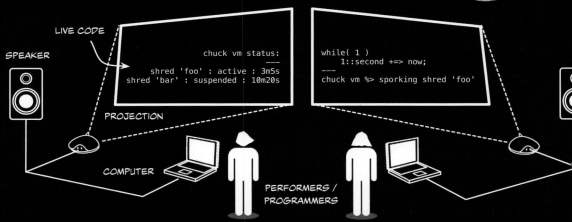

LIVE CODING: THE ACT OF MODIFYING THE LOGIC AND STRUCTURE OF A PROGRAM *DURING RUNTIME*, FOR THE PURPOSE OF EXPRESSIVE CONTROL AND *RAPID EXPERIMENTATION.*

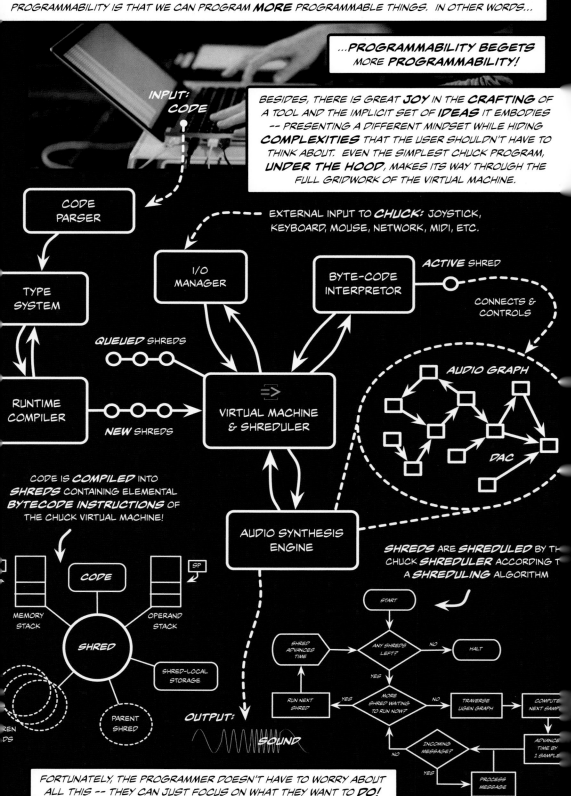

AS INTENTIONALITY DRIVES METHODOLOGY, AESTHETIC AGENDAS DRIVE THE DESIGN OF NEW **TOOLS.** LIVE CODING AND A **TEMPORALLY DETERMINISTIC** WAY OF PROGRAMMING SOUND MOTIVATED THE DESIGN OF **CHUCK.** RETURNING TO WHERE WE STARTED, THE MAGIC OF PROGRAMMABILITY IS THAT WE CAN PROGRAM **MORE** PROGRAMMABLE THINGS. IN OTHER WORDS...

...**PROGRAMMABILITY BEGETS** MORE **PROGRAMMABILITY!**

BESIDES, THERE IS GREAT **JOY** IN THE **CRAFTING** OF A TOOL AND THE IMPLICIT SET OF **IDEAS** IT EMBODIES -- PRESENTING A DIFFERENT MINDSET WHILE HIDING **COMPLEXITIES** THAT THE USER SHOULDN'T HAVE TO THINK ABOUT. EVEN THE SIMPLEST CHUCK PROGRAM, **UNDER THE HOOD,** MAKES ITS WAY THROUGH THE FULL GRIDWORK OF THE VIRTUAL MACHINE.

INPUT: CODE

CODE PARSER

EXTERNAL INPUT TO **CHUCK:** JOYSTICK, KEYBOARD, MOUSE, NETWORK, MIDI, ETC.

TYPE SYSTEM

I/O MANAGER

BYTE-CODE INTERPRETOR

ACTIVE SHRED

CONNECTS & CONTROLS

QUEUED SHREDS

RUNTIME COMPILER

NEW SHREDS

VIRTUAL MACHINE & SHREDULER

AUDIO GRAPH

DAC

CODE IS **COMPILED** INTO **SHREDS** CONTAINING ELEMENTAL **BYTECODE INSTRUCTIONS** OF THE CHUCK VIRTUAL MACHINE!

AUDIO SYNTHESIS ENGINE

SHREDS ARE **SHREDULED** BY TH CHUCK **SHREDULER** ACCORDING T A **SHREDULING** ALGORITHM

CODE

SP

MEMORY STACK

OPERAND STACK

SHRED

SHRED-LOCAL STORAGE

PARENT SHRED

REN DS

OUTPUT: SOUND

START

SHRED ADVANCES TIME

ANY SHREDS LEFT?

NO

HALT

YES

RUN NEXT SHRED

YES

MORE SHRED WAITING TO RUN NOW?

NO

TRAVERSE UGEN GRAPH

COMPUTE NEXT SAMP

INCOMING MESSAGE?

NO

ADVANCE TIME BY 1 SAMPLE

YES

PROCESS MESSAGE

FORTUNATELY, THE PROGRAMMER DOESN'T HAVE TO WORRY ABOUT ALL THIS -- THEY CAN JUST FOCUS ON WHAT THEY WANT TO **DO!**

THERE ARE **MANY** COMPUTER MUSIC LANGUAGES, EACH **TAILORED** FOR DIFFERENT GOALS AND **AESTHETICS** OF WORKING AND THINKING ABOUT PROGRAMMABLE SOUND AND GRAPHICS!

MAX/MSP

GRAPHICAL PATCHING AUDIO LANGUAGE -- NAMED MAX AFTER **MAX MATHEWS**, MSP AFTER **MILLER S. PUCKETTE**, THE ORIGINAL DESIGNER OF **MAX**

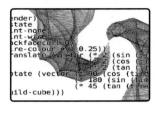

PROCESSING

VISUAL PROGRAMMING LANGUAGE, OFTEN USED IN CONJUNCTION WITH AUDIO LANGUAGES

FLUXUS

FUNCTIONAL VISUAL LIVE CODING LANGUAGE

CHUCK

STRONGLY-TIMED COMPUTER MUSIC LANGUAGE

IMPROMPTU

SCHEME-BASED LIVE CODING LANGUAGE

SUPERCOLLIDER

COMPLEX AND POWERFUL AUDIO SYNTHESIS LANGUAGE

OPENFRAMEWORKS

C++ FRAMEWORK FOR BUILDING AUDIOVISUAL SOFTWARE

OVERTONE

COLLABORATIVE LIVE CODING TOOLKIT

TIDAL

MINI LANGUAGE FOR LIVE CODING PATTERNS

Sonic Pi

LIVE CODING AUDIO SYNTHESIS LANGUAGE

CSOUND

OLD SCHOOL COMPUTER MUSIC LANGUAGE, THE MODERN DIRECT DESCENDANT OF MAX MATHEWS'S MUSIC-N FAMILY OF LANGUAGES

PURE DATA

OPEN-SOURCE GRAPHICAL PATCHING COMPUTER MUSIC LANGUAGE

SYNTHESIS TOOLKIT (STK)

CROSS-PLATFORM C++ TOOLKIT FOR PHYSICAL MODELING AND AUDIO SYNTHESIS; DIRECTLY INFLUENCED **CHUCK**

YET, AS **POWERFUL** AS PROGRAMMABILITY IS, IT IS **NOT** THE BE-ALL AND END-ALL. IT IS A **WAY** OF **THINKING** AND A **CREATIVE TOOL** TO CARRY IT OUT. NO MORE, NO LESS.

PRINCIPLE 4.9   AS **MARK WEISER*** SAID,

HE PURPOSE OF A COMPUTER IS DO SOMETHING ELS

CHNOLOGY IS **ALWAYS** A MEANS-TO-AN-END, AND **NEVER** AN END-IN-ITSELF.   * FROM CHAPT

OF THE MYRIAD WAYS IN WHICH DESIGNS CAN **FAIL**, THERE IS PERHAPS NO GREATER **PERIL** THAN LOSING SIGHT OF THE BIGGER PICTURE OF **WHAT** WE ARE CREATING, **WHY** WE ARE DOING IT IN THE FIRST PLACE, AND FOR **WHOM!** THERE IS A BALANCE TO BE STRUCK BETWEEN **TECHNOLOGY** AND **ART** IN EVERY ENDEAVOR. THE USE OF TECHNOLOGY SHOULD BE THOUGHTFUL AND **JUSTIFIED.** THE ARTFUL DESIGNER NEEDS A GREATER VISION THAN JUST THE TECHNOLOGY ITSELF. INDEED, THE PURPOSE OF A COMPUTER SHOULD BE TO DO **SOMETHING ELSE.**

PROGRAMMABILITY IS BOTH
BLESSING AND CURSE

PRINCIPLE 4.10

THE TRICK, I THINK, IS TO USE IT FREELY BUT **WISELY.** JUST BECAUSE SOMETHING IS **PROGRAMMABLE** DOESN'T MAKE IT **INTERESTING.** THERE IS AN INFINITE SPACE OF POSSIBILITIES FOR THE ARTFUL PROGRAMMER TO NEGOTIATE, AND IT'S POSSIBLE TO BUILD THE SOFTWARE AND **PROCRASTINATE** ON THE ART / PRODUCT. I AM REMINDED OF A JOHN WOODEN QUOTE: "PLAYERS WITH FIGHT NEVER LOSE A GAME, THEY JUST RUN OUT OF TIME." DESIGNERS DON'T FAIL. WE JUST UH... **RUN OUT OF TIME.**

YEAH, I CERTAINLY HAVE BEEN GUILTY OF THIS. I MEAN, I CREATED A PROGRAMMING LANGUAGE: A RATHER EXTREME FORM OF PROCRASTINATION! WAS IT THE **RIGHT** THING TO DO? **WHO KNOWS?** IT REMAINS AN UNENDING EFFORT OF DESIGN AND MAINTENANCE. NONETHELESS, OUR TEAM HAS WORKED ON **CHUCK** SINCE 2002, AN EXERCISE IN LONG-TERM TOOL-BUILDING THAT HAS SUPPORTED LAPTOP ORCHESTRAS, A **COMMUNITY** OF RESEARCHERS AND STUDENTS, AND TEN MILLION **OCARINA'S**. INTERNALLY, THE JOY OF **BUILDING THINGS TOGETHER** REMAINS.

THERE IS UNDENIABLE **POWER** AND POTENTIAL FOR ARTFULNESS IN **PROGRAMMING.** AS LONG AS WE REMAIN COGNIZANT OF ITS PITFALLS AND PERILS, PROGRAMMABILITY ALLOWS US TO CREATE, TRANSFORM, AND FASHION THINGS THAT CHANGE OUR **BEHAVIOR** AND THE WAY WE **THINK.**

☾ PRINCIPLE 4.11 — THERE IS AN *ART* TO PROGRAMMING

> IT LIES IN WHAT WE **DO** WITH IT.

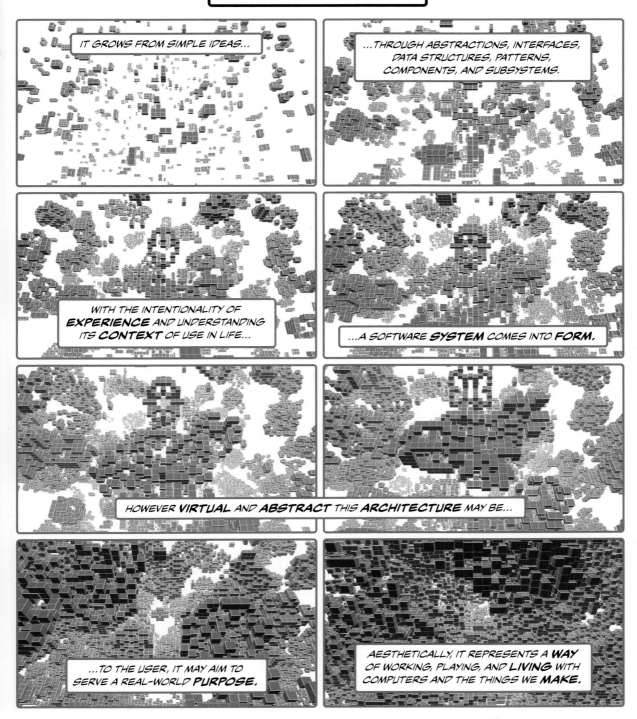

IT GROWS FROM SIMPLE IDEAS...

...THROUGH ABSTRACTIONS, INTERFACES, DATA STRUCTURES, PATTERNS, COMPONENTS, AND SUBSYSTEMS.

WITH THE INTENTIONALITY OF **EXPERIENCE** AND UNDERSTANDING ITS **CONTEXT** OF USE IN LIFE...

...A SOFTWARE **SYSTEM** COMES INTO **FORM.**

HOWEVER **VIRTUAL** AND **ABSTRACT** THIS **ARCHITECTURE** MAY BE...

...TO THE USER, IT MAY AIM TO SERVE A REAL-WORLD **PURPOSE.**

AESTHETICALLY, IT REPRESENTS A **WAY** OF WORKING, PLAYING, AND **LIVING** WITH COMPUTERS AND THE THINGS WE **MAKE.**

CHAPTER 4 DESIGN ETUDE

GOOD LUCK, HAVE FUN!

DESIGN A
SOUND LOGO!

DESIGN, USING **CHUCK** AND/OR ANOTHER PROGRAMMABLE TOOL, YOUR OWN COMPUTER-GENERATED **SOUND LOGO.** AIM FOR 15-30 SECONDS IN DURATION. THINK OF IT AS A "SHORT MUSICAL STATEMENT WITH A PURPOSE."

• PART 1: PRAGMATICS

UNDERSTAND AND ARTICULATE THE SOUND LOGO'S **PURPOSE** (WHO IS IT FOR, WHAT DO YOU WANT IT TO CONVEY?). IS IT FOR YOURSELF? OR AN ORGANIZATION YOU BELONG TO? YOUR COMMUNITY? A FAMILY MEMBER? YOUR CAT? PET SNAKE? WHAT DO YOU WANT IT TO **SAY** ABOUT THE ENTITY THAT THE SOUND LOGO IS TO REPRESENT?

• PART 2: AESTHETICS

AESTHETICALLY, ARTISTICALLY, WHERE DO YOU WANT YOUR SOUND LOGO TO END UP? WHAT KINDS OF SOUND DO YOU WANT TO USE? WILL YOU **RECORD** SOUNDS AND TRANSFORM THEM, OR WILL YOU **GENERATE** YOUR OWN? DO YOU WANT THE RESULT TO BE **RECOGNIZABLE** OR SYMBOLIC AND **SURREAL?** **EXPERIMENT** WITH HOW YOU WANT IT TO SOUND, THINKING ABOUT HOW THESE CHOICES INFLUENCE THE OVERALL **FEEL** OF THE SOUND LOGO AND HOW DIFFERENT CHOICES CAN CONVEY THE SAME GENERAL MESSAGE IN DIFFERENT WAYS. BE MINDFUL OF HAVING A **NARRATIVE** (E.G., BEGINNING, MIDDLE, END). WHAT **SENTIMENTS** AND **MEANINGS** DO YOU WANT THE LISTENER TO TAKE AWAY AS A RESULT OF **EXPERIENCING** THE SOUND LOGO?

• PART 3: DEPLOYMENT!

AFTER YOU CODE UP AND ASSEMBLE THE SOUND LOGO, TO THE EXTENT POSSIBLE, PLACE/TEST THE SOUND LOGO IN THE **INTENDED CONTEXT** OF ITS USE. WHEN AND WHERE WOULD YOU HAVE IT HEARD? IF, FOR EXAMPLE, IT'S A SOUND LOGO FOR YOURSELF, DO YOU WANT IT PLAYED ON YOUR WEBPAGE, OR PERHAPS...WHEN YOU **ENTER A ROOM?** (HMM, FIND SOME PORTABLE SPEAKERS AND TRY IT OUT!)

• PART 4: DOCUMENTATION

DOCUMENT THE **REACTION** AND **SENTIMENT** THE SOUND LOGO'S DEPLOYMENT GENERATES IN YOU AND IN OTHERS -- BY VIDEO OR IN WORDS. THEN **REFLECT** ON THE EXPERIENCE. DID YOU ACCOMPLISH WHAT YOU SET OUT TO ACHIEVE, **PRAGMATICALLY** AND **AESTHETICALLY?** FEEL FREE TO ITERATE AND RE-DEPLOY!

All media work us over completely. They are so pervasive in their personal, political, economic, aesthetic, psychological, moral, ethical, and social consequences that they leave no part of us untouched, unaffected, unaltered."

—Marshall McLuhan (1964)
Understanding Media: The Extensions of Man

THE **MEDIUM** AND THE **MESSAGE**

We can all appreciate that **how** we say something is not fully separable from **what** we say. The **medium** colors, transforms, and ultimately shapes the **message**. So it is with technology. If the message is a design's purpose and meaning, then the medium is how we get there.

The medium fundamentally and unmistakably leaves an aesthetic imprint on us. Our design choices pertaining to medium govern what is made, how we experience it, and what we get out of it.

Design **connects** the medium to the message. The art of design draws from the essential qualities of a medium, to create something that would not and could not exist as meaningfully in another medium. The medium should become the message so completely that the medium seemingly melts away, leaving only the essential message —and the illusion that the medium doesn't matter. However, this is only an illusion, for medium matters fundamentally; it is how we unfold the message.

INTERFACE:
THE MEMBRANE OF
INTERACTION BETWEEN
HUMAN AND TECHNOLOGY

CHAPTER 5
INTERFACE DESIGN

FROM THE PERSPECTIVE OF THE *USER*, AN *INTERFACE* IS HOW WE *USE* A DESIGN AND, THROUGH *INTERACTION*, HOW WE *EXPERIENCE* THAT DESIGN. THE INTERFACE GIVES RISE TO THE *NATURE* AND *AESTHETICS* OF ITS ENCOUNTER; IT ALLOWS *AFFORDANCES* TO EMERGE, WITH *NUANCES* IMBUED WITHIN. *INTERFACE DESIGN* IS AN ARTICULATION OF THE SPECIFIC MANNER IN WHICH THE HUMAN IS TO MEET THE *TECHNOLOGY.* MORE BROADLY, IT PROVIDES A FUNDAMENTAL MEANS BY WHICH HUMANS MAY *THINK* ABOUT AND *INTERACT* WITH OUR *WORLD.*

IN A TIME OF RAPIDLY INCREASING *AUTOMATION*, THE ARTFUL DESIGNER MUST BE COGNIZANT OF SITUATIONS IN WHICH IT IS *ESSENTIAL* TO DESIGN THE *HUMAN INTO THE LOOP.* INTERFACES OUGHT TO *EXTEND* US, MAKE US FEEL A SENSE OF *EMBODIMENT* IN THEIR USE, GIVING US *NEW HANDS* TO INTERACT WITH THE WORLD AROUND US.

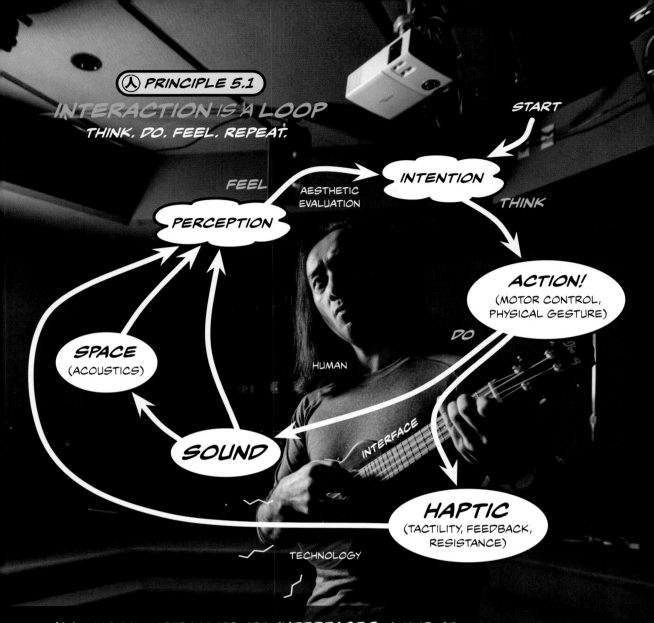

INTERACTION IS A LOOP
THINK. DO. FEEL. REPEAT.

START

FEEL

INTENTION

AESTHETIC
EVALUATION

THINK

PERCEPTION

ACTION!
(MOTOR CONTROL,
PHYSICAL GESTURE)

SPACE
(ACOUSTICS)

DO

HUMAN

INTERFACE

SOUND

HAPTIC
(TACTILITY, FEEDBACK,
RESISTANCE)

TECHNOLOGY

ALL MUSICAL INSTRUMENTS ARE **INTERFACES**, A KIND OF IMPLICIT **CONTRACT** OF **INTERACTION** BETWEEN A PLAYER AND AN UNDERLYING MECHANISM OF SOUND PRODUCTION. THE **INTERACTION** BETWEEN THEM MANIFESTS ITSELF AS AN ACTIVE, ONGOING **FEEDBACK LOOP** -- A DYNAMIC PROCESS, ONE IN WHICH WE ARE CONSTANTLY EVALUATING THE **RESULTS** OF OUR **ACTIONS**, EVER FINE-TUNING THE RELATIONSHIP.

AS THE WORD IMPLIES, **INTERACTION** IS ACTION BETWEEN TWO ENTITIES (E.G., OUR **PERSON** AND AN **OBJECT**). WHILE THIS CONSTITUTES A **BINARY** RELATIONSHIP, THE MOST EFFECTIVE AND ELEGANT INTERACTIONS ARE THE RESULT OF INTERFACES THAT MEDIATE AND SEAMLESSLY BIND THE USER AND ARTIFACT INTO A **SINGLE SYSTEM.**

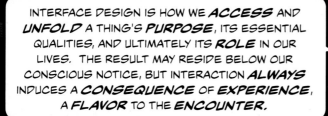

INTERFACE DESIGN IS HOW WE **ACCESS** AND **UNFOLD** A THING'S **PURPOSE**, ITS ESSENTIAL QUALITIES, AND ULTIMATELY ITS **ROLE** IN OUR LIVES. THE RESULT MAY RESIDE BELOW OUR CONSCIOUS NOTICE, BUT INTERACTION **ALWAYS** INDUCES A **CONSEQUENCE** OF **EXPERIENCE**, A **FLAVOR** TO THE **ENCOUNTER**.

FOR EXAMPLE, THE WAY A **MUSICAL INTERFACE** IS DESIGNED -- **HOW** IT PLAYS, HOW IT **FEELS** TO PLAY IT -- INSPIRES, CONSTRAINS, STIMULATES **WHAT** WE DO WITH IT, EXPRESSED THROUGH THE TYPES OF MUSIC WE MAKE. INTERACTION **IS** AN **EXPERIENCE** UNTO ITSELF...

...WHICH IS TO SAY:

PRINCIPLE 5.2

THERE IS AN *AESTHETIC* TO *INTERACTION*

IT'S **HOW** WE ENGAGE WITH AN INTERFACE, A SENSE THAT THE OBJECT WITH WHICH WE INTERACT BECOMES AN **EXTENSION** OF **OURSELVES** AND THAT WE BECOME PART OF **IT**. LIKE AN INSTRUMENT THAT FEELS **SATISFYING** IN OUR HANDS, A TOOL WE **LOVE** USING, A CAR THAT HANDLES LIKE IT **KNOWS** WHAT WE ARE **THINKING**, OR A VIDEO GAME WHOSE CONTROLS ARE SO NATURAL THAT WE FEEL **TRANSPORTED** INTO IT... THE AESTHETICS OF INTERACTION LIE IN THE **ELEGANCE** AND **EXPRESSIVENESS** OF USE.

INSTRUMENT DESIGN IS AN **EXTREME** FORM OF INTERFACE DESIGN. DECEPTIVELY **TRICKY**, IT DEMANDS BOTH **SIMPLICITY** OF INTERACTION AND THE POTENTIAL FOR **COMPLEXITY** AND **RICHNESS** IN ITS OUTPUT.

THEIR CHALLENGES MAKE **INSTRUMENTS** GREAT CASE STUDIES OF INTERFACE DESIGN.

IT IS **REMARKABLE** THAT A PIANO'S INTERFACE IS EXACTLY THE **SAME** FOR A BEGINNER AND AN EXPERIENCED PLAYER. THERE IS NO **POWER-USER MODE**. ITS **PREDICTABILITY** IS WHAT ENABLES US TO **LEARN** TO PLAY IT.

(MEANWHILE, BEHIND THE COVER, THE **TECHNOLOGY** OF ITS INNER WORKINGS IS **HIDDEN** FROM THE PLAYER)

ALL INSTRUMENTS ARE **DESIGNED.** ALL MAKE USE OF **TECHNOLOGY.**

GUITAR
DULCIMER MANDOLIN BANJO
UKULELE
HARPSICHORD KOTO
HARP GUZHEN SITAR
VIOLIN *PLUCKED* TANBURA
CELLO
BASS DOUBLE ZITHER
BOWED SAROD BASS PIPA PIANO
VIOLA OUD GUQIN CLAVICHORD
ERHU ĐÀN BẦU *STRUCK*
LUTE
LYRA LYRE
SARANGI HAMMERED
VIOL DULCIMER
HURDY- TALKING
GURDY *VIBRATING* DRUM
HAEGEUM *STRINGS*
NYCKELHARPA

VIBRAPHONE TUBULAR CHIMES
XYLOPHONE BELLS WASHBOARD
GLASS HARMONICA DJEMBE
TAMBOURINE GONG
HYDRAULOPHONE BIANZHONG
ZILL SHEKERE
TUNED TIMPANI TRIANGLE CONGA
STEEL GLOCKENSPIEL *UNTUNED*
DRUMS MBIRA MARACA BODHRÁN
MARIMBA TAIKO WOOD
TABLA SNARE BLOCK
ANGKLUNG CYMBAL
TAP SHOE TOMS
DHOLAK COWBELL BONGO
MEMBRANES. BARS, & SINGING
RESONANT SURFACES BOWL
STOMP BOX

A TAXONOMY OF MUSICAL INSTRUMENTS
(BY SOUND-PRODUCTION MECHANISM) ⚘ TAXONOMY 5.3

SINGING
BEATBOXING *AIR* PIPE ORGAN
ACCORDION
VOICE DIDGERIDOO
BOBBY HARMONICA
MCFERRIN *BODY* *BLOWN* BAGPIPES
CLAPPING
SNAPPING SLAPPING
WHISTLING
WOODWINDS
FLUTE TRUMPET
PICCOLO TROMBONE *HELMHOLTZ*
CLARINET SAXOPHONE *RESONATORS*
OBOE
DAEGEUM TUBA OCARINA (CLAY)
BASSOON XUN
ENGLISH HORN EUPHONIUM BLOWN BOTTLE
CHANTER FRENCH HORN
SHAKUHACHI BAZOOKA
BUGLE
PAN FLUTE MELODICA
RECORDER SOUZAPHONE
LEAF WHISTLE

ELECTRONIC &
COMPUTER-MEDIATED
OCARINA (PHONE)
LAPTOP
ORCHESTRA
THEREMIN INSTRUMENTS
RADIO BATON
ELECTRIC
ONDES
ELECTRIC GUITAR MARTENOT *COMPUTATIONAL*
ANALOG SYNTHESIZERS MUSICAL
YAMAHA TELHARMONIUM ROBOTS?
MOOG MELLOTRON
BUCHLA DRUM MACHINE **???**
HAMMOND SEQUENCER AUTO-TUNE
WURLITZER MONOME MPC **???**

AN **INSTRUMENT** IS A **TOOL:** IT **EXISTS**
BECAUSE IT OFFERS AT LEAST ONE CORE ASPECT,
HOWEVER SUBTLE, THAT IT DOES **BETTER** THAN
ANYTHING ELSE. THE DIFFERENCE BETWEEN
INSTRUMENTS JUSTIFIES THEIR **DIVERSITY --**
AND EVER INVITES NEW EXPLORATIONS.

INSTRUMENTS MAKE SOUND, TRANSLATING **INTENTIONALITY** INTO **EXPRESSION.** AS WITH ANY **USER INTERFACE,** AN INSTRUMENT FULFILLS ITS **FUNCTION** BY AFFORDING A SET OF **INTERACTIONS** BETWEEN **IT** AND OUR **BODIES.** THE SPECIFIC WAY IN WHICH THE INTERACTION ENGAGES US PHYSICALLY IS A CORE **EXPERIENCE** OF **USING** THAT INTERFACE. BODIES, FOR THIS REASON, ARE OF CENTRAL IMPORTANCE TO DESIGN.

(A) PRINCIPLE 5.4   BODIES MATTER!

"OUR **BODY** IS THE **ULTIMATE INSTRUMENT** OF ALL OUR EXTERNAL KNOWLEDGE, WHETHER INTELLECTUAL OR PRACTICAL. IN ALL OUR WAKING MOMENTS WE ARE RELYING ON OUR AWARENESS OF CONTACTS OF OUR BODY WITH THINGS OUTSIDE FOR **ATTENDING** TO THESE THINGS. OUR OWN BODY IS THE ONLY THING IN THE WORLD WHICH WE NORMALLY NEVER EXPERIENCE AS AN OBJECT, BUT EXPERIENCE ALWAYS IN TERMS OF THE WORLD IN WHICH WE ARE ATTENDING FROM OUR BODY. IT IS BY MAKING THIS INTELLIGENT USE OF OUR BODY THAT WE FEEL IT TO BE OUR BODY, AND NOT A THING OUTSIDE."

-- MICHAEL POLANYI, *THE TACIT DIMENSION*

IN A WAY, OUR BODIES ARE OUR **MIND'S INTERFACE** TO THE OUTSIDE WORLD, AND OUR HANDS AND ARMS ARE CAPABLE OF CARRYING OUT MANY NUANCED ACTIONS.

PRESS!

SLIDE

TWIST

SHAKE!

PLUCK!

BOW

PULL

STRIKE!

TAP!

SQUEEZE

HAMMER!

"...IT WAS THE **HANDS** THAT WERE THE WORKING SURFACE, THE HANDS THAT FELT AND **MANIPULATED THE UNIVERSE.** HUMAN BEINGS **THOUGHT WITH THEIR HANDS.** IT WAS THEIR HANDS THAT WERE THE ANSWER OF CURIOSITY, THAT FELT AND PINCHED AND TURNED AND LIFTED AND HEFTED. THERE WERE ANIMALS THAT HAD BRAINS OF RESPECTABLE SIZE, BUT THEY HAD NO HANDS AND THAT MADE ALL THE DIFFERENCE."

-- ISAAC ASIMOV, *FOUNDATION'S EDGE*

THE MANY TYPES OF **GESTURES** OUR BODIES ARE CAPABLE OF MAKING GIVE RISE TO **TECHNOLOGIES** TO **SENSE** THEM; THEY ARE THE **BUILDING BLOCKS** OF INTERFACE DESIGN.

BODY	ACTIONS	TECHNOLOGIES

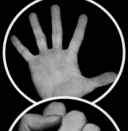

FINGERS

HIGHLY NUANCED, VERY ARTICULATED, SENSITIVE

PRESS, PUSH, PULL, HOLD, PLUCK, POKE, TOUCH, SWIPE, PINCH

BUTTON, KEY, STRING, KNOB, SLIDER, FORCE-SENSING RESISTOR (FSR), MULTITOUCH, SKELETAL-TRACKING

HAND

GESTURAL UNIT COMPRISING OF WRIST, PALM, FINGERS

OPEN, CLOSE, ROTATE, PUSH, PULL, TWIST, GRASP, STRIKE, POSITION, SYMBOL, GESTURE

ACCELEROMETERS, GYROSCOPES, SKELETAL-TRACKING

ARMS

CAPABLE OF MAKING BROAD STROKE GESTURES

EXTEND, CONTRACT, WAVE, FLAIL, POSE, HOLD, LIFT, GESTURE

ACCELEROMETERS, GYROSCOPES, SKELETAL-TRACKING

YEEHAW!

TORSO, LIMBS, HEAD

NUANCED AND EXPRESSIVE AT A FULL BODY SCALE

STAND, POSE, JUMP, SQUAT, CRAWL, BEND, KNEEL

ACCELEROMETERS, GYROSCOPES, SKELETAL-TRACKING

FACE

CAPABLE OF NUANCE AND SUBTLETY, TO WHICH HUMANS ARE EXTREMELY ATTUNED

MOUTH, EYES, BROWS, NOSE, EXPRESSION, SMILE, FROWN, TWITCH

COMPUTER VISION FACE TRACKING, EYE TRACKING

VOICE & BREATH

EXTREMELY EXPRESSIVE; BOTH VOICE AND BREATH CAN BE USED AS INPUT

VOCALIZE, BREATHE

MICROPHONE, SIGNAL PROCESSING, TRACKING

LESS OBVIOUSLY: BRAIN WAVES, HEART RATE, AND SKIN RESPONSE CAN ALSO BE SENSED AND TRACKED!

PARTLY DUE TO THE **EXPRESSIVE RICHNESS** OF OUR BODIES, INTERFACE DESIGN REMAINS A **SUBTLE** AND **TRICKY CRAFT**, ESPECIALLY IN THE DOMAIN OF THE **COMPUTER**, WHERE THE SAME **PHYSICAL ACTIONS** MUST BE TRANSLATED INTO MANY AND VASTLY DIFFERENT TASKS. IT IS PERHAPS NOT SURPRISING THAT INTERFACES LIKE THE **MOUSE** EMERGED AS A DOMINANT INTERACTION PARADIGM, ITS VALUE AND LIMITATIONS INHERENT IN ITS **SIMPLICITY** AND **ADAPTABILITY**. FLEXIBLE AND **ENDURING**, THE MOUSE MUST TRANSLATE A **SMALL** SUBSET OF WHAT THE BODY CAN DO INTO A **LARGE** RANGE OF APPLICATIONS. FOR EXAMPLE, THE **SAME** MOUSE IS USED FOR WORD PROCESSING, VIDEO EDITING, WEB BROWSING, AND COUNTLESS DIFFERENT VIDEO GAMES. THIS IS BOTH BLESSING AND CURSE, FOR THE DESIGN OF TRADITIONAL, DESKTOP-BASED **GRAPHICAL USER INTERFACES** (GUI'S) IS CONSTRAINED BY THE EXPRESSIVE RANGE OF THE MOUSE.

THE
GUI'S MENTAL MODEL OF A *USER*
FROM O'SULLIVAN AND IGOE'S **PHYSICAL COMPUTING** (2004)

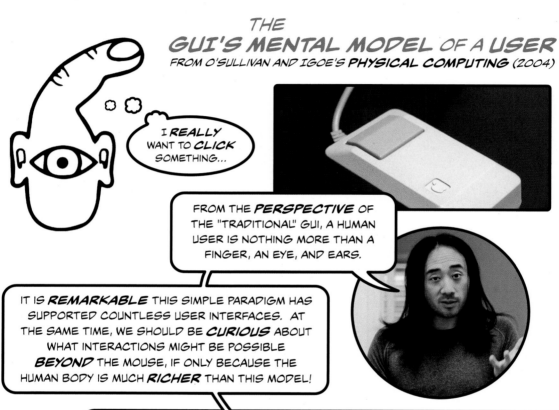

I **REALLY** WANT TO **CLICK** SOMETHING...

FROM THE **PERSPECTIVE** OF THE "TRADITIONAL" GUI, A HUMAN USER IS NOTHING MORE THAN A FINGER, AN EYE, AND EARS.

IT IS **REMARKABLE** THIS SIMPLE PARADIGM HAS SUPPORTED COUNTLESS USER INTERFACES. AT THE SAME TIME, WE SHOULD BE **CURIOUS** ABOUT WHAT INTERACTIONS MIGHT BE POSSIBLE **BEYOND** THE MOUSE, IF ONLY BECAUSE THE HUMAN BODY IS MUCH **RICHER** THAN THIS MODEL!

WHAT IF COMPUTER-BASED INTERFACES ARE DESIGNED TO BE AS RICH AND EXPRESSIVE AS TRADITIONAL **MUSICAL INSTRUMENTS,** OR AS CAPABLE OF NUANCE AS ARE SCULPTORS, SURGEONS, WATCHMAKERS, PUPPETEERS? WHAT NEW THINGS MIGHT WE THEN DO WITH COMPUTERS? WHAT NEW IDEAS MIGHT BE EXPRESSED? SHOULDN'T THE COMPUTER'S "MENTAL MODEL" OF US MORE CLOSELY RESEMBLE WHO WE ARE, WHAT WE MIGHT BE CAPABLE OF?

FOR EXAMPLE, IF OUR INSTRUMENTS HAD MENTAL MODELS OF **US**, WHAT WOULD WE LOOK LIKE TO THEM? APPLYING THE SAME LINE OF THINKING AS O'SULLIVAN AND IGOE, I'VE SKETCHED OUT A FEW **POSSIBILITIES**...

A PIANO'S MENTAL MODEL OF A PIANIST

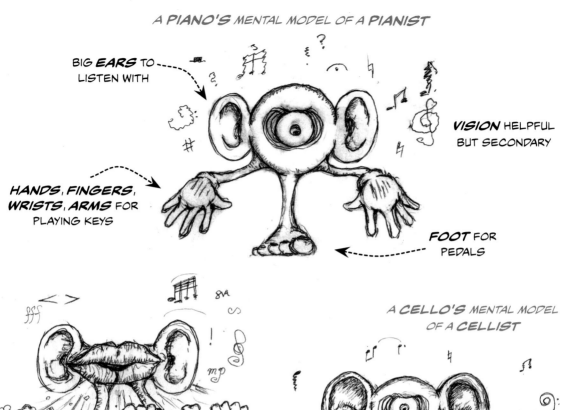

BIG **EARS** TO
LISTEN WITH

VISION HELPFUL
BUT SECONDARY

**HANDS, FINGERS,
WRISTS, ARMS** FOR
PLAYING KEYS

FOOT FOR
PEDALS

A **FLUTE'S** MENTAL MODEL
OF A **FLUTIST**

A CELLO'S MENTAL MODEL OF A CELLIST

A **DRUMSET'S** MENTAL MODEL
OF A **DRUMMER**

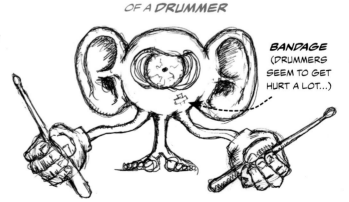

CHIN

BANDAGE
(DRUMMERS
SEEM TO GET
HURT A LOT...)

A **VIOLIN'S** MENTAL MODEL
OF A **VIOLINIST**

DRUMMERS ARE AN ACTIVE BUNCH: THEY
USE HANDS, ARMS, LEGS, AND FEET!

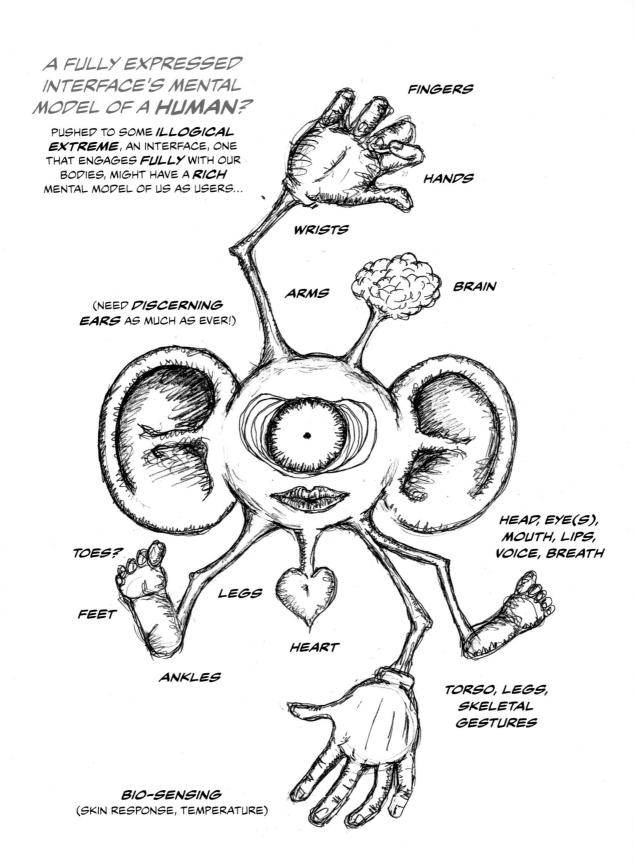

A FULLY EXPRESSED INTERFACE'S MENTAL MODEL OF A **HUMAN?**

PUSHED TO SOME *ILLOGICAL EXTREME*, AN INTERFACE, ONE THAT ENGAGES *FULLY* WITH OUR BODIES, MIGHT HAVE A *RICH* MENTAL MODEL OF US AS USERS...

FINGERS

HANDS

WRISTS

ARMS

BRAIN

(NEED *DISCERNING EARS* AS MUCH AS EVER!)

HEAD, EYE(S), MOUTH, LIPS, VOICE, BREATH

TOES?

LEGS

FEET

HEART

ANKLES

TORSO, LEGS, SKELETAL GESTURES

BIO-SENSING
(SKIN RESPONSE, TEMPERATURE)

THE EXPRESSIVE POTENTIAL OF INTERFACE DESIGN IS PERHAPS NONE MORE EVIDENT THAN IN THE WORLD OF **ELECTRONIC** AND **COMPUTER-BASED MUSICAL INSTRUMENTS.** ITS BRIEF HISTORY IS A VERITABLE KALEIDOSCOPE OF NEW FORMS AND NOVEL FUNCTIONS THAT EXTEND OUR BODIES, AMPLIFYING OUR MINDS.

THE **THEREMINOVOX**, DESIGNED BY **LEON THEREMIN** IN 1920, ONE OF THE EARLIEST ELECTRONIC MUSICAL INSTRUMENTS. NOTAB IT REQUIRES **NO PHYSICAL CONTACT** WITH THE INSTRUMEN INTERNAL **ANALOG CIRCUITS** OF VACUUM TUBES, COILS, AN OSCILLATORS CREATE **ELECTROMAGNETIC FIELDS** TH **SENSE** AND **SONIFY** THE **PROXIMITY** OF THE HANDS (OR MO GENERALLY, ANY PART OF THE BODY), PROVIDING SIMULTANEOUS AN CONTINUOUS CONTROL OVER TWO SONIC PARAMETERS: ONE HAI CONTROLS **PITCH**, WHILE THE OTHER CONTROLS **VOLUME.**

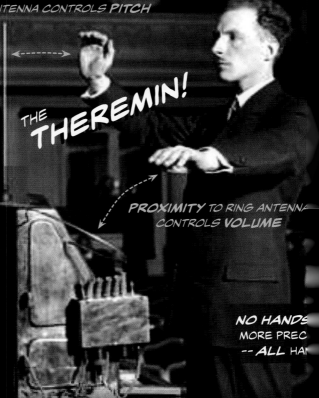

PROXIMITY TO UPRIGHT ANTENNA CONTROLS **PITCH**

W INTERFACES FOR **MUSICAL EXPRESSION!**

THE **THEREMIN!**

OUTLANDISH INSTRUMENT JCES AN EQUALLY OUTLANDISH TURE SOUND: AN EERIE, OTHER- DLY, **SIREN-LIKE WAIL** IN CLASSIC SCI-FI MOVIES, AS AS IN MANY LIVE PERFORMANCE EXTS SINCE.

PROXIMITY TO RING ANTENN CONTROLS **VOLUME**

NO HANDS MORE PREC -- **ALL** HAN

KNOWLEDGE OF THE **UNDERLYING TECHNOLOGY** DOES NOT DIMINISH THE SENSE OF **MAGIC** THE THEREMIN STILL INSPIRES TODAY. ALSO, IT'S FITTING THAT ONE OF THE EARLIEST ELECTRONIC MUSICAL INSTRUMENTS WAS SO INSISTENTLY DESIGNED FOR THE **HANDS!**

OWING UNDERSTANDING OF ELECTRONICS, CIRCUITS, AND SIGNALS GAVE RISE TO STRUMENT SYSTEMS SUCH AS **ANALOG MODULAR SYNTHESIZERS.**

WHILE MANY MODULAR SYNTHS WERE MANIPULATED USING ARRAYS OF **KNOBS** AND **BUTTONS**, MORE TRADITIONAL FORMS OF MUSICAL INPUT WERE USED TO DRIVE THE OUTPUT, SUCH AS IN THIS **MINI-MOOG.**

ANALOG VS. DIGITAL

IN THE DIGITAL DOMAIN, THE *YAMAHA DX7* WAS THE WORLD'S FIRST COMMERCIALLY VIABLE *DIGITAL SYNTHESIZER*, MADE POSSIBLE BY THE TECHNIQUE OF *FREQUENCY MODULATION SYNTHESIS.*

VAST NETWORKS OF ANALOG PATCH CABLES CONNECTED INPUTS TO OUTPUTS, DRIVING OSCILLATORS, GAINS, PULSE WIDTH MODULATIONS, STEP TIMERS, FREQUENCIES, LFO'S, FILTER CUTOFFS, RESONANCES... THIS CAN BE SEEN AS A DIRECT **ANCESTOR** TO DIGITAL GRAPHICAL PATCHING LANGUAGES SUCH AS **MAX/MSP** AND **PURE DATA**, AND EVEN TEXT-BASED LANGUAGES LIKE **CSOUND, SUPERCOLLIDER**, AND **CHUCK.**

(THE **DARK SIDE** OF GRAPHIC PATCHING -- GIVES NEW MEANING TO **SPAGHETTI CODE**)

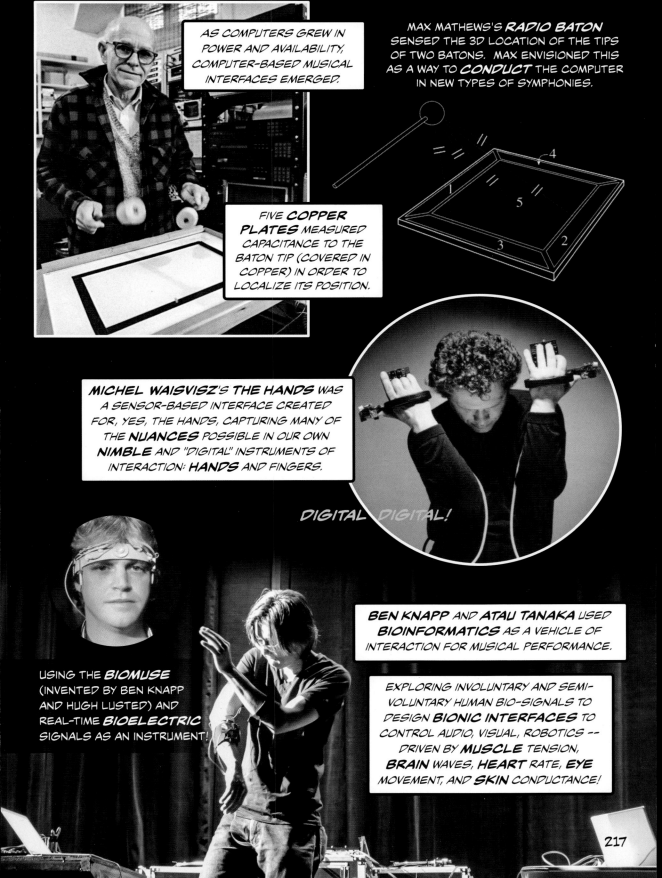

AS COMPUTERS GREW IN POWER AND AVAILABILITY, COMPUTER-BASED MUSICAL INTERFACES EMERGED.

MAX MATHEWS'S **RADIO BATON** SENSED THE 3D LOCATION OF THE TIPS OF TWO BATONS. MAX ENVISIONED THIS AS A WAY TO **CONDUCT** THE COMPUTER IN NEW TYPES OF SYMPHONIES.

FIVE **COPPER PLATES** MEASURED CAPACITANCE TO THE BATON TIP (COVERED IN COPPER) IN ORDER TO LOCALIZE ITS POSITION.

MICHEL WAISVISZ'S THE HANDS WAS A SENSOR-BASED INTERFACE CREATED FOR, YES, THE HANDS, CAPTURING MANY OF THE **NUANCES** POSSIBLE IN OUR OWN **NIMBLE** AND "DIGITAL" INSTRUMENTS OF INTERACTION: **HANDS** AND FINGERS.

DIGITAL DIGITAL!

USING THE **BIOMUSE** (INVENTED BY BEN KNAPP AND HUGH LUSTED) AND REAL-TIME **BIOELECTRIC** SIGNALS AS AN INSTRUMENT!

BEN KNAPP AND **ATAU TANAKA** USED **BIOINFORMATICS** AS A VEHICLE OF INTERACTION FOR MUSICAL PERFORMANCE.

EXPLORING INVOLUNTARY AND SEMI-VOLUNTARY HUMAN BIO-SIGNALS TO DESIGN **BIONIC INTERFACES** TO CONTROL AUDIO, VISUAL, ROBOTICS -- DRIVEN BY **MUSCLE** TENSION, **BRAIN** WAVES, **HEART** RATE, **EYE** MOVEMENT, AND **SKIN** CONDUCTANCE!

217

REBECCA FIEBRINK PIONEERED THE APPLICATION OF **MACHINE LEARNING** TO MUSICAL **HUMAN-COMPUTER-INTERACTION** AND **INSTRUMENT DESIGN.** SHE IS A MACHINE LEARNING EXPERT WHO EXPLORES COMPUTER-MEDIATED **SYSTEMS** INTO WHICH **HUMANS** ARE INTEGRATED AS PART OF THE CORE **INTERACTION LOOP.** SHE AUTHORED **WEKINATOR,** A FRAMEWORK FOR USING MACHINE LEARNING TO INFERENTIALLY **MAP** HUMAN **INPUT** (BROADLY SPEAKING, ANYTHING THAT CAN BE SENSED OR TRACKED) TO EXPRESSIVE **OUTPUT,** POTENTIALLY IN AN ENTIRELY DIFFERENT DOMAIN (E.G., SOUND, ANIMATION, TEXT, GAMES).

DON'T FORGET THE **HUMAN!**

REBECCA'S WORK EMBODIES A **UNIQUE** PARADIGM FOR DESIGNING INTERFACES: BY A HUMAN **EMPIRICALLY** AND **ITERATIVELY** **"SHOWING"** A COMPUTER HOW TO PERFORM THE MAPPING. A MAGNIFICENT INSTANCE OF **DESIGN-BY-EXAMPLE,** IT DEMONSTRATES HOW MACHINE LEARNING AND "HUMAN TEACHING" CAN WORK TOGETHER FOR DESIGN.

(꙰) PRINCIPLE 5.5 — *HAVE YOUR MACHINE LEARNING -- AND THE HUMAN IN THE LOOP!*

WE TEND TO THINK OF COMPUTERS, ALGORITHMS, MACHINE LEARNING, AND ARTIFICIAL INTELLIGENCE AS **PURE AUTOMATION** THAT PRODUCES SOME **OUTPUT** -- AND IT'S EASY TO OVERLOOK THEIR POTENTIAL FOR **HUMAN INTERACTION.** BUT MANY THINGS COMPUTERS CAN DO CAN BE FUNDAMENTALLY **IMPROVED** BY PLACING **HUMAN** INTENTIONALITY, TACIT KNOWLEDGE, INSTINCT, AND AESTHETICS INTO THE INTERACTION LOOP -- EMBODYING THE IDEA THAT COMPUTERS OUGHT NOT BE **REPLACEMENTS** BUT **EXTENSIONS** OF US. THIS IS THE ARTFUL DESIGN VERSION OF "HAVE YOUR CAKE AND EAT IT TOO": HAVE YOUR **MACHINE LEARNING** -- AND **THE HUMAN IN THE LOOP!**

...AND *DON'T FORGET* THE LAPTOP!

DESIGN WITH *PHYSICALITY!*

THINGS YOU CAN SENSE USING PHYSICAL LAPTOP:

	USING	
TILT		ACCELEROMETERS
SMACK		ACCELEROMETERS
PRESS		KEYBOARD
BOW		TRACKPAD
MOVE		CAMERA
SOUND		MICROPHONE

REBECCA HAS DONE SOME OF THE MOST DEEPLY INTERESTING WORK IN THE FIELD, AND SHE REMINDS US THAT AN *AUTHENTIC SYNTHESIS* OF TECHNOLOGY AND THE HUMAN IS NOT ONLY POSSIBLE, BUT TRULY INTERESTING.

ONCE UPON A TIME, REBECCA AND I ALSO DESIGNED MANY INSTRUMENTS USING THE *LAPTOP* AS A *PHYSICAL* OBJECT, LEADING US TO ISSUE A BATTLE CRY OF *"DON'T FORGET THE LAPTOP!"* IT'S A DESIGN *ETHOS* OF CONSIDERING COMPUTERS AS MORE THAN A *VIRTUAL MEDIUM*, OF TAKING WHATEVER PHYSICAL INPUTS AND SENSORS HAPPEN TO BE AVAILABLE AND THINKING AS BROADLY AS POSSIBLE ABOUT THEIR INTERACTIVE AFFORDANCES. THIS ETHOS RECOGNIZED THAT *PHYSICALITY* AND EXTREME *CONVENIENCE* CAN COEXIST, AND TOGETHER THEY CAN HAVE A TREMENDOUS POSITIVE EFFECT ON INTERFACE-BUILDING.

THIS WAY OF THINKING WAS LATER CARRIED INTO THE DESIGN OF THINGS LIKE *OCARINA* AS A *PHYSICAL ARTIFACT*, AS WELL AS...

THE
LAPTOP ACCORDION!

LAPTOP ACCORDION OPERATES BY OPENING AND CLOSING THE *SCREEN* (LIKE THE BELLOWS OF AN ACCORDION) TO MAKE SOUND. THE MOVEMENT IS TRACKED BY COMPUTER VISION (USING OPTICAL FLOW), WHILE THE KEYS ARE MAPPED TO PITCH.

IS IT A *LAPTOP...*

AIDAN MEACHAM

...OR AN *ACCORDION?!*

SANJAY KANNAN

THERE IS A HISTORY OF **AUGMENTING** EXISTING MUSICAL INSTRUMENTS. **AJAY KAPUR** DID SO WITH MANY **CLASSICAL INDIAN** INSTRUMENTS, INCLUDING THE **TABLA**, THE **SITAR**, AND THE **DHOLAK**.

AJAY WENT ON TO BUILD ORCHESTRAS OF **MUSICAL ROBOTS**, INSPIRED BY THE WORKS OF DESIGNERS AND INSTRUMENT INVENTORS SUCH AS **TRIMPIN** AND **ERIC SINGER**.

MAHADEVIBOT A ROBOTIC DRUMMER THAT OPERATES FROM CIRCUITS, SOLENOIDS, AND ALGORITHMS! CONTROLLABLE FROM CODE, IT EXTENDS BEYOND HUMAN DEXTERITY.

LET'S START A **REVOLUTION!**

AJAY WAS A **LAB MATE** AT PRINCETON. NOWADAYS HE IS THE ASSOCIATE DEAN OF MUSIC TECHNOLOGY AT **CALARTS** AND THE CO-FOUNDER OF **KADENZE.** ONE OF OUR FAVORITE PASTIMES IS QUESTIONING THE VALUE OF THE COMPUTER MUSIC WE MAKE...

AJAY DESIGNED THE **E-SITAR** BY **AUGMENTING** A SITAR WITH SENSORS (LIKE CIRCUITS THAT TRACK PARDA, OR FRET POSITION). THE AESTHETIC WAS TO KEEP THE SITAR INTACT AS AN INSTRUMENT, WHILE **EXTENDING** WHAT IT CAN DO THROUGH THE COMPUTER.

SPENCER SALAZAR IS THE CREATOR OF THE **MINIAUDICLE** INTEGRATED DEVELOPMENT ENVIRONMENT (IDE) FOR **CHUCK**; HE HAS BEEN A CORE DEVELOPER OF **CHUCK** ITSELF SINCE 2005!

LET'S **SKETCH** SOME SOUND...

SPENCER'S PH.D. THESIS WAS THE DESIGN OF **AURAGLYPH**, A WAY TO USE **HANDWRITTEN** GESTURES TO CRAFT AUDIO PROGRAMS WHOSE ELEMENTS ARE ALSO THE **INTERFACE**, ON TABLET COMPUTERS THAT SERVE AS BOTH PROGRAMMING **SURFACE** AND PHYSICAL **CONTROLLER.**

THE INTERCONNECTED, MODULAR **NODES** IN **AURAGLYPH** SERVE BOTH **FUNCTIONAL** AND **AESTHETIC** ROLES. THEY SPECIFY THE UNDERLYING AUDIO SYNTHESIS WHILE **VISUALIZING** SOUND AS IT MAKES ITS WAY THROUGHOUT THE SYNTHESIS GRAPH. THIS IS A USEFUL LEARNING/DEBUGGING TOOL THAT ALSO GIVES **AURAGLYPH** ITS DISTINCTIVE **FEEL** AND A KIND OF ORGANIC BEAUTY, AS COMPLEX AUDIO PROGRAMS PULSE AND SWAY VISUALLY WITH ITS SOUNDS...

NODES OF **AURAGLYPH** -- ONE SIMULTANEOUSLY CREATES AN AUDIO ALGORITHM **AND** A VISUALIZATION OF IT.

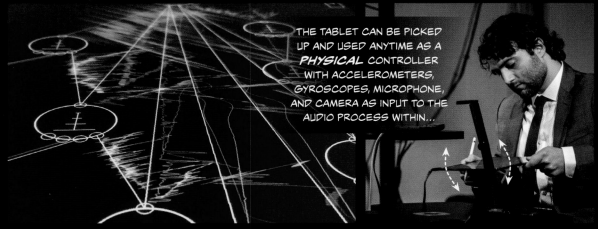

THE TABLET CAN BE PICKED UP AND USED ANYTIME AS A **PHYSICAL** CONTROLLER WITH ACCELEROMETERS, GYROSCOPES, MICROPHONE, AND CAMERA AS INPUT TO THE AUDIO PROCESS WITHIN...

221

INDEED, THE HISTORY AND COMMUNITY OF DESIGNING **NEW INTERFACES FOR MUSICAL EXPRESSION** (NIME) IS FILLED WITH WACKY CREATIONS AND **FASCINATING CHARACTERS** WHO DESIGN THEM!

MICHAEL LYONS'S **MOUTHESIZER** (COMPUTER VISION-TRACKED MOUTH SHAPE AS MUSICAL INPUT)

NIC COLLINS'S **SLED DOG** (HAND-SCRATCHABLE HACKED CD PLAYER)

MOBILE PHONE INSTRUMENTS

ERIC SINGER'S **GUITARBOT**

DAVID WESSEL'S **SLABS**

SID FELS'S **TOOKA** (A TWO-PERSON INSTRUMENT)

COOK / MORRILL TRUMPET

HAND / FINGERS TRACKING

JONGHYUN KIM

GINA GU

VIRTUAL AND AUGMENTED REALITY INTERFACES

ROGER DANNENBERG, COMPUTER ACCOMPANIMENT

ROMAIN MICHON'S **3D PRINT-STRUMENTS**

CURTIS BAHN'S **SENSOR BASS**

CHRIS CHAFE'S **CELLETTO**

TOMIE HAHN'S **PIKAPIKA** *(INTERACTIVE DANCE SYSTEM, MAPPING MOVEMENT TO SOUND)*

LAPTOP PERFORMANCE & LIVE CODING

MARK APPLEBAUM'S **MOUSEKETIER** (SCULPTED FROM FOUND OBJECTS LIKE SQUEAKY WHEELS, SPRINGS, THREADED RODS, A TOILET TANK FLOATING BULB, AND ELECTRONICS)

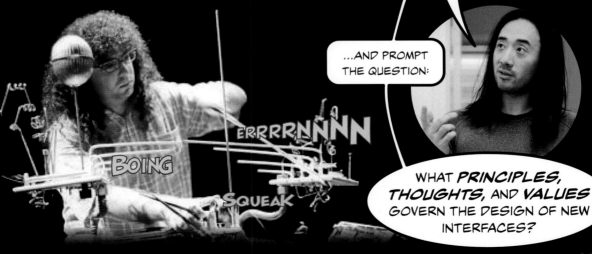

THESE ARE JUST THE **TIP** OF THE **ICEBERG** IN A WORLD OF NEW INTERFACES FOR MUSICAL EXPRESSION. THEIR DESIGNS EMBODY A UNIQUE SET OF QUESTIONS ABOUT EXPRESSION, MUSIC, AND OUR RELATIONSHIP TO OUR TECHNOLOGY...

...AND PROMPT THE QUESTION:

WHAT **PRINCIPLES, THOUGHTS,** AND **VALUES** GOVERN THE DESIGN OF NEW INTERFACES?

TO ANSWER THIS AND OTHER QUESTIONS, WE SHOULD FIRST RECOGNIZE A FUNDAMENTAL **DIFFERENCE** OF **MEDIUM** BETWEEN THE DESIGN OF **ACOUSTIC** INSTRUMENTS AND **COMPUTER-BASED** INSTRUMENTS. IN PARTICULAR, THE DESIGNS HAVE MUCH TO DO WITH THEIR **INPUT** AND THE **TECHNOLOGY** OF THEIR **SOUND PRODUCTION.** ON ONE HAND...

ACOUSTIC INSTRUMENT DESIGN
MUTUALIZES!
FORM FOLLOWS PHYSICS

THE **INTERFACE** OF AN **ACOUSTIC INSTRUMENT** IS INEXTRICABLY RELATED TO THE **MECHANISM** BY WHICH **SOUND IS PRODUCED.**

FRET

INPUT
INTERFACE

PLUCK!
STRUM!

FOLLOWS

SOUND PRODUCTION
MECHANISM

IT IS A CLASSIC EXAMPLE OF **FORM FOLLOWING FUNCTION**, OF **INPUT** GROUNDED IN THE **PHYSICAL CONSTRAINTS** OF THE UNDERLYING TECHNOLOGY.

VIBRATING STRINGS

OUTPUT
SOUND

FOR EXAMPLE, IF THE SOUND IS PRODUCED BY THE PHYSICAL PRINCIPLES OF **VIBRATING STRINGS**, THEN THE INTERFACE WOULD INVOLVE **EXCITING** THEM (PLUCKING, STRUMMING, OR BOWING) AND **CONTROLLING** FREQUENCY (FRETTING, FINGERING). THIS GIVES RISE TO THINGS LIKE FRETBOARDS, NECKS, SOUNDING CHAMBERS, AS WELL AS THE **MATERIALITY** OF STRING AND BODY.

ON THE OTHER HAND, THE **COMPUTER**, AS A DESIGN MEDIUM, IS PROFOUNDLY **UNDER-CONSTRAINED** -- A VIRTUAL SPACE WITHOUT PHYSICAL RULES.

COMPUTER INSTRUMENT DESIGN
DE-MUTUALIZES!
FORM DECOUPLED FROM FUNCTION

LAYER OF **ABSTRACTION**, SEPARATING INPUT FROM OUTPUT

```
    ...elay order
   ...> float L;
   ...set delay
 ::samp => delay.delay;
// set dissipation factor
Math.pow( R, L ) => delay.gain;
// place zero
-1 => lowpass.zero;

// fire excitation
1 => imp.gain;
// for one delay round trip
::samp => now;
...ase fire
...e.gain;
```

INPUT
CONTROLLER

THE INPUT INTERFACE CAN, IN THEORY, BE DESIGNED COMPLETELY INDEPENDENTLY OF UNDERLYING SOUND.

PROGRAMMABLE
MAPPING

MAPPING **BINDS** INPUT (INTERFACE) TO OUTPUT (SOUND SYNTHESIS), MAKING IT EASY TO CHANGE!

OUTPUT
SOUND

SYNTHESIZED IN SOFTWARE, SOUND IS NO LONGER BOUND BY **PHYSICAL** LAWS.

BEFORE WE GET INTO THE **ADVANTAGES** AND **PERILS** ASSOCIATED WITH THIS DE-MUTUALIZATION, WE SHOULD RECOGNIZE THAT THIS IS THE SORT OF THING THAT COMPUTERS ARE **GOOD** AT...

...IT GOES **BEYOND** INSTRUMENTS AND IS RELEVANT FOR COMPUTER-BASED INTERFACES IN GENERAL. THE **DECOUPLING** OF INPUT FROM UNDERLYING MECHANISM AFFORDS A **DE-MUTUALIZATION** OF FORM FROM FUNCTION: BOTH ARE PRESENT, BUT NEITHER **HAS** TO FOLLOW THE OTHER.

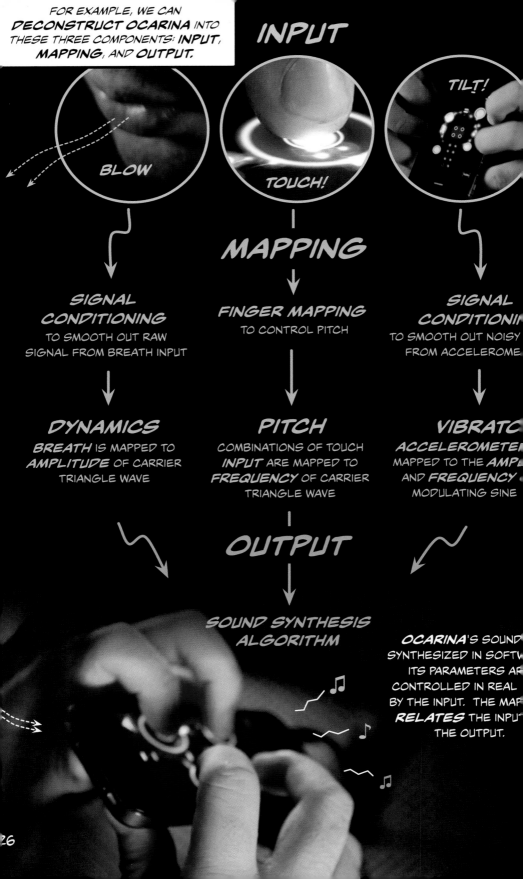

FOR EXAMPLE, WE CAN **DECONSTRUCT OCARINA** INTO THESE THREE COMPONENTS: **INPUT, MAPPING,** AND **OUTPUT.**

INPUT

BLOW

TOUCH!

TILT!

MAPPING

SIGNAL CONDITIONING
TO SMOOTH OUT RAW SIGNAL FROM BREATH INPUT

FINGER MAPPING
TO CONTROL PITCH

SIGNAL CONDITIONIN
TO SMOOTH OUT NOISY FROM ACCELEROME

DYNAMICS
BREATH IS MAPPED TO **AMPLITUDE** OF CARRIER TRIANGLE WAVE

PITCH
COMBINATIONS OF TOUCH **INPUT** ARE MAPPED TO **FREQUENCY** OF CARRIER TRIANGLE WAVE

VIBRATO
ACCELEROMETER
MAPPED TO THE **AMP** AND **FREQUENCY** MODULATING SINE

OUTPUT

SOUND SYNTHESIS ALGORITHM

OCARINA'S SOUND SYNTHESIZED IN SOFTW ITS PARAMETERS AR CONTROLLED IN REAL BY THE INPUT. THE MAF **RELATES** THE INPUT THE OUTPUT.

THIS DE-MUTUALIZATION MEANS THAT, IN THEORY, THE INPUT OF **OCARINA** (BREATH, TOUCH, TILT) CAN STAY EXACTLY THE **SAME**, WHILE WE **MAP** IT TO ENTIRELY **DIFFERENT** OUTPUT SOUNDS...

...LIKE, A STRANGE **CAT-OCARINA**

IF, INSTEAD OF TRIANGLE WAVES, WE MAP THE INPUT TO, SAY, **TUNED KITTENS** -- WE'D HAVE AN ENTIRELY DIFFERENT INSTRUMENT!

ME-OW! #♫

MEOW! ♪

ME-OW! ♫

?

DAT SOUNDZ LIKE MEEEE...

THIS **CHARACTERISTIC** OF COMPUTER INTERFACES SUGGESTS, LOGICALLY, A GENERAL METHOD TO DESIGN A MUSICAL INTERFACE AS **THREE SUB-TASKS**...

PRINCIPLE 5.6

HOW TO DESIGN A NEW INTERFACE
(A LOGICAL APPROXIMATION)

- **DETERMINE THE INPUT (ACTION)**
  (CONTROLLER, SENSORS, GESTURE)

- **CRAFT THE OUTPUT (SOUND)**
  (SYNTHESIS ALGORITHM, SOUND DESIGN)

- **CREATE THE MAPPING (FROM ACTION TO SOUND)**
  (CONNECTING INPUT TO OUTPUT PARAMETERS)

A **MEDIUM** OF **ABSTRACTION**, THE **COMPUTER** FACILITATES **STRATIFICATION** OF AN INTERFACE INTO **INPUT** (SENSORS), **OUTPUT** (SYNTHESIZER), AND **ACTION-TO-SOUND MAPPING** (LOGIC). IT ALSO IMPLIES A POSSIBILITY -- AT LEAST IN THEORY -- THAT THE THREE COMPONENTS ARE **MODULAR** AND **INTERCHANGEABLE.** THE SAME INPUT, FOR EXAMPLE, CAN BE MAPPED TO DIFFERENT SOUNDS, AND VICE VERSA. THE PHYSICAL CONSTRAINTS THAT **COUPLE** PLAYER MECHANICS TO ACOUSTIC MECHANICS ARE NO LONGER APPLICABLE. **FORM** FOLLOWS **IMAGINATION** AND **AVAILABILITY** OF SENSOR TECHNOLOGY, WHILE **FUNCTION** IS LARGELY ARTICULATED THROUGH THE RESULTING INPUT-TO-OUTPUT MAPPING.

WHILE THIS OFFERS US A **LOGICAL** PATH TO COMPARTMENTALIZE INTERFACE DESIGN, IT HAS ITS **DEFICIENCIES**, FOR THIS MODULAR APPROACH DOESN'T NECESSARILY ADDRESS THE AGGREGATE **EXPERIENCE** OF THE INTERFACE, OR THE **HUMAN** THAT INTERACTS WITH IT. FOR THAT WE NEED SOMETHING MORE...

IN DESIGN, PURE LOGIC IS **NECESSARY** BUT OFTEN NOT **SUFFICIENT** -- WE ALSO NEED SOMETHING THAT CONSIDERS THE RESULT AS A **WHOLE.** FOR THIS WE LOOK TO THE THINKING OF A UNIQUE INDIVIDUAL: **PERRY R. COOK!**

IN 2001, PERRY WROTE A PAPER CALLED "PRINCIPLES FOR DESIGNING COMPUTER MUSIC CONTROLLERS," IN WHICH HE ARTICULATED HIS APPROACH TO COMPUTER MUSIC INTERFACE DESIGN. IT GAVE **VOICE** AND **SUBSTANCE** TO AN **EMERGING COMMUNITY** OF NEW INTERFACES FOR MUSICAL EXPRESSION...

PERRY COOK'S

PRINCIPLES FOR DESIGNING COMPUTER MUSIC CONTROLLERS!

PRINCIPLE 5.7

INSTANT MUSIC, **SUBTLETY** LATER

PRINCIPLE 5.8

PROGRAMMABILITY IS A **CURSE**

PRINCIPLE 5.9

SMART INSTRUMENTS ARE OFTEN **NOT SMART**

PRINCIPLE 5.10

COPYING AN INSTRUMENT IS **DUMB,** LEVERAGING EXPERT TECHNIQUE IS **SMART**

PRINCIPLE 5.11

SOME PLAYERS HAVE **SPARE BANDWIDTH,** SOME DO NOT

≈AHEM≈

MUSIC INTERFACES ARE GREATLY *INFLUENCED* BY THE *MUSIC* WE LIKE AND SET OUT TO MAKE, THE INSTRUMENTS WE *ALREADY KNOW* HOW TO PLAY, AS WELL AS *AVAILABLE* SENSORS, COMPUTERS, TECHNOLOGY. BUT THE MUSIC WE CREATE AND *ENABLE* WITH OUR NEW INSTRUMENTS CAN BE EVEN MORE GREATLY INFLUENCED BY OUR INITIAL *DESIGN DECISIONS* AND *TECHNIQUES!*

(λ) PRINCIPLE 5.12

MAKE A *PIECE*, NOT AN INSTRUMENT OR CONTROLLER

(λ) PRINCIPLE 5.13

NEW *ALGORITHMS* SUGGEST NEW *CONTROLLERS*

(λ) PRINCIPLE 5.14

NEW *CONTROLLERS* SUGGEST NEW *ALGORITHMS*

(λ) PRINCIPLE 5.15

EXISTING INSTRUMENTS SUGGEST NEW CONTROLLERS

(λ) PRINCIPLE 5.16

EVERYDAY OBJECTS SUGGEST AMUSING CONTROLLERS

THESE HUMAN-BASED PRINCIPLES REMAIN AS *RELEVANT* AS EVER, AND MANY EASILY *EXTEND* BEYOND COMPUTER MUSIC. SIMPLE AND QUIRKY, DEEP AND TRUE, THERE IS A CERTAIN *ZEN* TO THEM. LET'S SEE 'EM IN ACTION!

PERRY COOK'S COFFEEMUG!

AN ICONIC EMBODIMENT OF THE PRINCIPLE "INSTANT MUSIC, SUBTLETY LATER"

POTENTIOMETER
AKA: A **KNOB** (CAN BE MAPPED TO ANY SINGLE CONTINUOUS PARAMETER; E.G., VOLUME)

DATA CABLE
(SENSOR DATA SENT TO A NEARBY COMPUTER AND MAPPED TO SOUND!)

2X BUTTONS
(SELECTS MODE BY STATE WHEN MUG IS TURNED ON; SITUATIONAL)

SWITCH
(ON/OFF)

4X FSRs
FORCE-SENSING RESISTORS (SENSES **PRESSURE**; GIVES FOUR INDEPENDENT AND **CONTINUOUS** INPUTS FOR FOUR FINGERS)

(INSIDE)

1-AXIS ACCELEROMETER
SENSES TILT (AND SHAKE)

COFFEE MUG
YEAH... IT REALLY IS A COFFEE MUG

INSTANT MUSIC, SUBTLETY LATER!

THERE ARE **FOUR MODES**, THE **FIRST** OF WHICH IS A PHYSICALLY-INSPIRED, COMPUTER-GENERATED **SHAKER**! MOTION IS SENSED BY THE ACCELEROMETERS AND MAPPED TO PERRY'S SHAKER SOUND SYNTHESIS ALGORITHM.

A **SECOND**, MORE **GENERATIVE**, MODE EMULATES AN OVER CAFFEINATED **LATIN FUSION BAND**, WHERE THE FOUR FSRS CONTROL THE **VOLUME** AND **ENTHUSIASM** LEVELS OF VARIOUS PARTS OF THE BAND (VOICES 1 & 2, BASS, PERCUSSION). ACCELEROMETER (TILT) CONTROLS GROSS PITCH RANGE AND DENSITY MAPPINGS OF THE ENSEMBLE.

SHAKE!

MODE 1: PHYSICALLY-INSPIRED SHAKER!

EASY TO **PICK UP** AND **IMMEDIATELY** MAKE SOUNDS. ASTUTE PLAYERS WILL DISCOVER DIFFERENT LEVELS OF MUSICAL DENSITY AND TEXTURE USING THE **FSRs** IN TANDEM WITH THE ACCELEROMETER.

MODE 2: BAND-IN-A-MUG
4X PRESSURE
(HOW HARD EACH FINGER PRESSES CONTROLS ONE OF FOUR PARTS OF THE BAND)

TEMPO CONTROL

TONAL CENTER & MODAL CONTROL

VOICE 1 & 2

BASS PLAYER

DRUMMER

TILT CONTROLS PITCH RANGE

DESPITE ITS APPARENT SIMPLICITY (INSTANT MUSIC), THIS ARTIFACT CONTINUES TO **REVEAL** ITSELF THROUGH DIFFERENT MODES AND AS ONE BEGINS TO LEARN ITS NUANCES (SUBTLETY LATER).

IT'S GOOFY, FUN, AND COMPLETELY DELIGHTFUL.

TOP FSR CONTROLS "BREATH PRESSURE"

THE **TILT** OF **COFFEEMUG** IN THIS MODE CONTROLS THE PITCH REGISTERS, TO BE USED IN TANDEM WITH THE "VALVES." IT AUGMENTS THE CONTROL OVER THE "BREATH PRESSURE" PARAMETER.

IN A **THIRD** MODE, **COFFEEMUG** CAN BE PLAYED LIKE A **TRUMPET** -- THE BOTTOM THREE FSRs SERVE AS VALVES ON A TRUMPET, WITH THE SAME MAPPINGS.

TRUMPET VIBRATO

MODE 3: MUG TRUMPET!

CONTROLS PITCH LIKE TRUMPET VALVES

THE BAND-IN-A-MUG IS **GENERATIVE**: THE PLAYER **INFLUENCES** HIGHER LEVEL MUSICAL PARAMETERS SUCH AS DENSITY AND TEXTURE. IN CONTRAST, PHYSICAL GESTURES IN THE **SHAKER** AND **TRUMPET** MODES **DIRECTLY** CAUSE AND CONTROL THE SOUND.

THE **FINAL** MODE: A ZEN JAPANESE FLUTE GENERATOR...

MODE 4: ZEN FLUTE GENERATOR

ORIGINALLY DESIGNED FOR A **TRANS-CONTINENTAL** INTERNET PERFORMANCE IN 1995 BETWEEN USA (COLUMBIA UNIVERSITY) AND JAPAN (UNIVERSITY OF TOKYO), THE JAPANESE FLUTE WAS ACTUALLY THE ORIGINAL MUG MODE. (PRINCIPLE: MAKE A PIECE, NOT A CONTROLLER!)

≈MMMMMMMMMM≈ SUCH ZENNNNNN

THE **INNARDS** OF **COFFEEMUG** ARE A SMALL NETWORK OF CIRCUITS AND A **MICRO-CONTROLLER** (E.G., ARDUINO OR BASIC STAMP), WHICH DIGITIZES THE DATA, COMPUTES MAPPING AND MUSICAL SEQUENCE, AND SENDS IT OUT SERIALLY!

THINGS I LEARNED FROM THE DESIGN OF COFFEEMUG

#1. IT'S OKAY TO USE A LOT OF **TAPE,**

#2. UNDERSTATE. A MUG IS A SIMPLE, EVERYDAY ARTIFACT, AND THUS PERHAPS MORE **INVITING**; WE ARE ALREADY FAMILIAR WITH THE FORM OF A COFFEE MUG.

#3. EVERYDAY ARTIFACTS MAKE **AMUSING** INSTRUMENTS. WE ARE NOT USED TO THINKING ABOUT THE MUG AS A MUSICAL THING -- THIS INCONGRUITY OF FAMILIARITY (MUG) VS. STRANGE (MUG AS INSTRUMENT) MAKES THEIR UNION A **DELIGHTFUL** ARTIFACT UNTO ITSELF.

#4. INSTANT AMUSEMENT, NUANCE AND LEARNING LATER. BOTH FOR THE PLAYER AND THE AUDIENCE, SOMETIMES YOU GOTTA WIN PEOPLE OVER FIRST, SO THEY'LL WANT TO EXPLORE FURTHER...

NEW **ALGORITHMS** SUGGEST NEW **CONTROLLERS!**

THE DIGITAL **SHAKERS**
(INTERFACES TO CONTROL **PH.I.S.E.M.** ALGORITHMS)

1-AXIS ACCELEROMETER
(SHAKE DETECTION)

POWER SUPPLY

PRESERVES TAMBOURINE FORM FACTOR

BELLS REMOVED; SOUND SYNTHESIZED BY COMPUTER

DATA OUT
(TO COMPUTER)

FSR
(PRESS TO DAMPEN)

BUTTON

SHAKE TO INJECT ENERGY INTO THE SYSTEM!

THE **FROG MARACA** -- ONLY ONE MODE: PICK UP AND **SHAKE!** LEARN SUBTLETIES OVER TIME.

OF COURSE THERE IS A **SECOND** MODE: A SEMI-AUTOMATED PERCUSSION ENSEMBLE!

ANOTHER: **"SPUTNIK"**

THE SHAKING INTERACTION IS IDENTICAL, BUT THE **SOUND** CAN BE VERY **DIFFERENT.**

MARACA

WATERDROPS   BAMBOO CHIMES

SLEIGH BELLS   GUIRO   TAMBOURINE   CRUNCH   **NEXT** MUG

WRENCH / RATCHET   STICKS   SEKERE   CABASA   PENNY IN MUG

MODELS   SAND PAPER   NICKEL IN MUG

BIG ROCKS   LITTLE ROCKS   DIME IN MUG

QUARTER IN MUG

THIS SHAKER IS DISGUISED AS A **PLASTIC LETTUCE,** NAMED **"ROMAINERACA,"** THIS CONTROLLER WAS INSPIRED BOTH BY THE PH.I.S.E.M. ALGORITHM AND AN EVERYDAY OBJECT...

FRANC IN MUG

PESO IN MUG

SYNTHESIS MODEL: PHISEM
(PHYSICALLY INSPIRED STOCHASTIC EVENT MODELING)

PERRY'S **PHISEM** MODELS THE SOUND OF **PARTICLE SYSTEMS** (E.G., BEANS IN GOURD, TAMBOURINE, SLEIGH BELLS, EVEN COINS ROLLING IN A MUG). THE SOUNDS OF MANY PARTICLE-GOURD COLLISIONS ARE COMPUTED FROM **STATISTICS** DERIVED FROM REAL-WORLD MEASUREMENTS OF ACTUAL SHAKERS, PRODUCING A **SYNTHESIS ALGORITHM** WITH FOUR REAL-TIME CONTROLLABLE CORE **PARAMETERS.** THE ALGORITHM IS **PHYSICALLY INSPIRED** (AS OPPOSED TO PHYSICALLY SIMULATED) AND IS MUCH MORE **EFFICIENT** THAN A **BRUTE-FORCE, EXHAUSTIVE** PARTICLE SIMULATION. AND, IT SOUNDS **GOOD!**

MODELING A SHAKER SYSTEM

INPUT PARAMETER #1
ENERGY INJECTED INTO SYSTEM MODELS **FORCEFULNESS** OF SHAKING; AFFECTS OVERALL INTENSITY AND AMOUNT OF SOUND

INPUT PARAMETER #2
RATE OF **DECAY** OF ENERGY (MODELS OVERALL ENERGY LOSS OVER TIME IN SYSTEM; AFFECTS THE DURATION OF SONIC TAIL)

INPUT PARAMETER #3
NUMBER OF PARTICLES (AFFECTS SONIC DENSITY; **PHISEM** MODELS THIS SCENARIO AND IS **STATISTICAL**, NOT CONCERNED WITH SIMULATING INDIVIDUAL PARTICLES)

GOURD

TIME **BETWEEN** SUCCESSIVE COLLISIONS

128 PARTICLES
64 PARTICLES
32 PARTICLES

PROBABILITY

TIME

PARTICLES

PARTICLE-GOURD COLLISIONS
THE STATISTICS OF THESE COLLISIONS (WHICH ACCOUNT FOR THE BULK OF THE SOUND) ARE MODELED ON REPEATED MEASUREMENTS OF ACTUAL PARTICLE COLLISIONS (WHICH EXHIBITS A **POISSON** DISTRIBUTION THAT ASSUMES THAT ALL COLLISIONS ARE **INDEPENDENT STOCHASTIC** EVENTS). THIS EFFICIENTLY MODELS OCCURRENCES OF COLLISION USING AN **EXPONENTIAL FUNCTION** TO COMPUTE TIME BETWEEN SUCCESSIVE COLLISIONS.

INPUT PARAMETER #4
RESONANCE FREQUENCIES OF DIFFERENT "GOURDS" CAN BE EXPRESSED THROUGH **STOCHASTIC MODAL SYNTHESIS**, ALLOWING THE USER TO "TUNE" THE GENERAL FREQUENCY CONTENT AND TO SYNTHESIZE OSCILLATORY, RINGING SOUNDS -- SUCH AS A **FRANC** COIN ROLLING IN A **NEXT MUG!** THE **MODAL** RESONANCES OF EACH TYPE OF SHAKER WERE DERIVED EMPIRICALLY FROM SPECTRAL ANALYSIS OF REAL-WORLD RECORDINGS.

PERRY HAS A SIMPLE WAY TO **CLASSIFY** COMPUTER MUSIC INTERFACES INTO **THREE** CATEGORIES.

SQUEEZEVOX

BOSSA

NUKULELE

DIGITALDOO

E-SITAR

OCARINA

INTERFACES BASED ON *EXISTING INSTRUMENTS*

FOWL-HARMONIC

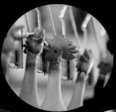

COFFEEMUG

ROMAINERACA

SONIC BANANA

FILLUP GLASS

P-RAY'S CAFE

SLED DOG

INTERFACES BASED ON "NONMUSICAL" THINGS

THEREMIN

THE HANDS

TOOKA

RADIO BATON

TWILIGHT

BIOMUSE

GAMETRAK

...AND THE *FANTASTICAL!*

INTERACTION DESIGN DEALS WITH HUMAN ACTIONS AS **INPUT** ON ONE SIDE AND COMPUTER **OUTPUT** ON THE OTHER. **MAPPING** IS WHAT CONNECTS THE TWO SIDES -- FOR EXAMPLE, AN **ACTION-TO-SOUND** MAPPING.

SOME WOULD SAY **MUCH** OF COMPUTER INTERFACE DESIGN IS ABOUT THE **MAPPING** -- SENSIBLY ARRIVING AT THE DESIRED OUTPUT FROM THE INPUT.

PERRY, HOW DO **YOU** THINK ABOUT MAPPING?

A GNARLY PROBLEM OF MAPPING?

HMM... LET'S SEE: FEW-TO-MANY, MANY-TO-FEW, LOTS OF INPUTS, OUTPUTS, AND THEY INTERACT IN SOME COMPLEX WAY! BUT TO BE HONEST, I AM NOT A BIG FAN OF THINKING ABOUT **MAPPING** LIKE IT'S SOME **BIG THING.** TO ME, MAPPING IS REALLY JUST WHAT INPUT GOES TO WHAT OUTPUT... AS SUCH, **MAPPING** ISN'T THIS **HOLY GRAIL** OF INTERFACE DESIGN.

I MEAN, YOU CAN HAVE A **PROVABLY** "GOOD" MAPPING AND STILL HAVE AN OVERALL CRAPPY, UNSATISFYING INTERFACE.

EXACTLY... YOU CAN GO PRETTY FAR ARMED WITH **COMMON SENSE** AND **EXPERIMENTATION.** I MEAN, SOME THINGS ARE FAIRLY **OBVIOUS**...

RIGHT, **RULES OF THUMB** LIKE **SIZE** OF INPUT **PROPORTIONAL** TO SIZE OF OUTPUT (SMALL GESTURES YIELD SUBTLE RESULTS; LARGE GESTURES YIELD SWEEPING CHANGES) AND CONSISTENCY OF EXPECTATION...

RIGHT.

MAPPING IS NECESSARY, BUT THERE IS, AS YOU KNOW, A **BIGGER PICTURE** HERE...

MUCH EFFORT IN *INTERFACE DESIGN* (MUSICAL AND OTHERWISE) HAS WORKED TO IMPROVE THE FIDELITY AND FLEXIBILITY OF A LINEARIZED, *FORWARD* LOGIC PATH, BY IMPROVING THE INPUT SENSORS, OUTPUT QUALITY, SYSTEM LATENCY, AND BY FINDING THE "RIGHT" MAPPING BETWEEN THEM.

LIKE *THIS*

A TRADITIONAL ENGINEERING APPROACH TO INTERFACE DESIGN

INPUT
E.G., CONTROLLER

SENSOR DATA

LOGIC
WHAT AND HOW INPUT CONTROLS WHAT OUTPUT

PARAMETER VALUES

SIGNAL CONDITIONING
CLEAN UP, AMPLIFY, AND OTHERWISE PREPARE DATA FOR PROCESSING

PREPARED SENSOR DATA

OUTPUT
E.G., SYNTHESIZER

THIS *TRADITIONAL ENGINEERING APPROACH* OF BUILDING A CONTROLLER (A BOX) AND CONNECTING IT TO A SYNTHESIZER (ANOTHER BOX) IS USEFUL BECAUSE IT'S *LOGICAL* AND *CONVENIENT*. *HOWEVER*, THERE ARE SITUATIONS WHERE SUCH AN APPROACH WOULD *NEVER* HAVE YIELDED THE SAME FINAL PRODUCT AS WOULD HAVE RESULTED (AND BENEFITED) FROM A *TIGHTLY COUPLED* CONSIDERATION OF A COMPLETE SYSTEM ALL AT THE SAME TIME.

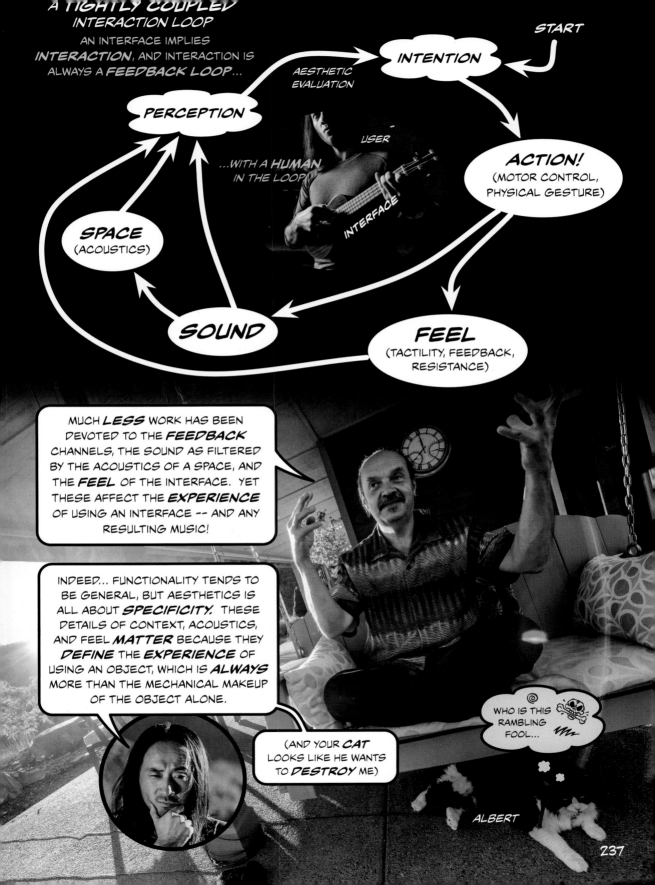

A TIGHTLY COUPLED INTERACTION LOOP

AN INTERFACE IMPLIES **INTERACTION**, AND INTERACTION IS ALWAYS A **FEEDBACK LOOP**...

START

INTENTION

AESTHETIC EVALUATION

PERCEPTION

...WITH A **HUMAN** IN THE LOOP

USER

INTERFACE

ACTION!
(MOTOR CONTROL, PHYSICAL GESTURE)

SPACE
(ACOUSTICS)

SOUND

FEEL
(TACTILITY, FEEDBACK, RESISTANCE)

MUCH **LESS** WORK HAS BEEN DEVOTED TO THE **FEEDBACK** CHANNELS, THE SOUND AS FILTERED BY THE ACOUSTICS OF A SPACE, AND THE **FEEL** OF THE INTERFACE. YET THESE AFFECT THE **EXPERIENCE** OF USING AN INTERFACE -- AND ANY RESULTING MUSIC!

INDEED... FUNCTIONALITY TENDS TO BE GENERAL, BUT AESTHETICS IS ALL ABOUT **SPECIFICITY**. THESE DETAILS OF CONTEXT, ACOUSTICS, AND FEEL **MATTER** BECAUSE THEY **DEFINE** THE **EXPERIENCE** OF USING AN OBJECT, WHICH IS **ALWAYS** MORE THAN THE MECHANICAL MAKEUP OF THE OBJECT ALONE.

(AND YOUR **CAT** LOOKS LIKE HE WANTS TO **DESTROY** ME)

WHO IS THIS RAMBLING FOOL...

ALBERT

237

YES! SPECIFICALLY, I WOULD SAY ONE MAJOR *FLAW* IN THE CONTROLLER/SYNTHESIZER PARADIGM IS THE *LOSS OF INTIMACY* BETWEEN *HUMAN* PLAYER AND *INSTRUMENT.*

I'D POSIT *THREE* REASONS FOR THIS INTIMACY LOSS:

1. LACK OF *HAPTIC FEEDBACK* FROM *CONTROLLER* TO *PLAYER*, INCLUDING THE COMBINED SENSES OF *TOUCH* (E.G., SKIN VIBRATION AND PRESSURE) AND THE MUSCLE SENSE OF MOTION, POSITION, AND FORCE.

2. LACK OF *FIDELITY* IN CONNECTION FROM *CONTROLLER* TO *SOUND GENERATOR* (*DELAY* AND *DISTORTION* IN RESPONSE TO GESTURES). "DISTORTION" HERE REFERS TO ANY RESPONSE THAT DOESN'T MEET SOME USUAL, LEARNABLE, OR REPEATABLE EXPECTATIONS!

3. LACK OF *ANY SENSE* THAT *SOUND COMES FROM THE INSTRUMENT ITSELF.* MORE GENERALLY, THIS IS A SUBSET OF A LARGER FEELING THAT NO MEANINGFUL PHYSICS GOES ON IN THE CONTROLLER. IN SHORT, SUCH INTERACTIONS LACK A SENSE OF *EMBODIMENT*, A TANGIBLE CONNECTION BETWEEN INSTRUMENT AND OUR IDEA OR FEELING OF IT AS AN OBJECT.

THESE THINGS MATTER.

☥ PRINCIPLE 5.17 *EMBODY!*

WE HUMANS ARE *EMBODIED* CREATURES; WE OPERATE MORE EFFICIENTLY, SATISFYINGLY WHEN WE "*FEEL AS ONE*" WITH THE INTERFACE WE ARE USING! SIMILAR TO USING OUR HANDS, AN EMBODIED INTERFACE ALLOWS US TO THINK LESS ABOUT HOW TO CONTROL IT AND MORE ABOUT WHAT WE'D WANT TO *DO* WITH IT -- LIKE HOW WELL A CAR HANDLES OR HOW AN INSTRUMENT OR A VIDEO GAME RESPONDS TO OUR INTENTIONS. OUR SENSE OF *EMBODIMENT* MATTERS, FOR IT SHAPES HOW WE THINK ABOUT AND *INTERACT* WITH THE WORLD.

HERE IS AN EXAMPLE OF AN **EMBODIED INTERFACE!** THE **SQUEEZEVOX** PROJECT WITH **COLBY LEIDER** WAS INSPIRED BY THE DESIRE TO CONTROL **VOCAL** AND **SINGING** SYNTHESIS ALGORITHMS, AS WELL AS THE NATURAL GESTURES OF **ACCORDIONS** AND **SQUEEZEBOXES.**

MAPPING-WISE, SOME THINGS ARE KIND OF **OBVIOUS,** LIKE THE **BELLOWS** ARE AN ANALOGY TO **BREATHING,** THE **KEYBOARD** IS USED TO CONTROL **PITCH...**

...OTHERS ARE **LESS** OBVIOUS.

LINEAR FSR (FOR CONTINUOUS PITCH CONTROL)

ELASTIC "SQUEEZER" (LITERALLY FOR SQUEEZING; FOUR CHANNELS OF CONTINUOUS CONTROL)

EMBEDDED BREATH SENSOR (CLOSE TO EXHALE; OPEN TO INHALE)

BELLOWS == BREATHING

63 RETROFITTED SWITCHES (63 DISCRETE TRIGGERS)

RETROFITTED KEYBOARD (FOR DISCRETE CONTROL)

MAPPING
SQUEEZEVOX: LISA
AN EMBODIED, ACCORDION-INSPIRED
VOICE CONTROLLER!

AND THAT'S JUST THE **BEGINNING** OF THE **MAPPINGS** IN **LISA...**

⸮ZZZ⸮ WHAT... UH? GE ARE YOU **STILL** HERE... I SUPPOSE YOU'D WANT TO SAY "**LISA** ILLUSTRATES THE PRINCIPLE THAT EXISTING INSTRUMENTS SUGGEST NEW BLAH BLAH BLAH"

OH BY THE WAY, I WILL BE ADDING THIS HERE **BAG** OF YOURS TO MY **PILLOW COLLECTION** -- IT'S INFINITELY MORE USEFUL THAN YOU ARE... ⸮SNORE⸮ ⸮SNORE⸮

SO THIS IS *LISA* -- WHICH *WAS* AN ACCORDION I GOT AT A THRIFT STORE AND SUBSEQUENTLY HAD ALL ITS ESSENTIAL INNARDS, REEDS AND ALL, EXTRACTED!

PUSH!

THE ORIGINAL *KEYBOARD* WAS *GUTTED* AND *REPLACED* WITH A SECOND KEYBOARD SEPARATELY *GUTTED* FROM A SYNTHESIZER. ALL OF THE *MECHANICAL* BUTTONS AND STOPS WERE REPLACED WITH *ELECTRICAL* SWITCHES.

FROM BELLOWS TO KEYBOARD, BUTTONS TO SWITCHES, THE DESIGN *PRESERVES*, WHEN POSSIBLE, THE *PHYSICAL INTERACTION* FROM THE ORIGINAL ACCORDION, WHILE AUGMENTING IT WITH NEW INPUT MECHANISMS.

SLIDE

NEXT TO THE KEYBOARD IS A *LINEAR FORCE SENSING RESISTOR* THAT CAN SENSE BOTH *POSITION* ALONG THE STRIP AS WELL AS THE AMOUNT OF *PRESSURE!*

THIS WOULD BE AN EXAMPLE OF *LEVERAGING EXPERT TECHNIQUE* AND HOW PEOPLE HAVE LEARNED TO PLAY THE ACCORDION OVER THE AGES.

ALSO, SOME PLAYERS HAVE *ADDITIONAL BANDWIDTH* TO WORK WITH ADDITIONAL CHANNELS OF EXPRESSION. OTHERS DO NOT. I MYSELF HAVE NO BANDWIDTH FOR THINGS NOT RELATED TO FOOD OR NAPPING.

WHEREAS THE **KEYBOARD** CAN BE OBVIOUSLY MAPPED TO **DISCRETE** PITCHES, THE **LINEAR FSR** CAN BE MAPPED TO **CONTINUOUS PITCH** -- MORE ANALOGOUS TO THE **HUMAN VOICE**, CAPABLE OF MORE CONTINUOUS EXPRESSIVE GESTURES SUCH AS **GLISSANDO** AND **VIBRATO.**

CONTINUOUS CONTROL
(LIKE THAT OF THE **VOICE** OR **FRETLESS** STRING INSTRUMENT)

VS.

DISCRETE CONTROL
(LIKE **KEYS** ON **PIANO**)

FSR MAPPING #1
THE **LINEAR FSR** IS MAPPED LENGTH-WISE TO **CONTINUOUS PITCH**, RANGE IDENTICAL TO ADJACENT KEYBOARD!

FSR MAPPING #2
ALTERNATELY, THE LINEAR FSR CAN BE MAPPED TO **MICRO-TONAL** CONTINUOUS PITCH, **RELATIVE** TO THE CURRENT PITCH SELECTED ON THE KEYBOARD!

FSR MAPPING #3
THE LINEAR FSR CAN ALSO BE MAPPED TO **TIMBRE** (NOT PITCH), TO SHAPE THE **SPECTRUM** OF THE VOCAL SYNTHESIS FOR **OVERTONE SINGING!**

PRESS!

BUTTON / SWITCHES
(CAN BE MAPPED TO ANY DISCRETE **TRIGGERS**)

THE ORIGINAL BUTTONS ON THE LEFT-HAND SIDE HAVE BEEN REPLACED WITH 63 **ELECTRICAL SWITCHES.** IN THE VOCAL SYNTHESIS MODEL MAPPING, THESE ARE MAPPED TO DIFFERENT **VOWELS** AND **CONSONANTS** TO CONTROL WHAT IS **SUNG** -- FOR EXAMPLE "AHH," "EEE," "OOO," OR "TSS," "SH," "K." WITH ENOUGH PRACTICE, ONE CAN USE THIS TO SING "A BICYCLE BUILT FOR TWO"...

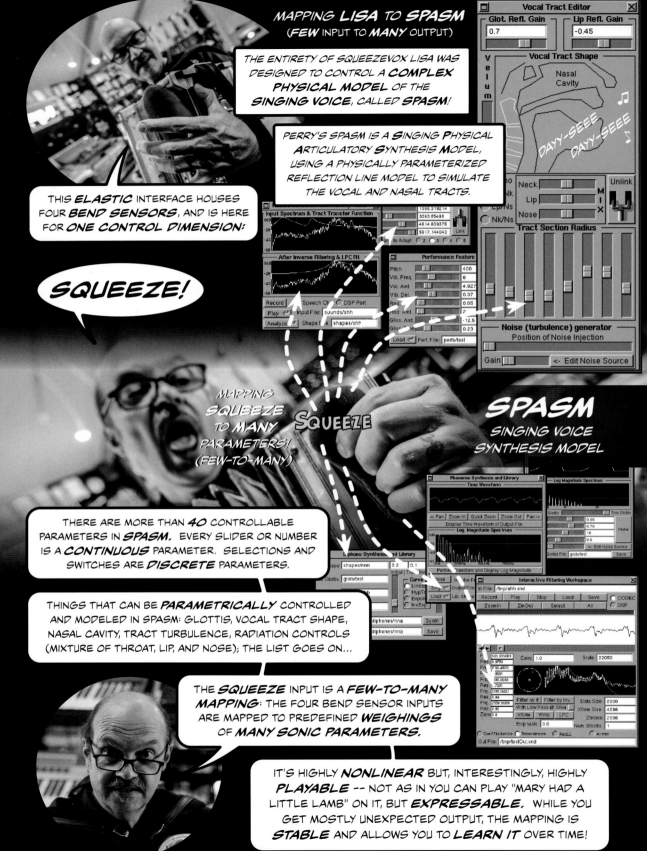

MAPPING **LISA** TO **SPASM**
(*FEW* INPUT TO **MANY** OUTPUT)

THE ENTIRETY OF SQUEEZEVOX LISA WAS DESIGNED TO CONTROL A **COMPLEX PHYSICAL MODEL** OF THE **SINGING VOICE**, CALLED **SPASM**!

PERRY'S SPASM IS A **S**INGING **P**HYSICAL **A**RTICULATORY **S**YNTHESIS **M**ODEL, USING A PHYSICALLY PARAMETERIZED REFLECTION LINE MODEL TO SIMULATE THE VOCAL AND NASAL TRACTS.

THIS **ELASTIC** INTERFACE HOUSES FOUR **BEND SENSORS**, AND IS HERE FOR **ONE CONTROL DIMENSION**:

SQUEEZE!

MAPPING **SQUEEZE** TO **MANY** PARAMETERS! (FEW-TO-MANY)

SQUEEZE

SPASM
SINGING VOICE SYNTHESIS MODEL

THERE ARE MORE THAN **40** CONTROLLABLE PARAMETERS IN **SPASM.** EVERY SLIDER OR NUMBER IS A **CONTINUOUS** PARAMETER. SELECTIONS AND SWITCHES ARE **DISCRETE** PARAMETERS.

THINGS THAT CAN BE **PARAMETRICALLY** CONTROLLED AND MODELED IN SPASM: GLOTTIS, VOCAL TRACT SHAPE, NASAL CAVITY, TRACT TURBULENCE, RADIATION CONTROLS (MIXTURE OF THROAT, LIP, AND NOSE); THE LIST GOES ON...

THE **SQUEEZE** INPUT IS A **FEW-TO-MANY MAPPING**: THE FOUR BEND SENSOR INPUTS ARE MAPPED TO PREDEFINED **WEIGHINGS** OF **MANY SONIC PARAMETERS.**

IT'S HIGHLY **NONLINEAR** BUT, INTERESTINGLY, HIGHLY **PLAYABLE** -- NOT AS IN YOU CAN PLAY "MARY HAD A LITTLE LAMB" ON IT, BUT **EXPRESSABLE,** WHILE YOU GET MOSTLY UNEXPECTED OUTPUT, THE MAPPING IS **STABLE** AND ALLOWS YOU TO **LEARN IT** OVER TIME!

243

PRINCIPLE 5.18

RE-MUTUALIZE!

INPUT + OUTPUT + HUMAN

JUST BECAUSE WE **CAN** DESIGN AN INTERFACE BY COMPONENT (E.G., AS INPUT-MAPPING-OUTPUT) DOES NOT MEAN WE ALWAYS **SHOULD**. THE ETHOS OF **RE-MUTUALIZATION** IS A COMMITMENT TO DESIGNING THE INTERFACE AS A WHOLE -- AND WITH THE **HUMAN** AS AN INTEGRAL PART OF THE **SYSTEM**.

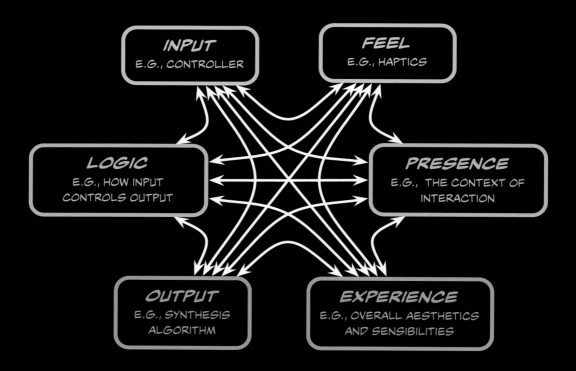

INPUT
E.G., CONTROLLER

FEEL
E.G., HAPTICS

LOGIC
E.G., HOW INPUT
CONTROLS OUTPUT

PRESENCE
E.G., THE CONTEXT OF
INTERACTION

OUTPUT
E.G., SYNTHESIS
ALGORITHM

EXPERIENCE
E.G., OVERALL AESTHETICS
AND SENSIBILITIES

THIS COMPLETELY **COMPLICATES** OUR INPUT-MAPPING-OUTPUT DESIGN APPROACH, BUT SOMETIMES THERE IS **NO WAY AROUND IT.**

245

SO, BACK TO YOUR QUESTION ABOUT MAPPING... I WILL **ADMIT**, MY MAPPINGS ARE **MUCH SIMPLER** THAN MOST PEOPLE WOULD CARE TO ACKNOWLEDGE, BECAUSE I THINK PEOPLE THINK **MAPPING** IS THE REALLY **HARD** AND **GNARLY** PART.

I THINK **LEARNING** TO **PLAY WHAT YOU'VE BUILT** AND **DARING** TO DO IT **IN FRONT** OF **PEOPLE** IS THE REALLY HARD AND GNARLY PART.

WOW, IMAGINE **THAT**...

...DOING SOMETHING **IN THE REAL WORLD** WITH THE DESIGN, AND AS A WAY TO ASSESS ITS ACTUAL **VALUE.**

THIS REMINDS ME OF SOMETHING PROMINENT HCI RESEARCHER (AND STANFORD PROFESSOR) **JAMES LANDAY** ONCE OBSERVED...

I MEAN THERE ARE MORE "**SCIENTIFIC**" WAYS TO GAUGE THE QUALITY OR EFFECTIVENESS OF AN INTERACTION. BUT THERE IS ALSO SOMETHING MORE **HOLISTIC** ABOUT DESIGN THAT SEEMS FUNDAMENTALLY DIFFICULT TO FORMALIZE.

IT'S MUCH EASIER (AND POSSIBLE IN A MATTER OF WEEKS) TO COME UP WITH **ISOLATED** INTERACTION TECHNIQUES AND RUN "SCIENTIFIC," TIGHTLY-CONTROLLED STUDIES WITH SMALL **ARTIFICIAL** TASKS -- IN CONTRAST TO DESIGNING AND BUILDING **REAL SYSTEMS** (WHICH CAN TAKE YEARS) AND EVALUATING THEM ON **REAL TASKS** WITH **REAL USERS.** WHILE THE FORMER IS INTERESTING IN ACADEMIC COMMUNITIES AND IS SELF-PERPETUATING, IT RARELY INVENTS ANYTHING TRULY NEW AND USEFUL.

COLOR ME CRAZY, BUT I THINK IT'S TIME WE DEVELOPED A DEEPER **AESTHETIC** UNDERSTANDING AND MORE SYSTEMIC, BALANCED CRITERIA FOR **EVALUATING** HUMAN-COMPUTER INTERACTIONS!

JAMES LANDAY
HCI RESEARCHER

YEAH, I'M WITH YA!

MORE IS **TRUE** THAN CAN BE **PROVEN**, AND WE KNOW MORE THAN WE CAN TELL. TO APPROACH DESIGN **PURELY SCIENTIFICALLY** WOULD BE TO MISS SOMETHING TACIT AND ESSENTIAL ABOUT DESIGN. THE QUALITY OF AN INTERFACE SHOULD ULTIMATELY BE JUDGED BY WHAT ONE **DOES WITH IT.**

NOW WHERE IS THAT TENNIS BALL...

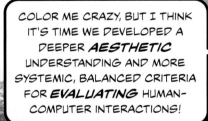

THIS LINE OF **RE-MUTUALIZING**, HUMAN-AWARE, **EMBODIED** THINKING ABOUT DESIGN LED TO SOME SUBTLE AND PROFOUND MUSICAL INTERFACES, PIONEERED BY FOLKS LIKE PERRY AND **DAN TRUEMAN**, A COMPOSER, PERFORMER, AND PIONEERING INSTRUMENT DESIGNER.

DAN TRUEMAN'S

BOSSA!

THE **VIOLIN** DECONSTRUCTED, RECONSTRUCTED!

THE **BOSSA** (**B**OWED **S**ENSOR **S**PEAKER **A**RRAY) IS A **DECONSTRUCTION** AND **RECONSTRUCTION** OF DAN'S PRIMARY ACOUSTIC INSTRUMENT: THE **HARDANGER FIDDLE.**

FANGERBOARD
FINGERING AND PRESSURE

R-BOW
FORCE, ACCELERATION, DISTANCE TO BRIDGE, TILT

NECK
ANGLES OF SWIVEL

BONGE
BOW PRESSURE ON FOUR SPONGES TO EMULATE STRINGS

THE CRITTER
12-SPEAKER **SPHERICAL SPEAKER ARRAY EMBODIED** SOUND OUTPUT, AKIN TO ACOUSTIC INSTRUMENTS

BOSSA'S COMPONENTS **LOOSELY ECHO** THE FUNCTIONAL ELEMENTS OF A VIOLIN, TRACKING BOW LOCATION, TILT, DISTANCE TO BRIDGE, "STRINGS" BEING BOWED, POSITION AND PRESSURE ON A **FANGERBOARD**, AND EVEN THE **SWIVEL** OF THE NECK. THE SENSORS SEND DATA TO A COMPUTER, WHICH INTERPRETS THE INPUT AND SYNTHESIZES THE SOUNDS, WHICH ARE THEN **SENT BACK** TO THE **SPHERICAL SPEAKER ARRAY.** ITS FULLY **EMBODIED** DESIGN **PRESERVES** A SENSE OF INTERACTIVE **INTIMACY** AND SONIC **PRESENCE**, GIVING RISE TO A NOTION OF **ELECTRONIC CHAMBER MUSIC** AND EVENTUALLY TO IDEAS LIKE THE **LAPTOP ORCHESTRA...**

THE BIRTH AND RISE OF THE LAPTOP ORCHESTRA!

DIRECTLY OUT OF PERRY AND DAN'S WORK ON EMBODIED INTERFACES AND SPEAKER ARRAYS, THE **LAPTOP ORCHESTRA** PUSHES THE IDEA OF ELECTRONIC CHAMBER MUSIC TO A NEW (IL)LOGICAL EXTREME: NOT ONE, BUT AN **ENSEMBLE** OF EMBODIED INTERFACES WITH HUMANS IN THE LOOP!

IN 2005, THE **FIRST** OF ITS KIND AND SCALE WAS BORN: THE **PRINCETON LAPTOP ORCHESTRA** (PLORK)!

LAPTOPS? ORCHESTRAS? NE'ER THE TWAIN SHALL MEET! AND **YET...**

HOW DOES IT CHANGE THE WAY WE **COMPOSE...**

...AND **DESIGN** INSTRUMENTS, AND CRAFT LIVE PERFORMANCES?

DAN TRUEMAN

...THIS **STRANGE PAIRING** MAKES IT ALL THE MORE INTRIGUING!

SCOTT SMALLWOOD

AND FIND A GOOD **BALANCE** BETWEEN **HUMAN** AND **TECHNOLOGY?**

PLORK'S WEST COAST SIBLING, STANFORD'S **SLORK** WAS FOUNDED THREE YEARS LATER IN 2008.

THE **LAPTOP ORCHESTRA** IS A LARGE-SCALE, COMPUTER-MEDIATED PERFORMANCE **ENSEMBLE, DESIGN LABORATORY,** AND **CLASSROOM,** EXPLORING A RADICAL INTERACTION OF **SCIENCE** AND **TECHNOLOGY** WITH **ART,** DRAWING FROM BOTH CONVENTIONAL AND CUTTING-EDGE PRACTICES.

SLORK IN BEIJING 2014

ENSEMBLES IN THIS MEDIUM CAN BE COMPRISED OF MORE THAN **20** LAPTOPS, HUMANS, AND...

...CUSTOM **MULTI-CHANNEL HEMISPHERICAL SPEAKER ARRAYS**...

...DESIGNED TO PROVIDE EACH COMPUTER **META-INSTRUMENT** AND **HUMAN PERFORMER** WITH THEIR OWN **SONIC IDENTITY** AND **PRESENCE.**

DOĞA

THE LAPTOP ORCHESTRA IS CAPABLE OF FUSING A POWERFUL SEA OF SOUND WITH THE **IMMEDIACY** OF HUMAN MUSIC-MAKING, ATTEMPTING TO CAPTURE THE **ENERGY** OF A LIVE ENSEMBLE AS WELL AS ITS **SONIC INTIMACY**...

...IN WHAT WE THINK OF AS A FORM OF **ELECTRONIC CHAMBER MUSIC.**

THE LAPTOP ORCHESTRA EMBODIES MANY OF THE IDEAS WE'VE ENCOUNTERED -- RE-MUTUALIZATION, BODY, CO-DESIGN, INTERFACES AS EXTENSIONS...

249

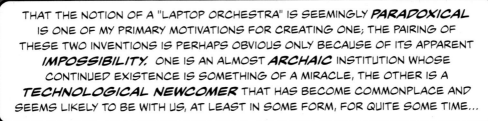

THAT THE NOTION OF A "LAPTOP ORCHESTRA" IS SEEMINGLY *PARADOXICAL* IS ONE OF MY PRIMARY MOTIVATIONS FOR CREATING ONE; THE PAIRING OF THESE TWO INVENTIONS IS PERHAPS OBVIOUS ONLY BECAUSE OF ITS APPARENT *IMPOSSIBILITY.* ONE IS AN ALMOST *ARCHAIC* INSTITUTION WHOSE CONTINUED EXISTENCE IS SOMETHING OF A MIRACLE, THE OTHER IS A *TECHNOLOGICAL NEWCOMER* THAT HAS BECOME COMMONPLACE AND SEEMS LIKELY TO BE WITH US, AT LEAST IN SOME FORM, FOR QUITE SOME TIME...

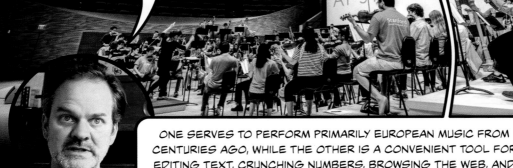

ONE SERVES TO PERFORM PRIMARILY EUROPEAN MUSIC FROM CENTURIES AGO, WHILE THE OTHER IS A CONVENIENT TOOL FOR EDITING TEXT, CRUNCHING NUMBERS, BROWSING THE WEB, AND CHECKING E-MAIL. *NEVER* THE TWAIN SHALL MEET.

THE *ORCHESTRA* VS. THE *LAPTOP* (IN PERFORMANCE)

THE ORCHESTRA

- IS *LARGE*
- TYPICALLY LIVES IN A REASONABLY LARGE PERFORMANCE HALL WITH GOOD MUSICAL ACOUSTICS
- *SOUND* IS NET SUM OF MANY RELATIVELY *PROXIMAL* INSTRUMENTS IN THIS HALL
- IS DIVIDED INTO *SECTIONS* ACCORDING TO THE NATURE OF THESE INSTRUMENTS
- INSTRUMENTS TYPICALLY TAKE DECADES TO MASTER, AND HAVE BEEN UNDER REFINEMENT FOR EVEN LONGER, SOMETIMES CENTURIES
- IS USUALLY *CONDUCTED*

THE LAPTOP

- IS TYPICALLY USED *ALONE*
- PLAYS IN ALL SORTS OF SPACES: BARS, CLUBS, SOMETIMES CONCERT HALLS
- SOUND IS TYPICALLY AMPLIFIED THROUGH A *CENTRALIZED* PA SYSTEM
- INSTRUMENT DESIGN IS CONSTANTLY *IN FLUX*, SOMETIMES EVEN GENERATED DURING THE ACTUAL PERFORMANCE (LIVE CODING); OFTEN CREATED BY THE *PLAYER*
- *MASTERY* OF INSTRUMENTS CAN TAKE A FEW MINUTES OR *MUCH LONGER*
- "WHAT? A CONDUCTOR?"

follow my lead frpm here

*"**BORROW** WHAT MAKES SENSE. **INVENT** THE REST."*

THE *LAPTOP ORCHESTRA*

- TYPICALLY BETWEEN 4 (QUARTET) AND 20 (FULL ENSEMBLE) PERFORMERS IN SIZE

- EACH *HUMAN PERFORMER* IS PAIRED WITH A *META-INSTRUMENT*, SO CALLED BECAUSE IT'S A LAPTOP STATION THAT CAN BE DESIGNED INTO DIFFERENT AND MORE *SPECIFIC* INSTRUMENTS

- A *META-INSTRUMENT* CONSISTING OF A *LAPTOP*, MULTI-CHANNEL AUDIO INTERFACE, AND -- CRUCIALLY -- A MULTI-CHANNEL HEMISPHERICAL *SPEAKER ARRAY*

- SOUND IS *LOCAL* AND *PROXIMAL* TO EACH INSTRUMENT AND PLAYER

- INSTRUMENTS ARE OFTEN DESIGNED ON A CASE-BY-CASE BASIS, TIGHTLY TAILORED TO EACH *WORK* IN QUESTION, AS BESPOKE EXPERIENCES FOR EACH PIECE

- THE NOTION OF *PLAYING* THE INSTRUMENTS ARE AS VARIED AS THE INSTRUMENTS

- *FORMATS* OF PIECES RANGE FROM FREE-FORM OR STRUCTURED IMPROVISATION TO RIGIDLY SCORED PIECES; NO PRESCRIBED LIMITATION ON TYPES OF MUSIC (E.G., GENRE)

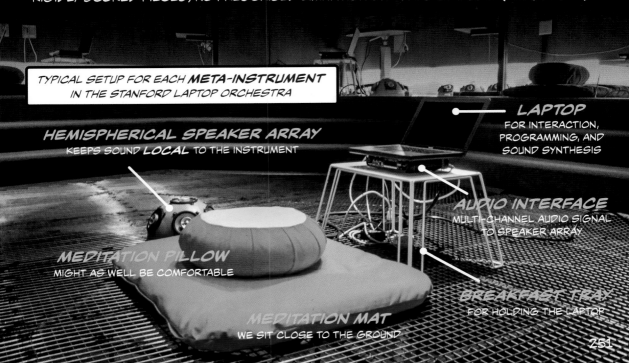

TYPICAL SETUP FOR EACH *META-INSTRUMENT* IN THE STANFORD LAPTOP ORCHESTRA

LAPTOP
FOR INTERACTION, PROGRAMMING, AND SOUND SYNTHESIS

HEMISPHERICAL SPEAKER ARRAY
KEEPS SOUND *LOCAL* TO THE INSTRUMENT

AUDIO INTERFACE
MULTI-CHANNEL AUDIO SIGNAL TO SPEAKER ARRAY

MEDITATION PILLOW
MIGHT AS WELL BE COMFORTABLE

BREAKFAST TRAY
FOR HOLDING THE LAPTOP

MEDITATION MAT
WE SIT CLOSE TO THE GROUND

FOR ME, THE ALLURE OF THE LAPTOP ORCHESTRA IS A **DESIGN LAB** WHERE THE INSTRUMENTS WE BUILD AND THE WORKS WE COMPOSE ARE "DEPLOYED" IN CONCERT SETTINGS WHERE, AESTHETICALLY, THE RUBBER MEETS THE ROAD.

IT IS IN THIS REAL-WORLD CONTEXT THAT SUCCESS IS MEASURED. IT'S ABOUT FINDING THE **RIGHT INTERPLAY** BETWEEN WHAT PEOPLE ARE GOOD AT DOING AND WHAT COMPUTERS ARE GOOD AT DOING (WHILE EMBRACING THEIR RESPECTIVE LIMITATIONS) TO ACHIEVE AND EXPLORE AN AESTHETIC GOAL.

THAT WE HAVE TO DESIGN THE INSTRUMENTS FROM THE **GROUND UP** FOR EACH NEW WORK IS BOTH A **BLESSING** AND **CURSE.**

FOR EACH DESIGN, A FIRST-ORDER **SANITY CHECK**

DOES THE END PRODUCT **JUSTIFY** THE TECHNOLOGY?

DOES IT DO **AT LEAST ONE THING** THAT CAN BE ACHIEVED BY NO OTHER MEANS?

DOES THE **DESIGN** USE THE **MEDIUM** TO SUPPORT THE RIGHT **INTERPLAY** BETWEEN **TECHNOLOGY** AND **HUMANS**?

IF IT **FAILS** ANY OF THESE CHECKS, THEN PERHAPS THE DESIGN, AS IT STANDS, ISN'T THAT INTERESTING OR SHOULDN'T USE THE TECHNOLOGY. DESIGN IS COGNIZANT OF SPECIFICITIES OF THE MEDIUM, AND WE TRY TO SEE **HUMANS** AND **COMPUTERS** AS TWO FUNDAMENTALLY **DIFFERENT** TYPES OF ENTITIES, EACH WITH BUILT-IN ADVANTAGES AND LIMITATIONS.

COMPUTERS VS. HUMANS

COMPUTERS

- NO INHERENT NOTION OF **INTENTION** OR **AESTHETICS**
- FOLLOW **CLEARLY DEFINED** INSTRUCTIONS AND LOGIC
- CAPABLE OF PRECISELY CARRYING OUT SEQUENCES OF SIMPLE OPERATIONS
- CAN SYNTHESIZE SOUNDS TO SPECIFICATION
- CAN BE NETWORKED

HUMANS

- INHERENT DESIRE TO **EXPRESS**
- CANNOT HELP BUT **INTEND**
- NATURALLY **SOCIAL**
- CAPABLE OF **REASON**
- CAPABLE OF AESTHETIC **JUDGMENT**
- REMARKABLY **ADAPTABLE** (WE ARE JACKS-OF-ALL-TRADES); SPECIALIZATION TAKES TRAINING

GOOD DESIGN EMBRACES EACH SIDE FOR WHAT IT IS. HERE, THE **MEDIUM** IS THE **MIXTURE** OF COMPUTERS AND HUMANS.

CENTRAL TO THE LAPTOP ORCHESTRA IS THE IDEA OF DESIGNING DIFFERENT TYPES OF MUSICAL INTERACTIONS THAT **BRIDGE** THE TRADITIONAL HUMAN-CENTRIC ASPECTS OF MUSIC-MAKING AND THE UNIQUE (AND LESS UNDERSTOOD) POSSIBILITIES OF TECHNOLOGY.

THE **SOUND** OF A **VIOLIN** DOES **NOT** NATURALLY COME OUT OF SPEAKERS AROUND YOU, BUT RATHER FROM THE **ARTIFACT ITSELF**...

OUR **SPEAKER ARRAYS** ARE DIRECT DESCENDANTS OF RESEARCH THAT PERRY AND DAN CONDUCTED IN THE 1990s. MUCH LIKE THE **BOSSA, HEMISPHERICAL SPEAKER ARRAYS** APPROXIMATE **OUTWARD-RADIATING** SOUND SOURCES, EMULATING THE WAY ACOUSTIC INSTRUMENTS RADIATE SOUND.

THEY PROVIDE A **SONIC PRESENCE** AND THE IMPRESSION OF A **PHYSICAL ARTIFACT** MAKING THE SOUND IN PROXIMITY TO YOU, IN STARK CONTRAST TO THE **DISEMBODIED** SOUND FROM SPEAKERS THAT SURROUND YOU.

INDEPENDENTLY ADDRESSABLE

MEANING WE CAN SEND **DIFFERENT** SOUND TO EACH SPEAKER, MAKING POSSIBLE TECHNIQUES FOR SPATIALIZATION AND EFFECTS

6 SPEAKERS
5 AROUND, 1 FACING UP

MULTIPLIED OUT TO AN ENSEMBLE OF SUCH SOUND SOURCES, THIS EMBODIED APPROACH CHANGES THE WAY WE DESIGN INTERFACES AND WRITE MUSIC

ENCLOSURE == IKEA SALAD BOWL
(THAT'S RIGHT, SALAD BOWLS)

THESE THINGS WEREN'T EXACTLY **OFF-THE-SHELF**, SO WE HAD TO **DESIGN** AND **BUILD** THEM FROM SCRATCH. HERE IS HOW WE BUILT OURS FOR SLORK...

6-CHANNEL INPUT AND **POWER**

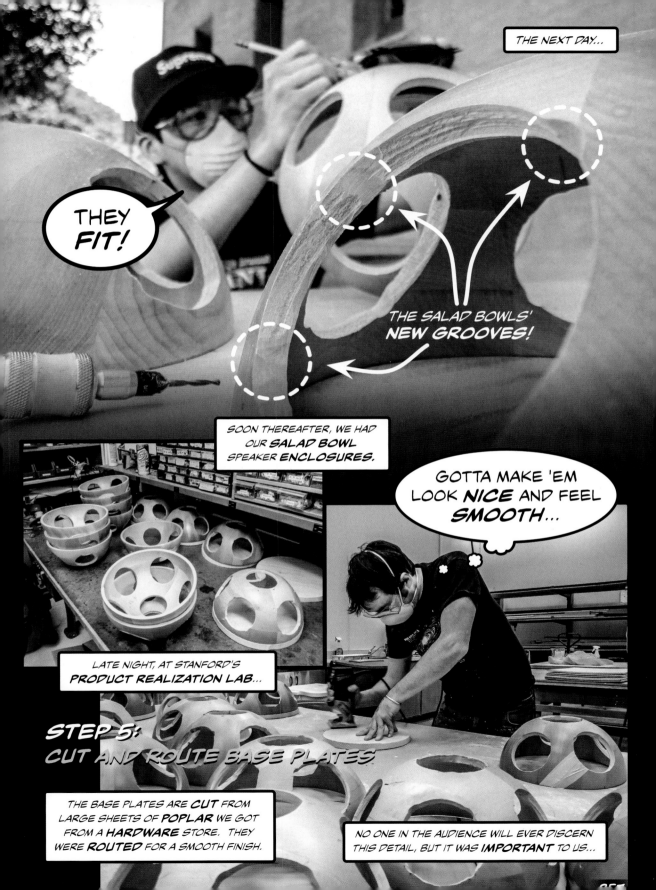

STEINUNN

MEANWHILE, A PARALLEL TEAM WORKED ON THE **ELECTRONICS** AND **CIRCUITRY!**

TURNER

TINKER, WIRE, SOLDER, PI...

HMM...

PROTOTYPES OF VARIOUS FORM FACTORS AND SIZES WERE **BUILT!**

NOPE...

BAEK

...**HEAR** ANYTHING?

LET'S CHECK THE WIRING **AGAIN.**

ADNAN

WHERE IS THAT OTHER AMP CIRCUIT?

HAYDEN

MAYBE **THIS** SWITCH WILL FIT BETTER...

HOW COULD THIS **NOT** WORK?

STEP 6:
ELECTRONICS

FOR THE NEXT TWO WEEKS, CCRMA'S **MAX LAB** (NAMED AFTER MAX MATHEWS) TRANSFORMED INTO AN AROUND-THE-CLOCK VENUE FOR **SOLDERING, CIRCUIT BENDING, DRILLING, CUTTING, GLUING, EXPERIMENTING, ASSEMBLING** THE LAPTOP ORCHESTRA.

EACH HEMISPHERICAL **SPEAKER ARRAY** (OR "HEMI") HOUSES THREE **STEREO** CLASS-D "T-AMP" AMPLIFIERS WIRED TOGETHER, FOR A TOTAL OF **SIX CHANNELS** EACH.

MULTIPLIED BY **20 HEMI**s FOR A TOTAL OF **120 CHANNELS.**

WE DRILLED HOLES IN **METAL STRIPS** THROUGH WHICH WE INSTALLED **AUDIO JACKS.**

A VIEW **BEHIND** THE **STRIP**...

DRAWER HANDLE FOR CARRYING THE HEMI

STEP 7: ASSEMBLE!

ENCLOSURES, SPEAKERS, AMPLIFIERS, WIRES, SWITCHES, JACK PANELS: ALL **READY** TO BE PUT TOGETHER!

UP AND DOWN, IN AND OUT OF THE BUILDING, CCRMA WAS **BURSTING** WITH ACTIVITY!

I CAN'T BELIEVE THIS ENSEMBLE **GOES LIVE** IN **TWO WEEKS!**

JIEUN

JUHAN

MAX

I BETTER CUT AND STRIP THESE WIRES **FASTER...**

JASON

GE! THIS PLACE IS **HOPPIN'!**

HEY JOHN!

I'D LIKE TO ORDER **SIX** "KING KONG" PIZZAS!

GINA

CONNECT EVERYTHING UP, AND **SEAL** THE ENCLOSURE!

LUKE

4" CAR SPEAKER DRIVERS **INSTALLED** AND **WIRED**

260

STEP 9: TEST-DRIVE

BEST WAY TO TEST A LAPTOP ORCHESTRA: **MAKE MUSIC** WITH IT!

FROM **CONCERT HALLS**...

TO AN OUTDOOR **SCULPTURE GARDEN**...

HEH THIS LAPTOP GOES TO **ELEVEN.**

TO INTIMATE **CHAMBER MUSIC** SETTINGS...

...WHERE WE DEPLOY **FEWER STATIONS**...

...AND THE **AUDIENCE** SITS **AMONG** THE **ENSEMBLE.**

ZZZ

Now, start cha
parameter. Q-P
as well increment

Change those para

Now, mess with the pi

262

STEP 10: DESIGN NEW WORKS!

THE **INSTRUMENTS** IN THE LAPTOP ORCHESTRA ARE AS DIVERSE AS THE WORKS THEMSELVES. THEY ARE TRULY A **TESTAMENT** TO PERRY'S "MAKE A PIECE, NOT AN INSTRUMENT" PRINCIPLE. COMPUTERS AS A DESIGN MEDIUM SEEM TO UNIQUELY SUPPORT THIS MODEL OF **PIECES** IN SEARCH OF A **CUSTOM INSTRUMENT**, WHERE INTERFACES ARE DESIGNED SPECIFICALLY TO SUPPORT EACH PIECE.

INDEED, WE USUALLY DON'T DESIGN **GENERAL-PURPOSE** INSTRUMENTS AND **THEN** WRITE MUSIC FOR THEM, BUT RATHER WE START WITH AN **IDEA** FOR A PIECE, AND WE WORK **BACKWARDS** TO INVENT THE INSTRUMENT(S) **SPECIFICALLY** FOR THAT PIECE, OR WE **CO-DESIGN** THE PIECE WITH THE INSTRUMENT(S). IT'S A GOOD WAY TO DISCOVER WHAT FEATURES THE INTERFACE ACTUALLY NEEDS!

NON-SPECIFIC GAMELAN TAIKO FUSION (2005)
BY PERRY R. COOK & GE WANG

THIS PIECE IS AN EXPERIMENT IN **HUMAN-CONTROLLED** BUT **MACHINE-SYNCHRONIZED** PERCUSSION ENSEMBLE PERFORMANCE. VARIOUS PERCUSSIVE SOUNDS ARE TEMPORALLY POSITIONED IN **PATTERNS** BY EACH PLAYER (AND ARE **SYNCHRONIZED** BY NETWORK ACROSS THE ENSEMBLE), AND THE PIECE GRADUALLY TRANSITIONS FROM **TUNED BELL** TIMBRES TO **DRUMS** AS THE TEXTURE AND DENSITY GROWS AND FLOWS ACCORDING TO INSTRUCTIONS CONVEYED BY A CONDUCTOR.

EACH INSTRUMENT IS PART OF A **NETWORKED STEP SEQUENCER** THAT **PRECISELY SYNCHRONIZES** ALL THE MACHINES, LEAVING THE PLAYER TO CONSTRUCT AND EVOLVE THE **MUSICAL PATTERNS** ON A DISCRETE TEMPORAL GRID.

A **CONDUCTOR** SIGNALS THE **DENSITY** ("WICKED SPARSE" TO "VERY DENSE") AND **TIMBRE** (WHICH COLORS TO USE) IN THE **PATTERNS** EACH PLAYER IS CONSTRUING.

CONDUCTOR
HOLDS UP PIECES OF THE SCORE

INKJET PRINTER
THE SCORE SHEETS ARE **PRINTED** LIVE ON-THE-FLY DURING THE PERFORMANCE, FURTHER UNDERSCORING THE **IMPROVISATIONAL** NATURE OF THE PIECE

ENSEMBLE
OCCASIONALLY ALSO INCLUDES **ACOUSTIC** BELLS AND DRUMS

A TOTALLY DIFFERENT PIECE, AND INSTRUMENT

ON THE FLOOR (2005)
BY SCOTT SMALLWOOD

THE **INSTRUMENT INTERFACE** IS A **MOCK SLOT MACHINE**, WHERE THE PLAYERS MAKE **WAGERS** OF ONE, TWO, OR THREE VIRTUAL COINS (CHOICES REPRESENTED AND VISUALIZED BY COLORED SPHERES). BY PLAYING THIS GAMBLING SIMULATION, THE ENSEMBLE RECREATES THE **SOUNDSCAPE** OF A **CASINO.**

A C-MAJOR CHORD **DRONE** CARPETS THE SOUNDSCAPE AND RAMPS UP INTENSITY OVER THE COURSE OF THE PERFORMANCE

REMAINING COINS FOR PLAYER

CHOICES OF WAGER: ONE, TWO, OR THREE, RESULTING IN ALGORITHMICALLY GENERATED MELODY FRAGMENTS

ready...   [30]

YOU WILL NOTICE WHEN YOU WALK INTO A CASINO THAT THE MACHINES ARE ALL TUNED TO THE **SAME KEY**: A C-MAJOR CHORD. THIS CHORD FLOATS AROUND THE SPACE, IN AND OUT OF EVERY CREVICE, CONSTANTLY ARPEGGIATING, HUMMING, DRONING, TWITTERING, ECHOING, SOMETIMES INCORPORATING SNIPPETS OF MELODY. THIS HAPPY DRONE **SOOTHES** THE NERVOUS CUSTOMERS AS THEY SLOWLY DROP THEIR MONEY INTO THE MACHINES. THEY CREATE A SEA OF C-MAJOR, EACH AND EVERY ONE OF THEM, PRESSING BUTTONS ON THE MACHINES, CREDIT AFTER CREDIT, ALL DAY AND ALL NIGHT.

AS PART OF THE THEATRIC GESTURE OF THIS PERFORMANCE, PLAYERS CONTINUE PLAYING UNTIL THEY **LOSE ALL THEIR CREDITS**, AT WHICH POINT THEY PHYSICALLY GET UP AND SLOWLY **WALK OFF** THE STAGE...

I AM **VIRTUALLY** BROKE!

THE PIECE **ENDS** WHEN **EVERYONE** LOSES THEIR VIRTUAL MONEY!

THE CONDUCTOR (A.K.A. "THE HOUSE") **SURVEILS** ALL THE PLAYERS FROM A CENTRAL MACHINE AND CAN REMOTELY **CHANGE THE ODDS** OVER THE COURSE OF THE PERFORMANCE (WHICH ALSO HELPS TO ENSURE THE PIECE **ENDS** ON TIME)!

DRONE (2005)
BY DAN TRUEMAN

TRACKPAD (CONTROLS **TIMBRE**)

NUMBER KEYS (CONTROLS BASE PITCH)

TILT: SIDE TO SIDE (CONTROLS **TUNING**)

TILT: FRONT TO BACK (CONTROLS **INTENSITY**)

SOME WORKS EMPLOY MORE **PHYSICAL GESTURES** THAN OTHERS...

USING **ACCELEROMETERS** (SUDDEN MOTION SENSORS INTENDED TO PROTECT MECHANICAL HARD DRIVES IN THE EVENT OF, WELL, SUDDEN MOTION) BUILT INTO THE LAPTOPS, THE PLAYERS INTRODUCE SUBTLE ADJUSTMENTS TO RICH (IF SIMPLE) ADDITIVE SYNTHESIS ALGORITHMS IN AN EFFORT TO CREATE RISSET-ARPEGGIO-LIKE PATTERNS. A RICH, PENETRATING DRONE ARISES WITH INTRICATELY SHIFTING **TIMBRES** AND **HARMONICS** CREATED BY THE SLIGHT CONTROLLED **DETUNING** BETWEEN ALL THE MACHINES.

THE CONDUCTOR SHAPES THE **TRAJECTORY** OF THE PERFORMANCE

CONDUCTING SIGNAL → RESULTING PLAYER ACTIONS

CONDUCTING SIGNAL		RESULTING PLAYER ACTIONS
SIGNAL NUMBER (1–8)	→	CHOOSE **BASE PITCH** OF DRONE
POINT DIRECTIONALLY	→	MOVE MOUSE CURSOR TO CHANGE TIMBRE
OPEN ARM FORWARD/BACK	→	TILT LAPTOP FORWARD/BACKWARD TO CONTROL INTENSITY
"CRADLE"; ROCK LEFT/RIGHT	→	TILT LAPTOP LEFT/RIGHT; **LISTEN**; MAKE INTERESTING BEATING PATTERNS
"SPRINKLE"	→	RANDOMIZE ONE OR MORE PARAMETERS; CONTINUE UNTIL NEXT GESTURE
"MIMIC"	→	ONCE THIS MODE IS SIGNALED, EACH PLAYER PLAYS WHEN CUED, **MIMICKING** AND **EMBELLISHING** ON GESTURE OF PREVIOUS PLAYER

 1   2   3   4   5   6   7   8

WE CONVEY **NUMBERS** USING **HAND GESTURES!**

SECTION A
FULL ENSEMBLE; EXPLORE VARIOUS TIMBRES; SHOULD SOUND **FULL** TO ALMOST (BUT NOT) OVERPOWERING; EVENTUALLY INCLUDE ALL CONDUCTING CUES EXCEPT FOR "MIMIC"; RANDOMIZATION SHOULD HAPPEN AT LEAST ONCE AND AT MOST TWICE; INCLUDE THREE TO SIX TOTAL CHANGES TO FUNDAMENTAL PITCH.

SECTION B
"MIMIC" ONLY; ONE PLAYER AT A TIME, AS CUED BY CONDUCTOR.

SECTION C (OR A')
RETURN TO SECTION A, AND (1) RIFF ON A YET-UNEXPLORED FUNDAMENTAL PITCH; (2) THE **LOUDEST** POINT IN THE PIECE OCCURS IN THIS SECTION; (3) COMMIT TO A PARTICULAR TIMBRE AND FUNDAMENTAL AND FADE OUT ON IT.

> UM, CAN YOU PLEASE PASS THE **LAPTOP?**

> BUT **OF COURSE.**

> MANY WORKS WERE COMPOSED FOR THE FULL ENSEMBLE. OTHERS -- LIKE THIS ONE -- WERE DESIGNED FOR A SMALLER, **CHAMBER-SIZED** SETTING...

20 (2008)

BY **ADNAN MARQUEZ-BORBON** AND **KYLE SPRATT**

IN THIS WORK FOR **20** "**UNPLUGGED**" LAPTOPS (AND ONLY 2 HUMAN PERFORMERS), EACH LAPTOP IS EQUIPPED WITH A PROGRAM THAT CAPTURES THE INCOMING SOUND FROM THE MICROPHONE AND PLAYS IT OUT ON THE ONBOARD LAPTOP SPEAKERS. THE LAPTOPS ARE INTRODUCED, ONE-BY-ONE, INTO A PHYSICAL CONFIGURATION OF CONTINUOUS **MUTUAL AUDIO FEEDBACK.** THE PHYSICAL LAPTOP **SCREENS** ARE USED AS A CRUDE LOW-PASS **FILTER** FOR THE SOUND, ADDING AN ADDITIONAL FUNCTIONAL AND VISUAL CONTROL ELEMENT.

> IT IS ABOUT THE **INDEPENDENCE** OF SOUND AND ITS BEHAVIOR ONCE **LIBERATED** FROM HUMAN CONTROL.

> THIS IS AN EXPERIMENT IN **EMERGENCE.**

> A SMALLER, MORE INTIMATE SETTING IS CRUCIAL TO PERCEIVE THE **SPATIALIZATION** OF SOUND.

> THE RELATIVE POSITIONS OF LAPTOPS AND THE **ANGLES** OF THEIR SCREENS (WHICH FILTER THE SOUND) HAVE A PRONOUNCED EFFECT ON THE OVERALL MIX.

267

IN ADDITION TO BEING AN ENSEMBLE AND DESIGN LAB FOR NEW INSTRUMENTS...

...THE LAPTOP ORCHESTRA IS ALSO A **CLASSROOM** THAT EXPLORES MUSIC, PROGRAMMING, INTERACTION DESIGN, COMPOSITION, AND LIVE PERFORMANCE AS PART OF A SINGLE CONTINUUM.

JIEUN

CHARLES

STUDENTS COME FROM MUSIC, COMPUTER SCIENCE, ARCHITECTURE, BIOLOGY, ECONOMICS... BUT AS PART OF OUR **TEACHING PHILOSOPHY**, WE **DO NOT** EXPLICITLY DIVIDE UP STUDENTS ACCORDING TO RESPECTIVE BACKGROUND (E.G., COMPUTER SCIENCE, MUSIC, DESIGN, ETC.); INSTEAD WE EXPECT EVERYONE TO NEGOTIATE THE **FULL** CREATIVE PIPELINE, EMPHASIZING THE CO-DESIGN OF ELEMENTS.

JIANFENG

AUDREY

ADNAN

MICHAEL

LIVE PERFORMANCE SERVES AS AN END GOAL, A **FORCING FUNCTION** TO BUILD SOMETHING THAT **WORKS** AND IS AESTHETICALLY **COMPLETE**.

269

TWILIGHT (2013)
BY GE WANG

INSPIRED BY THE CLASSIC SCIENCE FICTION SHORT STORY "TWILIGHT" BY *JOHN W. CAMPBELL* (PUBLISHED IN 1934, UNDER THE PSEUDONYM "DON A. STUART"), THIS PIECE RUMINATES NOT ON THE DAWN, ASCENSION, NOR TRIUMPH OF THE HUMAN RACE, BUT ON OUR POSSIBLE *DEMISE*, SET *SEVEN MILLION YEARS* IN THE FUTURE. THIS END IS NOT ONE OF ANNIHILATION THROUGH WAR, NOR DECIMATION FROM FAMINE OR DISEASE, BUT A GOLDEN *DECRESCENDO* OF DEFEAT BROUGHT ON BY THE GRADUAL, PEACEFUL, BUT UNSTOPPABLE USURPING OF *TECHNOLOGY* AND *MACHINES* -- AND THE LOSS OF HUMANKIND'S *CURIOSITY* AND SENSE OF WONDER. FROM THE ORIGINAL TEXT:

> "TWILIGHT--THE SUN HAS SET. THE DESERT OUT BEYOND, IN ITS MYSTIC, CHANGING COLORS. THE GREAT, METAL CITY RISING STRAIGHT-WALLED TO THE HUMAN CITY ABOVE, BROKEN BY SPIRES AND TOWERS AND GREAT TREES WITH SCENTED BLOSSOMS. THE SILVERY-ROSE GLOW IN THE PARADISE OF GARDENS ABOVE."

MOVEMENT ONE
THE DEAD CITY

"AND ALL THE GREAT CITY-STRUCTURE THROBBING AND HUMMING TO THE STEADY GENTLE BEAT OF PERFECT, DEATHLESS MACHINES BUILT MORE THAN THREE MILLION YEARS BEFORE -- AND NEVER TOUCHED SINCE THAT TIME BY HUMAN HANDS. AND THEY GO ON. THE DEAD CITY. THE MEN THAT HAVE LIVED, AND HOPED, AND BUILT -- AND DIED TO LEAVE BEHIND THEM THOSE LITTLE MEN WHO CAN ONLY WONDER AND LOOK AND LONG FOR A FORGOTTEN KIND OF COMPANIONSHIP. THEY WANDER THROUGH THE VAST CITIES THEIR ANCESTORS BUILT, KNOWING LESS OF THEM THAN THE MACHINES THEMSELVES."

THE DESIGN BRINGS TOGETHER A CLASSIC SCIENCE FICTION NARRATIVE, A *PHYSICAL METAPHOR* (PULLING A SOUND OUT OF THE GROUND), AND A *SYNTHESIS ALGORITHM* (GRANULAR SYNTHESIS).

"THE METAPHOR"
THE PRIMARY INTERACTION IN MOVEMENT ONE IS BASED ON THE *ABSTRACT* IDEA OF *PULLING* A *SOUND* OUT OF THE GROUND.

HEIGHT
CONTROLS PLAYBACK POSITION

SIMULTANEOUS CONDUCTING GESTURES

LEFT/RIGHT
CONTROLS PITCH DETUNING

GAMETRAK CONTROLLER

THE INTERACTION IS MAPPED ONTO *GRANULAR SYNTHESIS*, SUCH THAT THE VERTICAL POSITION DIRECTLY CONTROLS THE *PLAYBACK POSITION* OF ANY INPUT SOUND, EFFECTIVELY *SCRUBBING* THROUGH THE SOUND. IF THE MOTION STOPS HALFWAY, THE SOUND WILL *CONTINUE*, BUT IT IS *FROZEN* AT THE CURRENT PLAYBACK POSITION. GRANULAR SYNTHESIS MAKES THIS EFFECT SEEM *SMOOTH* AND TIMELESS.

GRANULAR SYNTHESIS
CHOPS UP AN INPUT SOUND INTO TINY (10-100::MS) WINDOWED PARTICLES (CALLED *GRAINS*), *TRANSFORMS* THEM (IN PITCH, DENSITY) AND *RECONSTITUTES* THEM INTO *IMPRESSIONISTIC* SOUND CLOUDS

INPUT SOUND

GRAINS

OVERLAP-ADD SYNTHESIS

GRANULATOR

TRANSFORMATION

RECONSTITUTED OUTPUT SOUND

THE RESULTING EFFECT IS A SENSE OF SOUND BECOMING *UNSTUCK* IN TIME, ALLOWING US TO "SCRUB" THROUGH IT WITH OUR *GESTURE.*

THE *INPUT* SOUNDS FED INTO THE *GRANULATOR* VARY FROM A METAL CHAIR BEING *DRAGGED* ACROSS A CONCRETE FLOOR IN A PARKING GARAGE, TO A METALLIC *RINGING* SOUND, TO A MAJOR-CHORD *DRONE*...

IT CREATES THE HAUNTING SOUND OF THE *DEAD CITY*, A SONIC MAELSTROM OF HUMMING, SCREECHING, AND DRONING MACHINERY, LONG LIBERATED FROM HUMAN DESIGN AND MAINTENANCE.

THE ENSEMBLE MIRRORS THE CONDUCTOR'S MOTION, MOVING IN UNITY, GIVING VOICE TO A *CITY* OF *MACHINES.*

271

DAY TURNS TO **DUSK.** THE ENSEMBLE -- WHO EARLIER ASSUMED THE ROLES OF MACHINES AND THE CITY SPIRES -- NOW REPRESENT **HUMANITY**, LYING DOWN TO SLEEP, PHYSICALLY, METAPHORICALLY...

TO **DREAM**...

...A **SONG OF LONGINGS,**

MOVEMENT TWO

A SONG OF LONGINGS

"AND THE SONGS. THOSE TELL THE STORY BEST, I THINK. LITTLE, HOPELESS, WONDERING MEN AMID VAST UNKNOWING, BLIND MACHINES THAT STARTED THREE MILLION YEARS BEFORE-- AND JUST NEVER KNEW HOW TO STOP. THEY ARE DEAD-- AND CAN'T DIE AND BE STILL."

THIS SONG OF LONGINGS IS RENDERED USING A SOLO INSTRUMENT INSPIRED BY THE **SOUND** AND **GESTURAL INTERACTION** OF THE **THEREMIN.**

PUSHING **FORWARD** SOUNDS THE PITCH, AND GRADUALLY ADDS **VIBRATO.**

PULLING **BACK** SILENCES THE INSTRUMENT.

PITCH IS CONTROLLED BY THE **HEIGHT** OF THE HAND.

G ------
E ------
D ------
C ------
B ------
A ------

TO HELP ACCURACY, PITCH IS SMOOTHLY **QUANTIZED** TOWARD THE NEAREST SCALE PITCH.

THE HIGHEST PITCH TRIGGERS A **SWELLING SEA** OF **CHIMES** THROUGHOUT ALL THE MACHINES...

THE SIMPLE **MELODY** RISES AND FALLS WITH EACH GESTURE...

WHILE A LOW DRONE **THROBS** ACROSS ALL THE MACHINES, SOMBERLY ACCOMPANYING THE SONG...

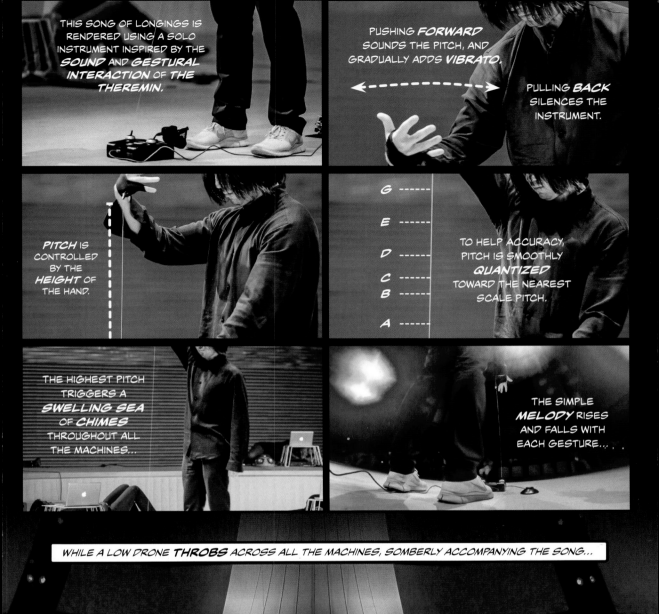

THE **FINAL BOW** OF HUMANKIND
IS A QUIET EXIT FROM THE STAGE.

AS THE MACHINES **DRONE ON**...

...BECAUSE **NO ONE** TOLD
THEM HOW TO **STOP**.

THERE IS SOMBER BEAUTY
EVEN IN **DESTRUCTION**
AND **DESOLATION**.
"TWILIGHT" IS AN IMAGINING
OF HUMANKIND'S SUNSET,
MAKING US EXAMINE OUR
PRESENT TIMES.

THIS WAS A PERFORMANCE DESIGNED TO EXPRESS THE PSYCHOLOGY,
LONGING, AND SADNESS OF A **TWILIGHT** OF HUMANITY ENDING NOT IN A
BANG, BUT AN IRREVERSIBLE **POWERDOWN**, BASKED IN THE GOLDEN,
LINGERING, DYING GLOW OF OUR DUSK. TOLD, FITTINGLY PERHAPS, THROUGH
THE **TECHNOLOGICAL MEDIUMS** OF OUR PRESENT.

PRINCIPLE 5.19

INTERFACES SHOULD *EXTEND* US

(AND NOT *REPLACE* US) WE WANT *TOOLS*, NOT *ORACLES!**

* ATTRIBUTED TO DYLAN FREEDMAN

SO THERE WE HAVE IT. AT LEAST IT'S A *START!* THE *WORLD* AND *ART* OF INTERFACE DESIGN IS *IMMENSE*, BUT HERE IS WHERE WE ALWAYS *BEGIN*, AND WHERE WE ALWAYS *END*...

...WITH THE *HUMAN* IN THE LOOP...

...AND WAYS IN WHICH *BODIES* MATTER...

...AND THE IDEA THAT *EMBODIMENT* CHANGES HOW WE THINK ABOUT INTERFACES -- NOT AS EXTERNAL OBJECTS, BUT AS *EXTENSIONS* TO OURSELVES, OUR BODIES, OUR MINDS!

IN SHORT, WE SHOULD THINK OF OUR INTERFACE WITH TECHNOLOGY EVER AS *TOOLS* -- NOT *ORACLES!* *NOT* EVERYTHING IS A *PROBLEM* TO BE *SOLVED*; *PROCESS* AND *EXPERIENCE* MATTER, BECAUSE THEY ARE *MEANINGFUL* TO WHO WE ARE.

THIS IS THE END OF CHAPTER 5, BUT BEFORE WE DIVE INTO CHAPTER 6 AND GAME DESIGN, WE ARE GOING TO VISIT AND SPEND A BIT MORE TIME WITH A *ZEN MASTER* OF COMPUTER MUSIC INSTRUMENT DESIGN...

CHAPTER 5 DESIGN ETUDE

> IT'S TIME TO DO SOME **THINKING** BY **MAKING!**

• PART 1: DECONSTRUCT

OBSERVE AND **DECONSTRUCT** AN INSTRUMENT INTO ITS INPUT, OUTPUT, AND MAPPING. DO THIS FOR BOTH A TRADITIONAL ACOUSTIC INSTRUMENT, AS WELL AS A COMPUTER-MEDIATED ONE. FOR THE LATTER, USE OUR **BROAD DEFINITION** OF A MUSICAL INSTRUMENT "THAT WHICH ALLOWS FOR INTENTIONED EXPRESSION OF SOUND." THIS INCLUDES EVERYTHING FROM SYNTHESIZERS TO CUSTOMER CONTROLLERS TO LIVE CODING AS INSTRUMENTS.

• PART 2: SKETCH

SKETCH OUT THE DESIGN FOR A **NEW INTERFACE** FOR **MUSICAL EXPRESSION** USING THE PRINCIPLES FROM THIS CHAPTER AS REFERENCES, BOTH TO FOLLOW AND FOR DEPARTURE. **WHO** IS IT FOR? EVERYONE? EXPERT? YOURSELF? FOR EXAMPLE, APPLY PERRY'S PRINCIPLES OF "DESIGN A PIECE, NOT AN INSTRUMENT" TO ENVISION THE **OUTPUT** (E.G., A MUSICAL STATEMENT) AND WORK BACKWARDS TO TAILOR AN INSTRUMENT SPECIFICALLY TO IT. WHAT DOES IT SOUND LIKE? HOW DOES IT PLAY? **OR**... APPLY THE "RE-MUTUALIZE" ETHOS TO CREATE A PHYSICAL MUSICAL ARTIFACT.

• PART 3: PROTOTYPE

PROTOTYPE THE INSTRUMENT, USING THE DESIGN CONSTRAINT OF LIMITING YOURSELF TO WHAT IS AVAILABLE TO YOU. THIS MAY REQUIRE YOU TO BACKTRACK AND RETHINK THE FUNDAMENTAL VISION OF YOUR INSTRUMENT (WHICH IS **OKAY**, AND OFTEN ESSENTIAL). DOES IT TAKE ADVANTAGE OF WHATEVER TECHNOLOGY YOU INTEND TO USE? HOW WOULD YOU MAP IT? HOW WOULD YOU CREATE A SENSE OF **EMBODIED** INTERACTION? HOW WOULD YOU MAKE THE INTERACTION **SATISFYING** AND **EXPRESSIVE** TO USE?

• PART 4: DEPLOY

DESIGN A **PERFORMANCE** WITH IT. IT DOES NOT HAVE TO BE COMPLEX -- JUST HAS TO DO THE INSTRUMENT JUSTICE! PERFORM IT LIVE FOR SOMEONE (E.G., YOUR FRIENDS, CO-WORKERS, STRANGERS, OR PETS -- ANY AUDIENCE) OR RECORD A VIDEO AND POST IT. PUT YOURSELF OUT THERE. THINK ABOUT THE WAYS IN WHICH IT WILL BE **EXPERIENCED** AND DESIGN AROUND THAT CONTEXT AND THE AESTHETICS OF THAT ENCOUNTER. WHAT ARE YOUR **METRICS** FOR **SUCCESS** FOR THE DESIGN OF THE INSTRUMENT AND THE PERFORMANCE?

AESTHETICS IS NOT A PROBLEM TO BE SOLVED

Music is precisely **more** than notes on the page, sound waves in the air. Music, like all art, is more than a product or an output: it is also a **process** with **intentionality**. In a way, art is what emerges from us earnestly trying (and often failing) to understand ourselves and our emotions. And through **experiencing** art, we derive its **meaning**.

We must not give in to the expedient notion that design and technology are about finding ever more clever algorithms, interchangeable modules that sit cleanly between an input and output, with the sole purpose of generating a **result** that solves a problem. The day that such a premise gives us everything we would want is the day we cease to be human.

This does not mean that we should shun technology. Rather, we should embrace technology always as a human medium—one that not only solves our problems but also enriches our lives and nourishes our sensibilities.

Aesthetics and the artful are **not** problems to be solved. They are values, and considerations of emotional and social consequences beyond the practical. Results matter, but no more than the process and the intrinsic experience.

PERRY ("RETIRED") AND WIFE **STACIE**, ALONG WITH THEIR CAT **ALBERT** AND DOG **GINA**, LIVE IN THE MOUNTAINS OF SOUTHERN OREGON...

AT DIFFERENT TIMES IN HIS LIFE, PERRY HAS BEEN A COMPUTER SCIENCE **PROFESSOR** AT PRINCETON, A **RESEARCHER** AT STANFORD AND *INTERVAL RESEARCH*, A ROCK 'N' ROLL **SOUND MAN**, AN AMUSEMENT PARK **AUDIO GUY**, A CONSERVATORY-TRAINED **VOCALIST** AND **TROMBONIST**, AN **ELECTRICAL ENGINEER** PH.D., A WORLD **EXPERT** ON THE TECHNOLOGY OF THE **HUMAN VOICE**, AND THE **CO-FOUNDER** OF ONLINE ARTS AND CREATIVE TECHNOLOGY EDUCATION COMPANY **KADENZE!**

HE ALSO PREVIOUSLY HELD PAYING JOBS AS DOG CATCHER, PICKER OF STRAWBERRIES, COUNTRY CLUB NIGHT WATCHMAN, BREAKFAST COOK, DISCO SOUND AND LIGHTING ENGINEER, PARKS AND RECREATION GARDENER, MATH/PHYSICS TUTOR, EARLY MUSIC SINGER, AND FASHION SHOW ORGANIST. INTERESTING GUY.

ALTHOUGH PERRY CLAIMS TO NOT REMEMBER, "WHATEVER YOU DO, DO IT WITH AESTHETICS" WAS ONE OF THE FIRST THINGS PERRY SAID TO ME WHEN I STARTED MY PH.D. AT PRINCETON IN 2001. PERRY WAS MY **ADVISOR** THEN; NOWADAYS, HE IS MORE OF A LIFE **SHERPA.**

PERRY HAS HAD A PROFOUND INFLUENCE ON HOW I THINK ABOUT COMPUTER MUSIC AND DESIGN. WHILE WE HAVE MANY ABSTRACT, PHILOSOPHICAL DISCUSSIONS ABOUT MUSIC, THE UNIVERSE, AND EVERYTHING, ONE THING I'VE LEARNED FROM PERRY IS THE WAY IN WHICH DESIGN IS ALWAYS **GROUNDED** IN THE **EVERYDAY**...

TO STEP INTO PERRY'S *STUDIO* IS TO ENTER A *PHYSICAL MANIFESTATION* OF HIS *MIND* -- WITH ITS SPECIAL BRAND OF *ORDER* AND *CHAOS*, FILLED WITH THINGS TO PLAY, BUILD, TAKE APART, PUT BACK TOGETHER... A BOTTOMLESS STORAGE PIT, A DATABASE OF MUSIC TECHNOLOGY ARTIFACTS ACROSS TIME, WHOSE RETRIEVAL SEEMS TO HINGE ON PRINCIPLES OF *INDETERMINISM* AND *ZEN*.

TECHNICAL LIBRARY

REVERSE DJEMBE / LAMP STAND

INSTRUMENT CASE STORAGE AREA

EUPHONIUM

MODELS OF HEADS

ZEN MASTER

TROMBONES

(TO STACIE'S ART STUDIO)

ROCKBAND DRUM SET

TRADITIONAL DRUM SET

DJEMBES

(TO MACHINE SHOP)

MULTI-CHANNEL RECORDING STUDIO

HEMISPHERICAL SPEAKER ARRAY

STUFF

STUFF

STUFF

IT'S A VERITABLE DESIGN HAVEN...

MONKEY WITH CYMBAL!

RUBBER CHICKEN MUSICAL INTERFACE

WHOA...

MINIMOOG

RESEARCH LAB...

HI!

Max V. Mathews and John R. Pierce

HAND-BUILT ELECTRO-ACOUSTIC MUSIC STUDIO...

...AND AD HOC MUSEUM...

⌇RRRRNN⌇ I AM THIS STUDIO'S UNCLE !

...BY VIRTUE OF THE FIRM *COMMITMENT* TO NEVER THROW ANYTHING AWAY.

IT'S THE **WORKSHOP** OF A DESIGNER, INVENTOR, AND CONSUMMATE MAKER...

ELECTRICAL ENGINEER BY TRAINING, **SOUND MAN** BY TRADE...

MUSICIAN BY DISPOSITION...

AHARHH

...AND ARTFUL **COMPUTER SCIENTIST** BY APPOINTMENT.

THESE SIDES, I'VE ALWAYS THOUGHT, REFLECT PERRY'S NATURAL COMFORT **ACROSS WORLDS**: ART AND SCIENCE, ENGINEERING AND HUMANITIES, PEOPLE AND MACHINE. WHEREAS MY DESIGN SPACE LIVES MOSTLY ON PAPER AND IN CODE, PERRY IS EQUALLY AT HOME **SOLDERING** AND **WIRING** HARDWARE AS HE IS PROGRAMMING.

VISITING PERRY IN THE MOUNTAINS IS A TIME FOR REFLECTION AND MEDITATION, TO WITNESS A ZEN MASTER OF MUSIC INTERFACE DESIGN IN HIS ELEMENT AND TO FURTHER SEE THE PRINCIPLES OF INTERFACE DESIGN IN MOTION...

COFFEEMUG HAS A **FRIEND**... WHO IS EVEN MORE "INSTANT MUSIC" AND **MINIMAL**...

HI, I AM **COFFEE MUG.**

MY NAME IS **FILLUP GLASS,** A MINIMALIST MUSICAL DRINKING UTENSIL.

6X FSRs (FORCE-SENSING RESISTORS) 6 CHANNELS OF CONTINUOUS CONTROL

— PLASTIC CUP THE CUP MAY BE PLASTIC, BUT IT'S **GLASS** IN SPIRIT!

COFFEEMUG INSTANT MUSICALLY EXPRESSIVE COFFEE MUG*

* FROM CHAPTER 5

FILLUP GLASS MINIMALIST MUSICAL CUP INTERFACE

INSPIRED BY COMPOSER **PHILIP GLASS** (WHO EITHER IS, OR ISN'T, OR PERHAPS NEVER WAS A **MINIMALIST**), THIS ARTIFACT GENERATES FOUR VOICES OF POLYPHONIC, MINIMAL MUSICAL PATTERNS (TWO OR THREE NOTES EACH). VOICE DENSITY IS CONTROLLED BY FOUR **FSRs**, TAKING ON THE TIMBRES OF PIANO, VIOLIN, VOICES, GUITAR.

THE TWO REMAINING **FSRs** CONTROL TEMPO AND TONAL CENTER / MODULATION.

LOOKING TOP DOWN INTO FILL-UP GLASS.

DEE-DOO DEE-DOO DEE-DOO DEE-DOO

DEW-DEW-DA! DEW-DEW-DA!

DOH-DAH-DUM DOH-DAH-DUM

AHHHHHHHHH AHHHHHHHHH

A STRATEGICALLY PLACED **BUTTON** ON THE BOTTOM OF **FILLUP GLASS** STARTS THE MUSIC GENERATION THE MOMENT YOU **PICK IT UP!**

PLACEMENT OF THE SIX FSRs ARE, BY DESIGN, **DISORIENTING:** AS YOU ROTATE THE ARTIFACT, IT IS **UNCLEAR** WHERE TO BEGIN OR END -- THE SENSORS HAVE NO ORDER OR PRIORITY. IN A WAY THIS IS FITTING, SINCE NO FSR/VOICE IS MORE IMPORTANT OR PROMINENT THAN ANY OTHER, AND THEY CAN BE USED IN ANY COMBINATION; IT ALSO HARKENS TO THE **CYCLIC** NATURE OF GLASS'S MUSIC.

285

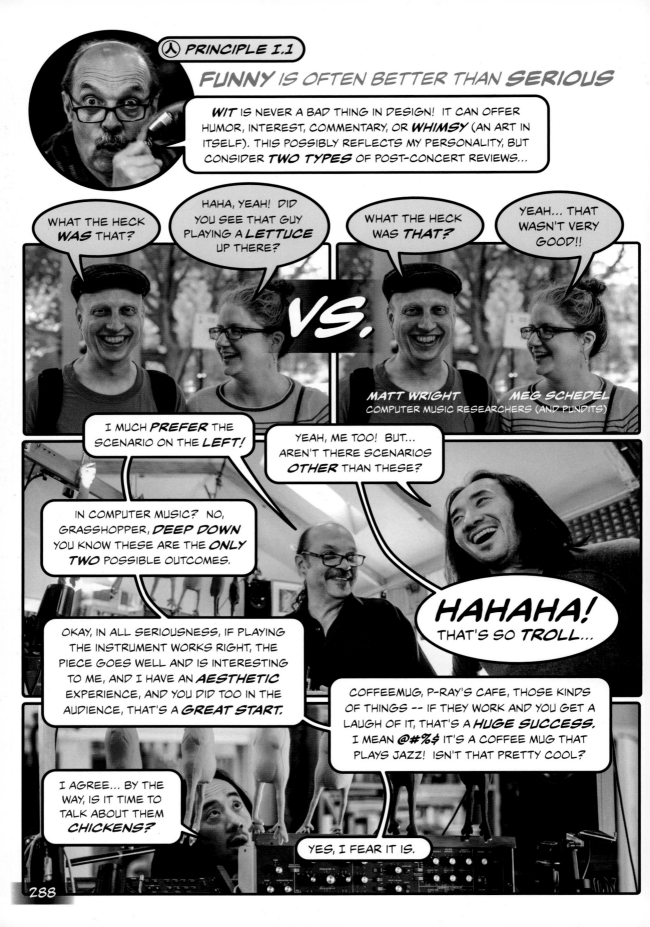

PRINCIPLE I.1

FUNNY IS OFTEN BETTER THAN SERIOUS

289

SENSEI, HOW DO *YOU* THINK ABOUT USING *PROGRAMMABILITY* AND *DESIGN?*

AS YOU KNOW, I DON'T BELIEVE IN *TOP-DOWN* DESIGN; I *TINKER*, I *MAKE*, I TRY TO CRAFT A *PIECE* -- NOT AN *INSTRUMENT* (THE LATTER NATURALLY EMERGES OUT OF NECESSITY).

AS YOU KNOW, I'M ALWAYS CODING, HACKING, BUILDING. BUT THERE IS A "DANGER" IN THERE...

YEAH, YOU'VE ALWAYS HAD A *LOVE-HATE* RELATIONSHIP WITH PROGRAMMABILITY...

THAT'S SO ZEN...

INDEED, THE PROGRAMMABILITY OF COMPUTER-BASED MUSIC SYSTEMS OFTEN MAKES THEM *TOO EASY* TO CONFIGURE, REDEFINE, REMAP, ETC. FOR PROGRAMMERS AND COMPOSERS, THIS PROVIDES AN *INFINITE LANDSCAPE* FOR EXPERIMENTATION, CREATIVITY, WRITING PAPERS, WASTING TIME, AND *NEVER* ACTUALLY COMPLETING ANY ART PROJECTS OR COMPOSITIONS!

AH YES. "PROGRAMMABILITY IS A *CURSE*," I'VE ALWAYS THOUGHT THAT'S PRETTY *DEEP*, IF ONLY BECAUSE WE RELY SO MUCH ON PROGRAMMING AS A MEDIUM!

THERE IS ALWAYS THE TEMPTATION AND DANGER OF *OVER-CUSTOMIZABILITY* THAT NATURALLY COME WITH PROGRAMMABILITY.

PRECISELY, GRASSHOPPER! AND FROM A PLAYER'S POINT OF VIEW, THE *WORST* THING IN THE WORLD WOULD BE AN INSTRUMENT WHOSE *MAPPING CHANGES A LOT!* (WHICH, BY THE WAY, IS REALLY EASY WITH A COMPUTER)

MOST OF MY SHAKERS HAVE A SINGLE PRIMARY AND *FIXED FUNCTION* WHERE IT'S JUST A SHAKER, WHERE THE PLAYER IMMEDIATELY MAKES SOUNDS AND CAN GET A FEEL FOR IT. OVER TIME, YOU *ADAPT* TO *IT* AND CAN PLAY IT CONTROLLABLY, EXPRESSIVELY.

THIS GOES WITH THE NOTION THAT *SMART* INSTRUMENTS OFTEN *AREN'T.*

291

LIKE, FOR EXAMPLE, *THE COWE?*

HUH HUH, I SUPPOSE SO!

OKAY, WHAT DID THE *ZEN MASTER* SAY TO THE *HOT DOG VENDOR?*

UHH... "MAKE ME *ONE WITH EVERYTHING*"?

I BASICALLY GOT A *TRACHEOTOMY* AND HAVE A TUBE STICKING OUT OF MY NECK! MY PURPOSE IN LIFE IS A *BOVINE AIRBAG* TO DRIVE THE BREATH SENSORS.

PRECISELY. YOU HAVE LEARNED WELL, GRASSHOPPER.

ORIGINALLY, THE *COWE* WAS TO BE A *SUPER SHAKER CONTROLLER!* BUT, I PUT MORE AND MORE SENSORS ON IT, AND I *UNSTUFFED* A STUFFED *COW*, AND THEN PUT AN *EXERCISE BALL* IN THE CAVITY. AT THAT MOMENT IT BECAME A *STRANGE BAGPIPE*, WHERE THE INTERFACE IS LIKE A BAGPIPE *CHANTER...*

THE COWE
CONTROLLER: ONE WITH EVERYTHING

AIR TUBE
COW TO COWE AIRDUCT

INFLATABLE COW
THE "BAG" IN BAGPIPE

BREATH SENSOR
CAN DETECT *BLOW* AND *SUCK*

BUTTON ARRAY
MULTIPLE *DISCRETE* INPUTS

LINEAR FSR
FORCE-SENSING RESISTOR
CAN SENSE *POSITION* AND
PRESSURE

(INSIDE)

- - - - - **2-AXIS ACCELEROMETERS**
SENSES *TILT* AND *SHAKE*

LINEAR SLIDER
CONTINUOUS CONTROL

ROTARY KNOBS
CONTINUOUS CONTROL

OF COURSE, THIS IS ALL TO SAY I *DON'T* HAVE A "BLUEPRINT" FOR DESIGNING THINGS. IT *DEPENDS* GREATLY ON THE *CONTEXT* AND THE *MEDIUM.* THAT'S WHY IT'S A *CREATIVE* ENDEAVOR...

...IN FACT, I'D CONTEND THAT DESIGN *SHOULDN'T* HAPPEN "STAGE BY STAGE" -- FOR EXAMPLE, YOU CAN'T SAY "NOW IT'S TIME TO DESIGN THE INPUT, THEN THE OUTPUT" AND SIMPLY HOPE IT'LL WORK. INSTEAD, CO-DESIGN *BOUNCES* BACK AND FORTH BETWEEN INPUT AND OUTPUT, SHAPING ONE TO THE OTHER. DESIGN IS NOT *FOLLOWING* A BLUEPRINT -- IT IS *MAKING* THE BLUEPRINT IN THE FIRST PLACE, OR EVEN *DISCOVERING* WHAT THE BLUEPRINT IS FOR!

PRINCIPLE I.2

DESIGN == CO-DESIGN

CONVERGE WAS THE CO-DESIGN OF *GRAPHICS* AND *AUDIO* AROUND THE METAPHOR OF MEMORY.

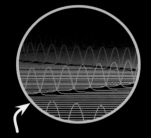

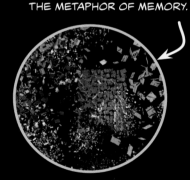

SNDPEEK WAS THE CO-DESIGN OF REAL-TIME *VISUALS* IN RESPONSE TO *AUDIO.*

OCARINA WAS THE CO-DESIGN OF AUDIO, VISUAL, INTERACTIVE, AND SOCIAL -- DESIGNED *BACKWARDS* FROM AVAILABLE TECHNOLOGY IN SEARCH OF AN EXPRESSIVE TOY!

DESIGN IS *CO-EVOLUTION.* A COFFEE MUG SUGGESTS A PARTICULAR *FORM* AS WELL AS THE POTENTIAL *TYPES* AND NUMBER OF *SENSORS* THAT MIGHT GO ON IT. HOW ONE MIGHT HOLD THE MUG SUGGESTS WHAT THOSE SENSORS MIGHT DO, THUS DERIVING THE INTERFACE'S MUSICAL *FUNCTIONS.*

IN MY ENCOUNTER WITH THE *PENCIL BAG*, THAT THE ENTIRE BAG IS ALSO A *ZIPPER* CONNECTS ITS FUNCTION TO ITS FORM -- AND MAKES THE WHOLE THING WORK AS AN *ARTIFACT* AND *EXPERIENCE,* AND QUITE *DELIGHTFULLY.* THAT'S CO-DESIGN IN ACTION.

NOT AGAIN...

OF COURSE, THIS ETHOS OF **CO-DESIGN** TRANSLATES TO TIME AND EFFORT. BUT THEN AGAIN, **CREATIVITY IS NOT AN EFFICIENT PROCESS**, AND I'D OFFER THAT SUCH CO-DESIGN STANDS TO PRODUCE MORE INTERESTING, AUTHENTIC, AND ORIGINAL DESIGNS. I MAINTAIN THIS IS TRUE EVEN IF IT HAS AN EXTERNAL PURPOSE OR DEADLINE (E.G., FOR WORK OR SCHOOL), AS LONG AS THE ESSENCE OF THE DESIGN COMES FROM **WITHIN**...

THAT SOUNDS PRETTY **ZEN!**

WELL, SOMEHOW WE'VE COME TO THINK OF DESIGN AS A MOSTLY **ENGINEERING** THING, WHEN IT IS **ALSO** AN ARTISTIC AND **HUMANISTIC** ENDEAVOR...

AND IF THAT IS SO, THEN ENGINEERING **ITSELF** IS BY DEFINITION AN ARTISTIC AND HUMANISTIC THING.

HISTORY WOULD SUPPORT THAT, AS MANY DISCIPLINES ORIGINALLY CAME FROM PURE CURIOSITY. ARISTOTLE, FOR EXAMPLE, WAS A PHILOSOPHER, LOGICIAN, SCIENTIST, ETHICIST, AND AESTHETE. ADA LOVELACE A POETICAL SCIENTIST, LEONARDO DA VINCI AN ARTFUL INVENTOR...

YES, OUR PRESENT-DAY **SEPARATION** BETWEEN ENGINEERING AND ARTS/HUMANITIES SEEMS LARGELY TO BE AN **ARTIFICIAL** ONE; THERE IS NO **ESSENTIAL** REASON FOR IT, WHILE MANY GOOD REASONS EXIST FOR **RE-MUTUALIZING** THEM. PERHAPS THE DISTINCTION AROSE OUT OF PRACTICAL CONSIDERATIONS AS DISCIPLINES FORMED AND GREW, AND OUT OF **EFFICACY** FOR TRAINING **SPECIALISTS** TO WORK IN SOCIETY...

...BUT WE MAY HAVE GONE **TOO FAR**, AS WE NO LONGER TRAIN PEOPLE TO THINK **ACROSS** THE CONTINUUM BETWEEN THE TECHNICAL AND THE ARTISTIC AND HUMANISTIC. WE'VE EXCELLED AT EQUIPPING OURSELVES, OUR STUDENTS, WITH **TOOLS**, BUT NOT THE **SENSIBILITIES** TO USE THEM -- OR TO **CREATE** THEM IN THE FIRST PLACE.

YES GRASSHOPPER...

I THINK WE NEED **AESTHETIC** EDUCATION FOR BUILDERS. MUCH LIKE HOW **YOU** TRAINED ME! WE NEED ARTFUL DESIGNERS, AND **HUMANIST ENGINEERS!**

LET ME GUESS, YOU'RE GOING TO TALK ABOUT THIS SUBJECT MORE, LATER IN THIS BOOK -- AMIRITE?

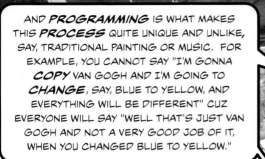

AND **PROGRAMMING** IS WHAT MAKES THIS **PROCESS** QUITE UNIQUE AND UNLIKE, SAY, TRADITIONAL PAINTING OR MUSIC. FOR EXAMPLE, YOU CANNOT SAY "I'M GONNA **COPY** VAN GOGH AND I'M GOING TO **CHANGE**, SAY, BLUE TO YELLOW, AND EVERYTHING WILL BE DIFFERENT" CUZ EVERYONE WILL SAY "WELL THAT'S JUST VAN GOGH AND NOT A VERY GOOD JOB OF IT, WHEN YOU CHANGED BLUE TO YELLOW."

POTWEET!

BUT IN THE **DIGITAL, PROGRAMMABLE** ART WORLD, TAKING A SOFTWARE PROGRAM THAT MAKES SOUND OR GENERATES GRAPHICS AND **CHANGING ONE LINE** OF CODE (OR EVEN **ONE NUMBER** IN A **CRITICAL SPOT**) CAN MAKE A **COMPLETELY NEW** PIECE OF ART.

IF YOU TAKE OUR "NON-SPECIFIC GAMELAN TAIKO FUSION" AND **SWAP** OUT THE SOUND, IT'S A DIFFERENT PIECE! OR CHANGE THE RANDOM NUMBER GENERATOR SEED!

OR TAKE YOUR GRAPHICAL FLARE THINGAMAJIGS -- YOU CHANGE THE LOOP VARIABLE TO 500, AND YOU GET A DIFFERENT **THING** ALTOGETHER!

```
// number of flares
for( int i = 0; i < 500; i++ )
{
    g_locations.push_back( vec3(rand2f
```

CHANGE FROM **ONE** TO **MANY** (BUILD COMPLEXITY FROM SIMPLICITY)

DRAWING A SIMPLE THING MANY TIMES IN A **LOOP** -- TRANSLATING, SCALING, AND COLORING **EACH INSTANCE** IN A SENSIBLE WAY. THE **SETUP** CAN TAKE **A LOT** OF WORK, BUT MOST CRITICAL PARAMETERS ARE MANIPULATED IN A SURPRISINGLY TINY AMOUNT OF CODE.

AND SO I ARGUE IT'S REALLY A **DIFFERENT WORLD** FOR DESIGN. RIGHT?

INDEED. FURTHERMORE, PROGRAMMING AS A DESIGN MEDIUM IS NOW INCREASINGLY **ACCESSIBLE** TO EVERYONE -- NOT NECESSARILY PART OF A CAREER, BUT MORE LIKE PLAYING AN INSTRUMENT FOR ITS SHEER **INTRINSIC JOY** AND EXPERIENCE. IT'S A **CREATIVE TOOL** -- NO MORE, NO LESS!

THE WAY MANY OF US **LEARN** SOUND PROGRAMMING WITH **CHUCK** OR GRAPHICS PROGRAMMING WITH **PROCESSING** IS THAT WE TAKE **EXISTING CODE** AND WE **TINKER** WITH IT -- AND **SEE** WHAT HAPPENS!

THEN WE LOOK STUFF UP WHEN WE CAN'T FIGURE OUT WHAT THE FUNCTIONS ACTUALLY DO. AND THEN EVENTUALLY WE GO BACK AND READ A LITTLE BIT ABOUT IT, IF WE HAVE TIME...

SO WHEN YOU TEACH SOFTWARE AND DESIGN AT STANFORD, YOUR COURSE, HOW DO YOU TEACH IT?

IT'S A **BRAVE NEW WORLD** OUT THERE!

I BELIEVE IN TEACHING **THINKING BY DOING.** I ASSIGN **DESIGN ETUDES**, PROJECTS, EXPRESSIVE SOFTWARE STATEMENTS -- WHERE A STUDENT IS TO DESIGN THROUGH CODING AND **AESTHETIC** EXPERIMENTATION / EVALUATION.

THE ASSIGNMENTS GET INCREASINGLY MORE **OPEN- ENDED.** I PROVIDE AS MUCH **STARTER CODE** AS POSSIBLE, AND I LIKE TO SAY: COPY WHAT YOU **LOVE**, AS LONG AS YOU EVENTUALLY **MAKE IT YOUR OWN.**

ULTIMATELY IT'S NOT ABOUT HOW MUCH CODE YOU'VE SHARED OR REUSED. IT'S ABOUT **HONING** ONE'S OWN **DESIGN SENSE** AND **AESTHETIC AWARENESS** -- AND **ARTICULATING** ONE'S OWN **DESIGN VOICE.** I ENCOURAGE **GROUP WORK** TO WORK TOGETHER THROUGH TECHNICAL PROBLEMS AS WELL AS AESTHETIC QUESTIONS. PROJECTS ARE HEAVILY **MILESTONE-BASED** AND **CRITIQUE-BASED.** IT'S LIKE A **DESIGN STUDIO** WHERE SOFTWARE IS THE **PRIMARY MEDIUM.**

RIGHT!

SO, WHAT COMPUTER SCIENCE DEPARTMENTS AROUND THE COUNTRY WOULD CALL **SOFTWARE DESIGN,** YOUR COURSE **ISN'T** THAT.

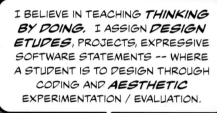

NOPE, I TEACH A LOT OF PROGRAMMING, ALGORITHMS, SOFTWARE, AND SYSTEM BUILDING... ALSO PRINCIPLES, SOME **PHILOSOPHY,** AND ABOVE ALL **AESTHETICS** -- BOTH ARTISTIC AND MORE BROADLY HUMANISTIC -- OF **DESIGN** AS A **HUMAN MEDIUM**...

IN A WAY, IT'S *YOUR* FAULT: ONE OF THE *FIRST* THINGS YOU, AS MY PH.D. ADVISOR, EVER SAID TO ME WAS "*WHATEVER* YOU DO, DO IT WITH *AESTHETICS!*"

AS A *COMPUTER SCIENCE* STUDENT, I WAS *INTRIGUED* BY THE STATEMENT. I DIDN'T REALLY KNOW WHAT IT MEANT (AND I STILL DON'T). BUT I FELT THERE WAS *SOMETHING* TO IT -- AND HAVE TRIED EVER SINCE TO *UNDERSTAND* IT... AND TO *ACT* ON IT.

I DON'T REMEMBER SAYING THAT.

PERHAPS, AS A ZEN MASTER, YOU JUST NEEDED TO GIVE ME A *NUDGE* AT A CRITICAL TIME, WHEN I WAS TRYING TO FIGURE OUT WHAT I WAS DOING... AND HAVING CONVEYED IT TO ME, YOU NO LONGER HAD ANY *NEED* TO REMEMBER IT, FOR IT HAS SERVED ITS PURPOSE AND I HAVE TAKEN IT TO HEART. AND NOW I TEACH IT TO *MY* STUDENTS...

OH GEEZ. HAH.

BUT A PART OF ME *KNEW* YOU WERE RIGHT ABOUT SOMETHING. THE DESIGNS THAT REALLY *MATTER* TO US ARE *USEFUL* BUT ALSO EMBODY A KIND OF *TRUTH*, AN AUTHENTICITY THAT SPEAKS TO SOMETHING IN US AND THAT MAKES US A LITTLE BIT MORE THOUGHTFUL, WITTY, PLAYFUL, EXPRESSIVE, EMPATHIC, AND *VULNERABLE* IN A GOOD WAY. THESE ARE CONSEQUENCES OF *AESTHETICS*, WHICH *ARISE FROM* -- BUT IS PRECISELY EVERYTHING *BEYOND* -- A THING'S *FUNCTION*...

WAIT. YOU GOT *ALL THAT* FROM SOME SENTENCE I MAY OR MAY NOT HAVE UTTERED TO YOU?

WELL YES. I MEAN, EVENTUALLY OVER, LIKE, *15 YEARS*...

HMM, EITHER I *AM* A ZEN MASTER, OR YOU HAVE GONE OUT OF YOUR GOURD, GRASSHOPPER...

...OR BOTH! AND I DO WANT TO ASK YOU ABOUT ONE MORE THING, BEFORE IT'S TIME FOR DINNER. THIS GOES BACK TO THE QUESTION, *WHY* DO WE DESIGN? WE SUSTAIN A *CONTINUUM* OF *INVENTION* THAT LEADS TO A *CONTINUUM* OF NEW THINGS. BUT *WHY* AND *WHERE* IS IT GOING? IS IT *BEAUTY* WE ARE AFTER? BEAUTY NOT SO MUCH IN MAKING SOMETHING "LOOK PRETTY" BUT AS IN THIS KIND OF *TRUTH*. OR IT IS JUST SOME INSATIABLE DRIVE FORWARD? *WHAT* ARE WE TRYING TO ACCOMPLISH? IS THERE AN ENDGAME? WHAT DOES IT ALL *MEAN*?!

DUDE, SHUT UP!

SOME THINGS HAVE NO *PRECISE* ANSWERS. YOU KNOW THIS.

YOU MIGHT AS WELL ASK WHAT IS THE PURPOSE OF EATING, WHICH IS BOTH *NECESSARY* FOR SURVIVAL BUT ALSO HAS A DIMENSION OF *TASTE.* THE QUESTION OF "WHY" NO LONGER HAS A CLEAR ANSWER THERE.

OR YOU MIGHT QUESTION IF THIS CONVERSATION IS ACTUALLY TAKING PLACE, OR JUST IN *YOUR MIND*, OR IN THE MIND OF THE *READER?*

AND IF I *CANNOT* TELL THE DIFFERENCE, DOES IT MATTER?

GRASSHOPPER, ONE HAS GOT TO FIGURE A FEW THINGS OUT FOR ONE'S SELF. I'VE NO DOUBT YOU WILL CONTINUE THIS POINTLESS EXPLORATION OF YOURS IN THE REST OF THIS BOOK...

I UNDERSTAND, UHH, SORT OF...

FOR NOW, LET ME LEAVE YOU WITH A THOUGHT TO CHEW ON, MAYBE FOR THE NEXT 15 YEARS... OR AT LEAST THE NEXT *15 SECONDS*...

STACIE AND I DO A LOT OF TRAVELING TO AFRICA, WHERE WE'VE SPENT A GOOD DEAL OF TIME JUST WATCHING ANIMALS IN THEIR HABITAT, INCLUDING A LOT OF *MONKEYS*...

THEY DO MANY OF THE *SAME* THINGS WE HUMANS DO: THEY EAT, THEY MATE, THEY *PLAY*, THEY THROW THEIR *POOP* AT EACH OTHER, THEY USE *TOOLS* (AS PRIMITIVE AS THOSE MAY SEEM TO US), AND THEY CARE FOR THEIR YOUNG.

SO... WE ARE, LIKE, MONKEYS, WITH MORE ADVANCED TECHNOLOGY?

302

303

AESTHETICS IS NOT LUXURY

Consumerism has perverted the notion of the "designer" object (e.g., "designer" handbag, "designer" watch) to signal you are paying for some extra, and often exterior, stylistic care to an otherwise purely utilitarian object. And while it at least recognizes that aesthetics is of emotional (and wallet-opening) appeal, it intentionally conveys an elitism around aesthetics, as if to say only those who can afford it deserve to have good, beautiful things.

But design is about what is inside, and its aesthetics are derived from experiencing the totality of a thing. Truth and beauty can be imbued into all design, but they must be deeper than the surface polish our consumeristic society equates with design. It is time to take back design to mean something more authentic—through the products and experiences we design, but also through an awareness of beauty and truth when we see it, regardless of the price tag or marketing jargon.

Elegance is not marked by lavishness, superficial sophistication, or opulence. Instead it is defined by authenticity and grace, plain and simple, deep and true.

≒RARRR≒ I'M GONNA **PWN** ALL THESE **NOOBS**!!

GLHF, BRING IT! I'LL TAKE ON YOUR **ENTIRE** TEAM -- 1 VS. 8, WITH NO MONITOR, AND WHILE EATING A **TACO**!

CHAPTER 6
GAME DESIGN
DESIGNING FOR **PLAY** AND **GAMIFICATION!**

PLAY IS A MASSIVELY **OVERLOADED** WORD... WE PLAY **GAMES** AND **TOYS**, WE PLAY **INSTRUMENTS**, WE EVEN PLAY **ROLES.** HOW DO WE **DESIGN** PLAY?

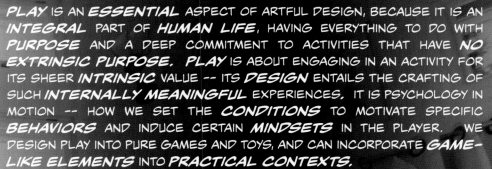

PLAY IS AN *ESSENTIAL* ASPECT OF ARTFUL DESIGN, BECAUSE IT IS AN *INTEGRAL* PART OF *HUMAN LIFE*, HAVING EVERYTHING TO DO WITH *PURPOSE* AND A DEEP COMMITMENT TO ACTIVITIES THAT HAVE *NO EXTRINSIC PURPOSE*. PLAY IS ABOUT ENGAGING IN AN ACTIVITY FOR ITS SHEER *INTRINSIC* VALUE -- ITS *DESIGN* ENTAILS THE CRAFTING OF SUCH *INTERNALLY MEANINGFUL* EXPERIENCES. IT IS PSYCHOLOGY IN MOTION -- HOW WE SET THE *CONDITIONS* TO MOTIVATE SPECIFIC *BEHAVIORS* AND INDUCE CERTAIN *MINDSETS* IN THE PLAYER. WE DESIGN PLAY INTO PURE GAMES AND TOYS, AND CAN INCORPORATE *GAME-LIKE ELEMENTS* INTO *PRACTICAL CONTEXTS*.

I AM THE *QUEEN OF BLADES!*

KERRIGAN

PYLON

I'M POWERING GE'S OFFICE!

YOU MUST CONSTRUCT *ADDITIONAL* PYLONS!

AL

ALL THE WORLD'S A STAGE, AND ALL THE MEN AND WOMEN MERELY *PLAYERS*: THEY HAVE THEIR EXITS AND ENTRANCES...

WHEREAS CHAPTERS 3, 4, AND 5 EXPLORED DESIGN THAT WE *SEE*, *HEAR*, AND *TOUCH*, THIS CHAPTER MARKS A TRANSITION INTO MORE *INVISIBLE DESIGN*...

BILL

> PLAY IS PRECISELY **MORE** THAN **ENTERTAINMENT**, IT IS AN **ACTIVE** AND **ENGAGED** EXPRESSION OF BEING **OURSELVES** -- AND BEING **FREE** IN THE CORE SENSE OF THE WORD. PLAY IS THE EPITOME OF AN **END-IN-ITSELF.**

PLAY IS...

"...ACTIVITY NOT CONSCIOUSLY PERFORMED FOR THE SAKE OF ANY RESULT BEYOND ITSELF." -- JOHN DEWEY

"...WHAT WE DO WHEN WE ARE FREE TO DO WHAT WE WILL." -- LUTHER GULICK

"...A PROFOUND MANIFESTATION OF CREATIVE ACTIVITIES." -- THOMAS PERCY NUNN

"...A KIND OF VOLUNTARY SELF-CONSTRAINED ACTIVITY." -- WILLIAM STERN

"...THE ACTIVITY IN WHICH A PERSON ENGAGES WHEN HE IS FREE TO DO WHAT HE WANTS TO DO." -- LESTER CROW & ALICE CROW

"...ANY PLEASURABLE ACTIVITY CARRIED ON FOR ITS OWN SAKE, WITHOUT REFERENCE TO THE ULTERIOR PURPOSE OR FUTURE SATISFACTION." -- CARTER VICTOR GOOD

"...A WAY, A MEANS WHICH IS USED BY THE SELF WHEN THE DIFFERENT INSTINCTIVE URGES ARE TRYING TO EXPRESS THEMSELVES." -- WILLIAM MORTON RYBURN

WORK **VS.** PLAY

• PERFORMED FOR **FUTURE** BENEFIT

• DONE FOR AN **EXTERNAL PURPOSE,** AN EXTERIOR "THAT FOR THE SAKE OF WHICH"

• IS **DELIBERATE**

• IS OFTEN GOVERNED BY RIGID **RULES** AND **CONDITIONS**

• DONE FOR THE **PRESENT**

• DONE PRIMARILY FOR **ITS OWN SAKE**

• IS **SPONTANEOUS**

• IS OFTEN **IMPROVISED,** LEAVING ROOM FOR **OPEN EXPRESSION**

IT IS A MATTER OF **PURPOSE** THAT DISTINGUISHES WORK FROM PLAY, THOUGH SUCH DISTINCTIONS ARE OFTEN **BLURRED** -- FOR EXAMPLE, WHEN WE ENJOY OUR WORK FOR ITS **OWN SAKE,** OR ARTISTS AND PRO-GAMERS WHO "PLAY" TO MAKE A LIVING.

PLAY SHARES KEY SIMILARITIES WITH **ART** AND **ART-MAKING** AND THEIR **IMPORTANCE** TO OUR **HUMANNESS**...

WE ARE NEVER SO AUTHENTICALLY **OURSELVES** AS WHEN AT **PLAY**.

OET AND PHILOSOPHER **FRIEDRICH SCHILLER**, WRITING N 1794*, WAS DRAWING A DEEP CONNECTION BETWEEN **PLAY**, **REEDOM**, AND **EXPRESSION**. THESE IDEAS WOULD ELP FUEL THE AESTHETIC REVOLUTION OF **ROMANTICISM**. Y THE WAY, THIS IS THE SAME SCHILLER WHO WROTE "AN DIE REUDE" ("ODE TO JOY"), AFTER WHICH **BEETHOVEN** ASHIONED THE SUBLIME CHORAL MOVEMENT OF HIS 9TH YMPHONY.

*FROM **LETTERS** ON THE AESTHETIC EDUCATION OF **MAN**

(⚙) PRINCIPLE 6.1

PLAY IS WHAT WE **DO** WHEN WE ARE **FREE**; PLAY IS WHAT WE DO **TO BE FREE**

LAY IS BOTH A **CONSEQUENCE** AND AN AUTHENTIC **EXPRESSION** OF **FREEDOM** T IS, BY DEFINITION, THE ABSENCE OF **EXTERNAL PURPOSE**; WE ENGAGE IN PLAY T OF **CHOICE**, AND WE PLAY FOR THE **SHEER INTRINSIC WORTH** OF THE ACTIVIT

TO SEE EXAMPLES OF **AUTHENTIC PLAY**, WE HAVE BUT TO WATCH CHILDREN AT PLAY OR FIND SOMEONE WITH A **PASSIONATE HOBBY** -- SOMEONE WHO SCULPTS OR IS OBSESSED WITH COOKING, OR SOMEONE WHO, OUT OF **SHEER INTEREST**, DESIGNS VIDEO GAMES, LEARNS AN INSTRUMENT, PAINTS, OR PLAYS PING PONG. PLAY IS A **PROTECTED** TIME AND PLACE SET ASIDE TO DO THAT WHICH IS PRECISELY NOT "**USEFUL**." IN OTHER WORDS, PLAY IS BOTH **PURPOSEFUL** AND **WITHOUT** PURPOSE.

THIS FORM OF PLAY IS NOT A **PASSIVE** OR **CASUAL** PROCESS, BUT A **CONSUMING** ACTIVITY, OFTEN TAKING UP SIGNIFICANT TIME AND ENERGY. THE ACTIVITY CARRIES **INTRINSIC MEANING** FOR THE **PERSON** ENGAGED IN IT -- WHO MIGHT GIVE UP LIFE'S ESSENTIALS TO PURSUE IT. IN THE AGGREGATE, SUCH ACTIVITIES HELP FORM THE **AESTHETIC DIMENSION** OF **LIFE**, OR WHAT ONE MIGHT CALL "WAY OF LIFE" -- AN ONGOING PURSUIT THAT IS **SELF-NOURISHING** AND **SELF-DEFINING**.

AS SUCH, PLAY IS **MEANINGFUL** AND **ESSENTIAL** FOR **ALL** OF US. IT'S NOT A **WASTE** OF TIME BUT A MEANINGFUL USE OF IT FOR NO USEFUL PURPOSE. THAT WE FIND SUCH **VALUE** IN PLAY, TO ME, IS **PROFOUND** EVIDENCE THAT WE ARE **CAPABLE** OF BEING TRULY FREE -- THAT WE **ARE** FREE.

IT IS THROUGH THIS **LENS** THAT WE LOOK AT AS WELL AS **GAMIFICATION**: A KIND OF PLA

WE HAVE DEVISED **INTRICATE** AND SOPHISTICATED WAYS TO PLAY...

LET'S START WITH **GAMES** -- IN PARTICULAR COMPUTER GAMES. WE **LOVE** THEM, WE **OBSESS** OVER THEM, WE STAY UP UNTIL 3 AM PLAYING THEM...

THEY MAKE US JOYFUL, HAPPY, FRUSTRATED, SAD...

Workers: 16/16

STARCRAFT 2
(BLIZZARD ENTERTAINMENT, 2010)

THEY POINT OUR IMAGINATION **OUTWARD** TOWARD THE **FANTASTICAL**...

Do burgers sound good?
How about Thai food?
I think tonight needs to be taco night.
Actually, I don't think I want to meet up tonight after all.

SAVE THE DATE
(CHRIS CORNELL, 2013)

FIREWATCH
(CAMPO SANTO, 2016)

...AND **INWARD** TO OUR **INNATE DESIRE** TO DISCOVER, PARTICIPATE, ACT, EXPRESS, COMPETE, EMPATHIZE, OR TO SIMPLY **BE.**

THAT DRAGON, CANCER
(NUMINOUS GAMES, 2016)

THE GAMES THAT STAY WITH US, THAT SEEM **TIMELESS**, LIKE GREAT ART, **UNDERSTAND** SOMETHING ABOUT **US**, EXPRESSED UNIQUELY THROUGH THE SPECIFICITIES AND NUANCES OF AN INTERACTIVE **MEDIUM.**

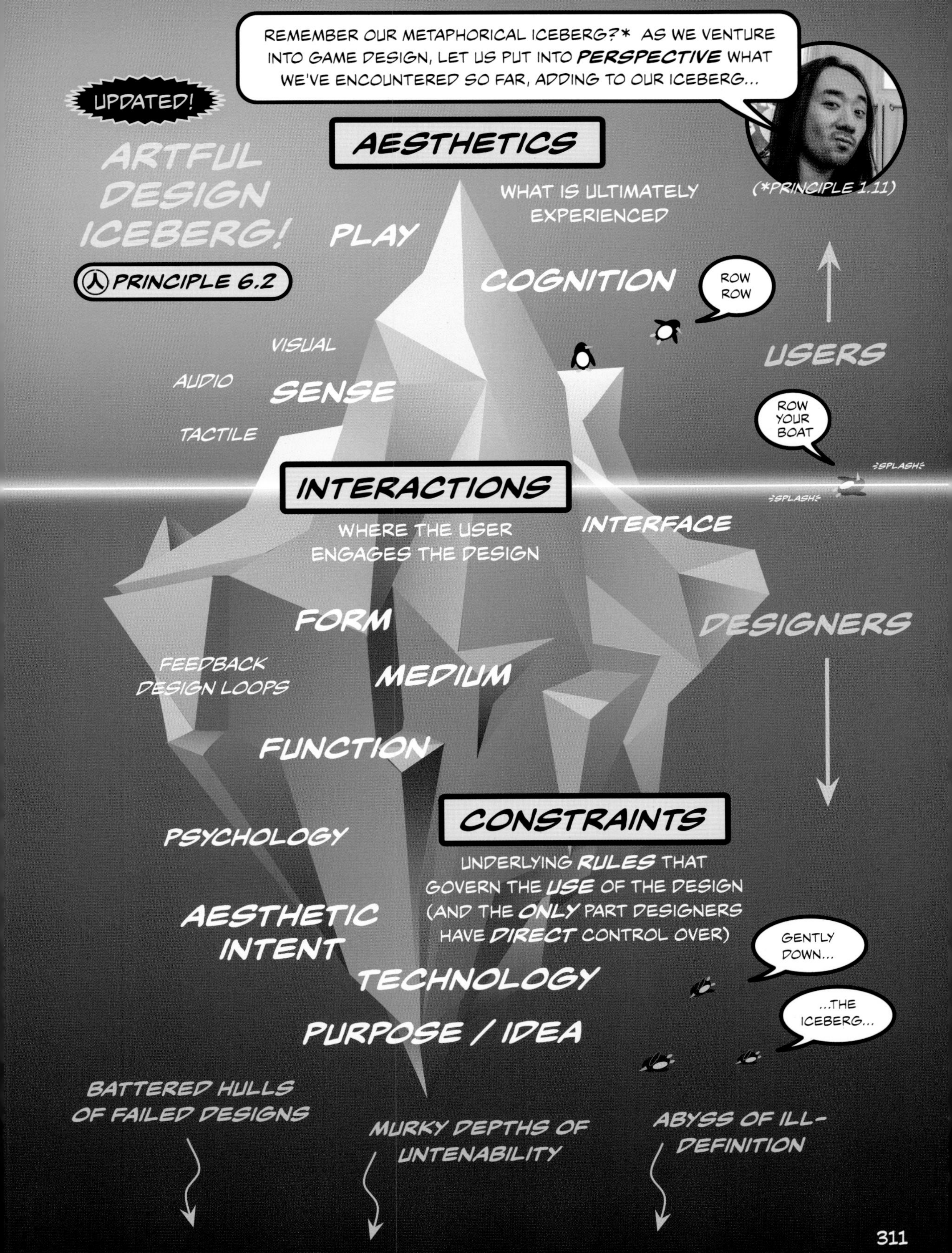

REMEMBER OUR METAPHORICAL ICEBERG?* AS WE VENTURE INTO GAME DESIGN, LET US PUT INTO **PERSPECTIVE** WHAT WE'VE ENCOUNTERED SO FAR, ADDING TO OUR ICEBERG...

(*PRINCIPLE 1.11)

UPDATED!

ARTFUL DESIGN ICEBERG!

PRINCIPLE 6.2

AESTHETICS

WHAT IS ULTIMATELY EXPERIENCED

PLAY

COGNITION

ROW ROW

VISUAL

AUDIO

SENSE

TACTILE

USERS

ROW YOUR BOAT

≈SPLASH≈

≈SPLASH≈

INTERACTIONS

WHERE THE USER ENGAGES THE DESIGN

INTERFACE

FORM

DESIGNERS

FEEDBACK DESIGN LOOPS

MEDIUM

FUNCTION

PSYCHOLOGY

CONSTRAINTS

UNDERLYING **RULES** THAT GOVERN THE **USE** OF THE DESIGN (AND THE **ONLY** PART DESIGNERS HAVE **DIRECT** CONTROL OVER)

AESTHETIC INTENT

GENTLY DOWN...

TECHNOLOGY

PURPOSE / IDEA

...THE ICEBERG...

BATTERED HULLS OF FAILED DESIGNS

MURKY DEPTHS OF UNTENABILITY

ABYSS OF ILL-DEFINITION

AWE! TERROR!

EXPRESSION! **FEELING** AND TACTILE JOY!
SENSIBILITY INVOKED
IN THE PLAYER

COMPETITIVE
STRIFE!

IMMERSION!

AESTHETICS

SOCIAL
CONNECTION! COOPERATION!

PLAYERS PLAY THE
GAME (DYNAMICS) AND
DERIVE **AESTHETICS**
FROM THE EXPERIENCE...

SAD.

OUR GENERALIZED DESIGN ICEBERG
METAPHOR HAS A CORRESPONDING
PARALLEL ICEBERG IN THE GAME
DESIGN AND GAME STUDIES COMMUNITY
-- THE **MECHANICS-DYNAMICS-
AESTHETICS (MDA)** FRAMEWORK!

HAPPY!

PLAYER

A

D

M

DYNAMICS

THE **RUN-TIME** BEHAVIOR,
ENVIRONMENT, AND CONTEXT;
THE **SYSTEM** THROUGH
WHICH MECHANICS ARE
EXPRESSED TO THE USER

DESIGNER

...**DESIGNERS** BUILD
GAMES FROM THE
MECHANICS ON UP.

AN **MDA**
ICEBERG

MECHANICS

BASE COMPONENTS OF THE GAME;
RULES, GAMUT OF POSSIBLE ACTIONS A
PLAYER CAN TAKE; THE UNDERLYING
ALGORITHMS THAT REALIZE THESE
RULES AND CONSTRAINTS

PRINCIPLE 6.3

FORMALIZED BY **HUNICKE, LEBLANC, AND ZUBEK** (2004), THE **MECHANICS,
DYNAMICS,** AND **AESTHETICS (MDA)** FRAMEWORK BREAKS DOWN A GAME INTO UNDERLYING
MECHANICS, RUN-TIME **DYNAMICS,** AND THE **AESTHETICS** ULTIMATELY EXPERIENCED BY
THE PLAYER. DRAWING CORRESPONDENCE TO OUR ARTFUL DESIGN FRAMEWORK, **MECHANICS**
CAN BE SEEN AS **FORMALIZED CONSTRAINTS, DYNAMICS** AS **BEHAVIORS** AND
INTERACTIONS, AND **AESTHETICS** AS, ULTIMATELY, THE **TAKEAWAY** FROM PLAYING.

ARTFUL DESIGN AND THE MDA FRAMEWORK TREAT AESTHETICS SIMILARLY -- AS **CONSEQUENCES** INVOKED BY THE **EXPERIENCE**, FOR WHICH THE MDA FRAMEWORK OUTLINES **EIGHT** CATEGORIES OF WHAT WE, AESTHETICALLY, GET FROM GAMES.

PRINCIPLE 6.4 THE AESTHETICS OF GAMES

SENSATION
GAMES AS **SENSE-PLEASURE**
TACTILE JOY. AUDIOVISUAL HAPPINESS. VISCERAL SATISFACTION OF SIGHT, SOUND, AND PHYSICAL INTERACTION

DISCOVERY
GAMES AS **UNCHARTED TERRITORY**
HUMAN DESIRE TO EXPLORE NEW WORLDS

NARRATIVE
GAMES AS **DRAMA**
GAMES AS UNIQUE MEDIUM FOR STORYTELLING

FELLOWSHIP
GAMES AS **SOCIAL** FRAMEWORK
A COMMUNITY OF CONNECTED PARTICIPANTS (E.G., MULTIPLAYER ENVIRONMENTS)

FANTASY
GAMES AS **MAKE-BELIEVE**
IMAGINARY WORLDS

EXPRESSION
GAMES AS **SELF-DISCOVERY**
CREATIVE SELF-EXPRESSION (A CORE AESTHETIC COMPONENT OF MUSIC-MAKING GAMES)

CHALLENGE
GAMES AS **OBSTACLES** TO OVERCOME
HUMAN URGE TO DEVELOP SKILL, OVERCOME PROBLEMS, MASTER SOMETHING

SUBMISSION
GAMES AS **PASTIME**
CONNECTION TO THE GAME AS A WHOLE EXPERIENCE, AS SOMETHING WE CHOOSE TO DO

IN THE CONTEXT OF **ARTFUL DESIGN**, I HEREBY PROPOSE **ONE MORE** TYPE OF AESTHETICS CATEGORY OF GAMES, NOT IN THE MDA FRAMEWORK...

REFLECTION
PRINCIPLE 6.5 GAMES AS **MIRROR** OF OUR HUMANNESS

GAMES CAN OFFER COMMENTARY, MAKE A STATEMENT, OR ARTICULATE LIFE AS WE ALL KNOW IT, ENRICHING IT AND ELEVATING US TO BE MORE **AUTHENTICALLY OURSELVES**. GAMES CAN INDUCE **SUBLIMITY** AND STRIVE TO BE **ARTFUL.** MOST GAMES ARE **NOT** THIS -- NOR DO THEY NEED TO BE. THE ONES THAT ARE, HOLD **MEANING** IN A WAY THAT **CANNOT** BE EXPRESSED IN A DIFFERENT MEDIUM. (AS WE LOOK AT CASE STUDIES OF GAME DESIGN IN MOTION, WE'LL HIGHLIGHT THESE!)

ANY DESIGN MIGHT BE **DECONSTRUCTED** INTO CONSTRAINTS, INTERACTIONS, AND AESTHETICS, AND WE CAN DECONSTRUCT A GAME INTO ITS **MECHANICS** (RULES), **DYNAMICS** (SYSTEM AND BEHAVIORS), AND **AESTHETIC** TAKEAWAYS. IT'S KINDA FUN TO DO -- THE EXERCISE ITSELF NOT UNLIKE A GAME. TAKE, FOR EXAMPLE...

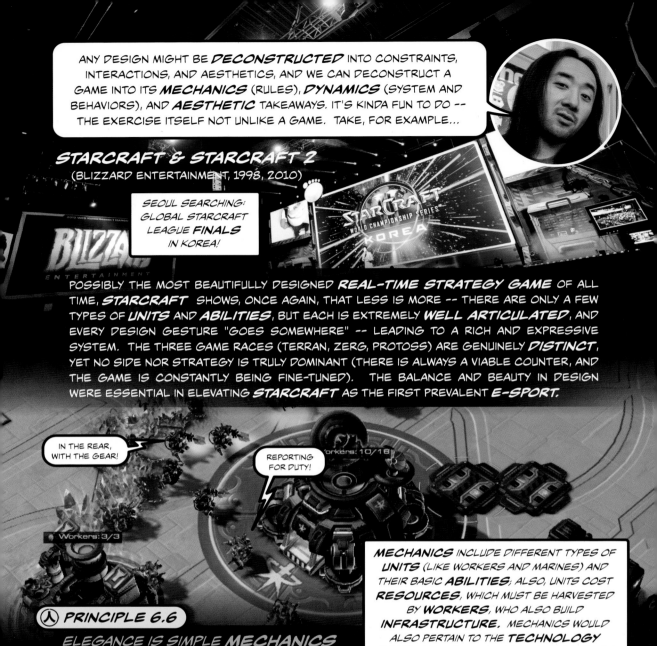

STARCRAFT & STARCRAFT 2
(BLIZZARD ENTERTAINMENT, 1998, 2010)

SEOUL SEARCHING: GLOBAL STARCRAFT LEAGUE **FINALS** IN KOREA!

POSSIBLY THE MOST BEAUTIFULLY DESIGNED **REAL-TIME STRATEGY GAME** OF ALL TIME, **STARCRAFT** SHOWS, ONCE AGAIN, THAT LESS IS MORE -- THERE ARE ONLY A FEW TYPES OF **UNITS** AND **ABILITIES**, BUT EACH IS EXTREMELY **WELL ARTICULATED**, AND EVERY DESIGN GESTURE "GOES SOMEWHERE" -- LEADING TO A RICH AND EXPRESSIVE SYSTEM. THE THREE GAME RACES (TERRAN, ZERG, PROTOSS) ARE GENUINELY **DISTINCT**, YET NO SIDE NOR STRATEGY IS TRULY DOMINANT (THERE IS ALWAYS A VIABLE COUNTER, AND THE GAME IS CONSTANTLY BEING FINE-TUNED). THE BALANCE AND BEAUTY IN DESIGN WERE ESSENTIAL IN ELEVATING **STARCRAFT** AS THE FIRST PREVALENT **E-SPORT**.

IN THE REAR, WITH THE GEAR!

REPORTING FOR DUTY!

Workers: 10/18

Workers: 3/3

MECHANICS INCLUDE DIFFERENT TYPES OF **UNITS** (LIKE WORKERS AND MARINES) AND THEIR BASIC **ABILITIES**; ALSO, UNITS COST **RESOURCES**, WHICH MUST BE HARVESTED BY **WORKERS**, WHO ALSO BUILD **INFRASTRUCTURE**. MECHANICS WOULD ALSO PERTAIN TO THE **TECHNOLOGY TREE** WHEREBY MORE ADVANCED BUILDINGS AND UNITS CAN BE ACCESSED, AND MORE!

PRINCIPLE 6.6

ELEGANCE IS SIMPLE MECHANICS GIVING RISE TO COMPLEX DYNAMICS

THIS IS **IT**, MAN! **GAME OVER**, MAN!

SHUT IT JENKINS, AND PULL YOURSELF TOGETHER!

MARINES
BASIC TERRAN INFANTRY UNIT, RANGED, BIOLOGICAL, LIGHT ARMORED
50 MINERALS
0 GAS

SIEGE TANKS
150 MINERALS
125 GAS

SIEGE MODE IS A MECHANIC THAT GIVES TANKS EXTRA RANGE AND EXTRA DAMAGE -- BUT RENDERS THEM IMMOBILE AND UNABLE TO TARGET CLOSE-PROXIMITY TARGETS...

MEANWHILE...

ROACHES
75 MINERALS
25 GAS

... A SMALL ZERG FORCE PUSHES ACROSS THE MAP...

BWAHAHAHA, THEY'LL NEVER SEE US COMING...

HYDRALISKS
100 MINERALS
50 GAS

SAY, DID YOU HEAR SOMETHING?

HMM, CALLING IN A COM-SAT SCAN -- FIRE!

OKAY.

KA-

...TRADE-OFFS SUCH AS SIEGE MODE COMPRISE THE MECHANICS, AND GIVE RISE TO THE GAME'S DYNAMICS...

OW!!

!

-BOOM

RUN AWAY!

+150 XP

BORN OUT OF MECHANICS, DYNAMICS IN STARCRAFT INCLUDE STRATEGIES, TACTICS, BUILD ORDERS, COUNTERS, ECONOMIC AND UNIT MANAGEMENT. THAT THERE IS NO SINGLE DOMINANT STRATEGY IS A TESTAMENT TO A DESIGN THAT VALUES BALANCE.

-BOOM

IN THE FIVE, FIVE BY FIVE.

I FEEL LIKE WE ARE IN A JAMES CAMERON MOVIE...

KA-

TACTICAL DYNAMICS: SIEGE TANKS, DUE TO THEIR MECHANICS, NEED CAREFUL POSITIONING (IN THE BACKLINE, USUALLY) TO ACHIEVE OPTIMAL UTILITY!

IT'S SUCH A **BEAUTIFUL** GAME, DAN.

ATTAINING **NERDVANA** WITH PRO-GAMERS TURNED LEGENDARY E-SPORTS **BROADCASTERS**...

NICK **"TASTELESS"** PLOTT

MOST DEF, NICK. GIVES ME **NERD-CHILLS** JUST THINKING ABOUT IT.

DAN **"ARTOSIS"** STEMKOSKY

MY HEROES! SIGN MY **MOUSEPAD**!

USE MY BACK TO WRITE ON!

MARK **"SIR CARRIER"** CERQUEIRA (ADORING FAN)

I MEAN, YOU GET TO BUILD MINI CITIES, BALANCE AN ECONOMY, RAISE AN ARMY, AND CONTROL THEIR EVERY ACTION... IT'S LIKE **PLAYING GOD.**

STARCRAFT INVOKES MULTIPLE **AESTHETIC** RESPONSES, INCLUDING **SENSATION** (TACTILE JOY), **FANTASY** (SCI-FI WORLD), **CHALLENGE** (EXTREMELY HIGH SKILL CEILING), AND **SOCIAL FRAMEWORK** (COMPETITION).

THE **LORE** OF THE **STARCRAFT** UNIVERSE PROVIDES **FANTASY** AND **NARRATIVE**: **KERRIGAN** -- A CENTRAL CHARACTER -- WAS ONCE A **TERRAN** DOMINION GHOST WITH **PSIONIC POTENTIAL**, ONLY TO BE BETRAYED BY HER OWN PEOPLE AND SACRIFICED TO THE BUG-LIKE, **BORG**-ESQUE, HIVE-MINDED **ZERG**, WHO **ASSIMILATE** AND **REBIRTH** HER INTO A FEARSOME INSTRUMENT OF **CARNAGE.** SHE EVENTUALLY BECOMES THE ALL-POWERFUL **QUEEN OF BLADES.** REMNANTS OF HER HUMANITY REMAIN, CHANNELED THROUGH RAGE AND VENGEANCE. KERRIGAN IS LIKE A MODERN-DAY (OR ACTUALLY 24TH-CENTURY) **MEDEA** FROM GREEK TRAGEDY AND IS THE VERY EMBODIMENT OF "HELL HATH NO FURY LIKE PSIONIC DOMINION OPERATIVE SCORNED."

NEVER CROSS THE QUEEN OF BLADES! ⁊RARRRRR⁊

HER OVERWHELMING **POWER** AND THE **FRAGILITY** OF HER HUMANITY MAKE KERRIGAN TRULY ONE OF THE ALL-TIME FAVORITE VIDEO GAME CHARACTERS.

⁊HEY⁊ WHO YOU CALLING A VIDEO GAME CHARACTER?!

AS WE **DECONSTRUCT** MORE VIDEO GAMES IN THE CONTEXT OF ARTFUL DESIGN, WE SHOULD NOTE SOME FUNDAMENTAL **COMMONALITIES** ACROSS ALL GAMES.

ALL GAMES REQUIRE INTERACTION AND *ACTIVE PARTICIPATION* ⚘ PRINCIPLE 6.7

*LIKE A MUSICAL INSTRUMENT, A GAME DOES NOT **MOVE FORWARD** WITHOUT YOUR ACTIONS AND DECISIONS!*

THIS **INTERACTIVE** QUALITY IS AT THE HEART OF **ALL** GAMES AND FUNDAMENTALLY DIFFERENTIATES GAMES AS A **MEDIUM** FROM MORE "FIXED" MEDIUMS SUCH AS FILM OR BOOKS (CHOOSE-YOUR-OWN-ADVENTURE BOOKS MIGHT BE AN EXCEPTION, THOUGH ONE MIGHT ARGUE THEY ARE ALSO GAMES). IN DESIGNING A GAME, THE DESIGNER MUST **RELINQUISH** SIGNIFICANT **CURATORIAL CONTROL** TO THE PLAYER. THE **EXPERIENCE** AT SOME LEVEL MUST UNFOLD **IN COOPERATION** WITH **HOW** THE PLAYER PLAYS THE GAME.

THIS **AGENCY** BESTOWED UPON THE PLAYERS IS WHAT MAKES GAMES, WELL, GAME-LIKE AND **PLAYFUL.** FROM THE PERSPECTIVE OF THE PLAYER, THIS PARTICIPATORY, AGENTIC **QUALITY** OF GAMES IMPARTS A SENSE OF EMBODIMENT.

⚘ PRINCIPLE 6.8

ALL GAMES ARE PLAYED IN *HYPER-1ST PERSON* RESULTING IN A SENSE OF *EMBODIMENT* THROUGH CONTROL

EVEN IF THEY ARE PLAYED FROM A "3RD-PERSON VISUAL PERSPECTIVE," THERE IS STILL THIS SENSE OF EMBODIMENT. IN **SUPER MARIO BROS.**, YOU **CONTROL** MARIO OR LUIGI, AND EVEN THOUGH YOU ARE VIEWING THEM FROM A 3RD-PERSON VANTAGE POINT, YOU NONETHELESS **PARTICIPATE** AND MAKE DECISIONS. THE GAME WORLD DOES NOT PROCEED MEANINGFULLY **WITHOUT** YOU. GAME DESIGN MUST, THEREFORE, GENUINELY **CEDE** CONTROL TO THE PLAYER AND CREATE A SENSE OF EMBODIMENT, NOT UNLIKE AN INSTRUMENT.

WITH THESE ADDED DIMENSIONS, LET'S ANALYZE GAMES AND THEIR DESIGN THROUGH A FEW MORE CASE STUDIES. WHILE **STARCRAFT** REPRESENTS A GARGANTUAN-SCALE EFFORT FROM A LEGENDARY COMMERCIAL GAME DEVELOPER, WE HAVE ALSO ENTERED A GOLDEN AGE OF **INDEPENDENT** "INDIE" GAMES FROM SMALL TEAMS (SOMETIMES JUST A SINGLE PERSON)...

MOVE... JUMP... FIRE...

...I **AM** MARIO!

A game by Chris Cornell

SAVE THE DATE!

- Start Game
- Load Game
- Preferences
- Credits
- Quit

WE CONTINUE OUR **DISSECTION** OF GAME DESIGN WITH A RATHER **UNCONVENTIONAL** GAME. **SAVE THE DATE**, MADE IN 2013 BY **CHRIS CORNELL**, IS PROBABLY THE BEST RETRO, EXISTENTIAL, SELF-DECONSTRUCTING GAME EVER MADE ABOUT **DINNER.**

FROM THE DESIGNER

"IT'S A PERFECTLY NORMAL EVENING, AND YOU HAVE A QUIET DINNER PLANNED WITH ONE OF YOUR FRIENDS. AND SO BEGINS ONE OF MY WEIRDER GAMES. **SAVE THE DATE** IS A GAME ABOUT A LOT OF THINGS. FRIENDSHIP. STORIES. HOPE. DESTINY. AND ABOVE ALL ELSE, DINNER."

INSPIRED BY THE **MECHANICS** AND **IDIOM** OF 1990s TEXT-BASED ADVENTURE GAMES, **SAVE THE DATE** IS A THROWBACK AND HOMAGE TO AN EARLIER AGE, BUT IT COMPLETELY USES THAT IDIOM TO CRAFT AN **EXPERIENCE** ALL ITS OWN.

BASIC PREMISE: GO ON A **DINNER DATE**, WITH SOMEONE NAMED **FELICIA.** BUT DINNER TURNS OUT TO BE ONLY ONE OBSTACLE. FIRST, WE MUST FIGURE OUT **WHERE** TO GO! ALSO, GOTTA LOVE GAMES ABOUT **FOOD.**

AND CHECK OUT THESE OLD-SCHOOL **MICROSOFT PAINT** STYLE 1990s BITMAP GRAPHICS!

WHOA... A **MOTOROLA RAZR** -- I USED TO HAVE THAT PHONE... BACK WHEN PHONES WERE **JUST PHONES!**

Felicia
Where are we having dinner?

LIKE MANY GAMES OF THAT ERA, THE MAIN MODE OF **INTERACTION** HERE: **MULTIPLE-CHOICE** SELECTION.

Do burgers sound good?
How about Thai food?
I think tonight needs to be taco night.
Actually, I don't think I want to meet up tonight after all.

I **LOVE** TACOS! BUT DON'T KNOW IF **TACOS** WOULD BE THE GREATEST IDEA... IS THIS A FIRST DATE? TACOS CAN BE... YOU KNOW, GASTRONOMICALLY **VOLATILE.**

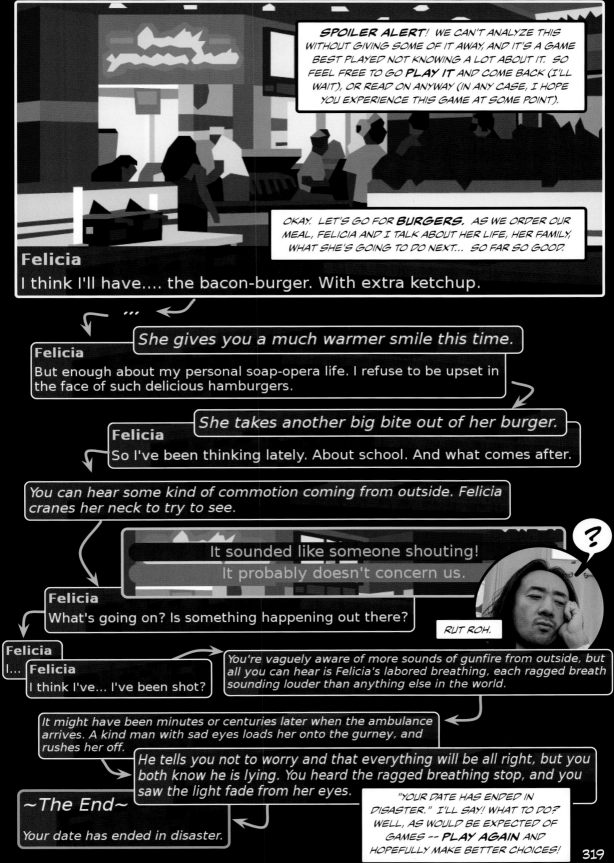

SPOILER ALERT! WE CAN'T ANALYZE THIS WITHOUT GIVING SOME OF IT AWAY, AND IT'S A GAME BEST PLAYED NOT KNOWING A LOT ABOUT IT. SO FEEL FREE TO GO PLAY IT AND COME BACK (I'LL WAIT), OR READ ON ANYWAY (IN ANY CASE, I HOPE YOU EXPERIENCE THIS GAME AT SOME POINT).

OKAY. LET'S GO FOR BURGERS, AS WE ORDER OUR MEAL, FELICIA AND I TALK ABOUT HER LIFE, HER FAMILY, WHAT SHE'S GOING TO DO NEXT... SO FAR SO GOOD.

Felicia
I think I'll have.... the bacon-burger. With extra ketchup.

...

She gives you a much warmer smile this time.

Felicia
But enough about my personal soap-opera life. I refuse to be upset in the face of such delicious hamburgers.

She takes another big bite out of her burger.

Felicia
So I've been thinking lately. About school. And what comes after.

You can hear some kind of commotion coming from outside. Felicia cranes her neck to try to see.

It sounded like someone shouting!
It probably doesn't concern us.

Felicia
What's going on? Is something happening out there?

RUT ROH.

Felicia
I...

Felicia
I think I've... I've been shot?

You're vaguely aware of more sounds of gunfire from outside, but all you can hear is Felicia's labored breathing, each ragged breath sounding louder than anything else in the world.

It might have been minutes or centuries later when the ambulance arrives. A kind man with sad eyes loads her onto the gurney, and rushes her off.

He tells you not to worry and that everything will be all right, but you both know he is lying. You heard the ragged breathing stop, and you saw the light fade from her eyes.

~The End~

Your date has ended in disaster.

"YOUR DATE HAS ENDED IN DISASTER." I'LL SAY! WHAT TO DO? WELL, AS WOULD BE EXPECTED OF GAMES -- PLAY AGAIN AND HOPEFULLY MAKE BETTER CHOICES!

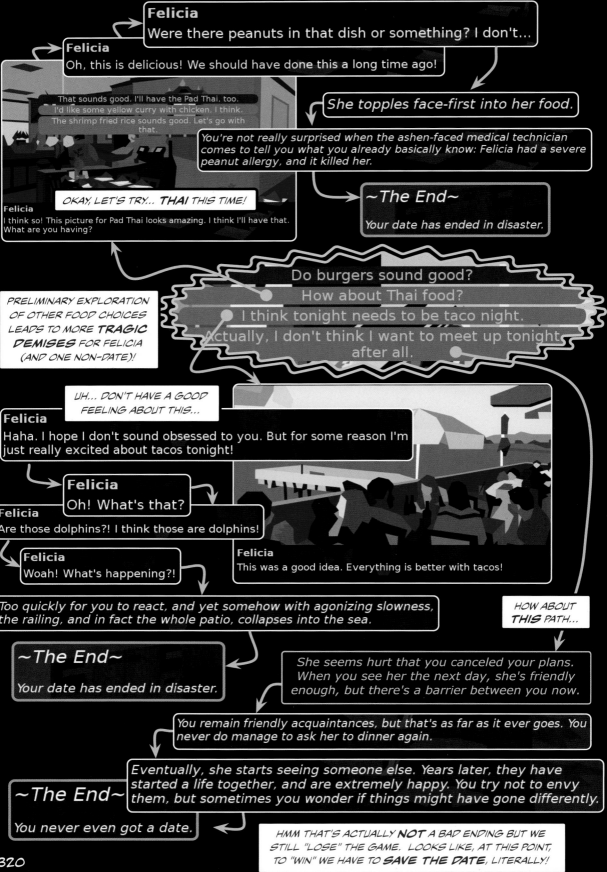

Felicia
Were there peanuts in that dish or something? I don't...

Felicia
Oh, this is delicious! We should have done this a long time ago!

That sounds good. I'll have the Pad Thai, too.
I'd like some yellow curry with chicken, I think.
The shrimp fried rice sounds good. Let's go with that.

She topples face-first into her food.

You're not really surprised when the ashen-faced medical technician comes to tell you what you already basically know: Felicia had a severe peanut allergy, and it killed her.

OKAY, LET'S TRY... **THAI** THIS TIME!

Felicia
I think so! This picture for Pad Thai looks amazing. I think I'll have that. What are you having?

~The End~
Your date has ended in disaster.

PRELIMINARY EXPLORATION
OF OTHER FOOD CHOICES
LEADS TO MORE **TRAGIC
DEMISES** FOR FELICIA
(AND ONE NON-DATE)!

Do burgers sound good?
How about Thai food?
I think tonight needs to be taco night.
Actually, I don't think I want to meet up tonight, after all.

UH... DON'T HAVE A GOOD
FEELING ABOUT THIS...

Felicia
Haha. I hope I don't sound obsessed to you. But for some reason I'm just really excited about tacos tonight!

Felicia
Oh! What's that?

Felicia
Are those dolphins?! I think those are dolphins!

Felicia
Woah! What's happening?!

Felicia
This was a good idea. Everything is better with tacos!

Too quickly for you to react, and yet somehow with agonizing slowness, the railing, and in fact the whole patio, collapses into the sea.

HOW ABOUT
THIS PATH...

~The End~
Your date has ended in disaster.

She seems hurt that you canceled your plans. When you see her the next day, she's friendly enough, but there's a barrier between you now.

You remain friendly acquaintances, but that's as far as it ever goes. You never do manage to ask her to dinner again.

Eventually, she starts seeing someone else. Years later, they have started a life together, and are extremely happy. You try not to envy them, but sometimes you wonder if things might have gone differently.

~The End~
You never even got a date.

HMM THAT'S ACTUALLY **NOT** A BAD ENDING BUT WE
STILL "LOSE" THE GAME. LOOKS LIKE, AT THIS POINT,
TO "WIN" WE HAVE TO **SAVE THE DATE**, LITERALLY!

QUICKLY, IT BECOMES CLEAR THAT WHATEVER YOU DO, YOUR DATE (PERSON-WISE AND SOCIALLY) IS **DOOMED.** FELICIA DIES. A LOT. AND -- IN VIDEO GAME AND **GROUNDHOG DAY** STYLE -- REVIVES, READY TO HAVE DINNER...

~The End~
Your date has ended in disaster.

~The End~
disaster.

ON THE OTHER HAND, AS **PLAYERS** (AND NOT UNLIKE BILL MURRAY IN **GROUNDHOG DAY**) WE **ACCUMULATE** INFORMATION , WHILE THE REST OF THE GAME RESETS. THAT'S THE MECHANIC, THE UNDERLYING, **UNSPOKEN PREMISE** OF GAMES LIKE THESE. UPON **REPLAY**, INFORMATION FROM PREVIOUS SESSIONS BECOMES USEFUL, AND NEW OPTIONS BECOME AVAILABLE.

~The End~
Your date has ended

~The End~
led in disaster.

~The End~
Your date has ended in disaster.

date has ended in disaster.

~The End~
Your date has ended in disaster.

~The End~
Your date has ended in disaster.

~The End~
Your date has ended in disaster.

~The End~
Your date has en

~The End~
Your date has ended in disaster.

FOR EXAMPLE, REPLAYING THE BURGER PATH, THERE IS NOW A **NEW CHOICE** TO TELL FELICIA TO "**DUCK**"!

It sounded like someone shouting!
It probably doesn't concern us.
Okay, this is important. I need you to duck. Now.

Felicia
I... Oh my god. Was that gunfire?

BUT, OF COURSE IN **THIS** GAME, SHE WONDERS **HOW** YOU KNEW THAT...

I didn't - I was just making a joke, and I guess we got lucky!
I'm actually a wizard.
I'm actually from the future.
I'm psychic.
I reloaded from a saved game.

...AND IN FACT, YOU HAVE AN OPTION TO TELL HER "THE ULTIMATE TRUTH" -- THAT YOU ARE PLAYING A VIDEO GAME AND JUST RELOADED THE GAME.

Felicia
Did you just save my life? What... How did you know that was going to happen?

THIS IS WHERE THINGS GET **EXISTENTIAL**, REFLECTING ON LIFE, ITS MEANING, AND THE GAME BEGINS TO **DECONSTRUCT** ITSELF AND THE **MEDIUM** OF VIDEO GAMES...

MECHANICS: TEXT-BASED, MULTIPLE-CHOICE, NONLINEAR, 1990s DIALOGUE-DRIVEN ADVENTURE "GENRE"

...SETTING IN MOTION A **FOURTH WALL BREAK** (IN THEATER PARLANCE) OF EPIC PROPORTIONS!

Hmm. Nothing, I guess. Sorry, I was making a joke. Ignore me.
I mean this is a video game. You died, I got a game-over, and so I hit reload so I could try different choices.

Felicia
What exactly do you mean, 'reloaded'...?

DYNAMICS: PLAYER TRIES ALL THE CHOICES OVER TIME TO SEE WHERE EACH ENDING GOES

A **STRANGE DESIGN LOOP**, SET IN MOTION -- A THING THAT **DECONSTRUCTS** ITSELF AND ITS VERY MEDIUM!

Felicia
Heh. That sounds kind of far-fetched. But wow, it would be nice to just be able to hit 'reload' on our mistakes, huh?

Felicia
I just need some time to sort things out. I'm a little too rattled for dinner now, anyway.

You finish your burgers and head out.

POOR FELICIA, OF COURSE, **STILL** DIES IN THIS PATH.

You haven't made it more than a foot from the door, though, when a car crashes into the front of the restaurant.

~The End~
Your date has ended in disaster.

You are flung to the side, but Felicia is not so lucky.

THE GAME IS *PERFECTLY* SELF-AWARE...

You're a character in a video game. I'm trying to win, but when you die, I get a game over.

This world is a computer simulation. I'm kind of like Neo.

You're not actually real.

Felicia

What is all this about?

WHOA, A *PHILOSOPHICAL* CHOICE WE GET TO MAKE. *WHAT* IS FELICIA, AND WHAT DO WE *CHOOSE* TO ASSUME ABOUT HER? THAT SHE IS "NOT REAL" -- OR DO WE *PLAY ALONG*, AS PART OF THE GAME? WHAT ABOUT OUR ROLE IN THIS ENTERPRISE? ARE WE "THE PLAYER" OR "THE CHARACTER" -- OR BOTH?! THIS VERY CONSIDERATION IS A *BRIDGE* TRANSCENDING THE GAME WORLD INTO OUR WORLD.

Felicia

What does that mean exactly?

Felicia

I mean, I've taken Philosophy 101. Sure, we could all be brains in jars, living simulations or whatever.

Felicia

Or I guess, as you're saying, computer programs that only think that they're people.

Felicia

OH DANG, AN EPIC *QUESTION* FOR THE AGES...

But if the illusion is good enough that I can't tell - how does that affect me?

Well, I just pushed this button on my screen to say this, so I'm definitely playing a game.

It's possible. I haven't almost died today, though, so my game is more boring than yours, at least.

No, I'm pretty sure you're the one in the game.

Felicia

For that matter, how do you know that *you* are even real?

NO LONGER JUST A GAME, *SAVE THE DATE* BRINGS US INTO A PHILOSOPHICAL DIMENSION ABOUT THE NATURE OF *BEING*. I MEAN IF WE CAN PLAY *THIS* AS A GAME, ARE WE ALSO *BEING "PLAYED"* BY A YET GREATER GAME WE ARE NOT AWARE OF (AND CANNOT GRASP EVEN IF WE KNEW)? *IS* ALL THE WORLD A *GAME*, AND WE *ARE* BUT PLAYERS, LIKE FELICIA IN OURS?

Felicia

What, your game has *buttons*?

BTW FELICIA HAS QUITE THE SENSE OF HUMOR.

AS WE PLAY, WITH THE ADVANTAGES OF BEING **OUTSIDE** THE SYSTEM OF THE GAME WORLD, YOU BEGIN TO CONVINCE FELICIA -- THAT SHE **IS** A CHARACTER IN A VIDEO GAME.

Felicia
Let's start with the big question. Just what exactly is going on? Tell me the truth.

You can't handle the truth!
Well, see, I'm a wizard...
I'm actually from the future...
As it turns out, I have psychic powers!
Everything here, including you, is a video game that I'm playing...

Exactly how many times have you had to watch me die?

Not that many, actually.
A lot.
I've lost count, honestly.
So far? 73 times, in 10 different ways.

THIS IS A BRILLIANT EXAMPLE OF **USING** THE MEDIUM TO **TRANSCEND** THE MEDIUM! AN EXQUISITELY CRAFTED STRANGE DESIGN LOOP!

AS YOU AND FELICIA TRY TO FIGURE OUT WHAT THE **PURPOSE** OF THIS GAME IS ALL ABOUT (MORE EXISTENTIALISM), YOU PONDER THE NATURE OF **STORYTELLING**...

Have you ever thought about how stories work?

Some, I guess?
A lot, actually.
Not really?

ULTIMATELY, **SAVE THE DATE** PULLS OFF AN **INTERACTIVE RUMINATION** ABOUT THE VERY **NATURE** OF VIDEO GAMES AND WHAT MAKES THEM INTRINSICALLY UNIQUE FROM OTHER **MEDIUMS** LIKE BOOKS AND MOVIES...

You can't just stop reading before you get to the last page and write your own.
Even if you don't like the ending, it's still the official ending, no matter what you want.
I can't just change the ending because I disagree with it.
Are you suggesting that I just, what, *imagine* a better ending?

Felicia
What would happen if the audience rebelled? And played a different story in their minds than the one that was being told? Would that story be any less valid or real?

...THE **MEDIUM** IS THE **MESSAGE** IS THE **MEDIUM** IS THE...

Felicia
This is a game, right? If regular storytelling wasn't collaborative enough already, game-storytelling kicks it up to a whole new level.

Felicia
In a book or movie, the author can at least know that you'll read everything exactly as they wrote it.

Felicia
Someone writing a game, though, already has to cede a ton of control to the player, just to make it a game.

WHOA. THAT'S **RIGHT ON**, I'M LIKIN' THIS FELICIA!

ME TOO! SHE'S ARTICULATING OUR **HYPER-1ST PERSON** PRINCIPLE ABOUT VIDEO GAMES: **WE** AS PLAYERS HAVE SIGNIFICANT CONTROL, OR THE ILLUSION THEREOF!

323

SOME TIME LATER...

DID I **FINISH** THE GAME? HMM... I AM NOT... SURE? IN ANY CASE, THE EXPERIENCE STAYED WITH ME.

OF ALL THE GAMES I'VE PLAYED, THIS IS AMONG THE MOST DEEPLY **ARTFUL**, FOR WHILE IT HAS A **PURPOSE** (DINNER, THEN "SAVING" THE DATE), IT IS **NOT** ABOUT THAT, BUT THE EXPERIENCE **ITSELF**, AND THE IDEAS AND SENTIMENTS IT LEAVES YOU WITH. IT IS WORTH **PLAYING** -- AND **REFLECTING** -- THROUGH.

A SELF-REFERENTIAL VIDEO GAME AS STORYTELLING!

ITS DESIGN **INVOKES** SOMETHING OF **THE SUBLIME.** IT MAKES THE PLAYER **THOUGHTFUL** AND **CURIOUS** ABOUT OUR OWN **CONDITION**, GIVING US A FRAMEWORK TO PONDER THE TRANSCENDENTAL QUESTION "WHAT IF **WE** ARE IN A SIMULATION?". FELICIA FEELS STRANGELY **HUMAN**, AND THE GAME'S INSISTENCE ON BREACHING THE **BOUNDARIES** OF THE GAME UNIVERSE INTO OUR UNIVERSE INVITES US TO CONSIDER **OUR** EXISTENCE, AS WELL AS THE **ROLE** OF STORIES, AND THE UNIQUE NATURE OF **GAMES** AS AN EXPRESSIVE **MEDIUM** -- ALL AT THE SAME TIME!

THE TOPICS OF EXPLORATION ARE **NOT NEW** -- HAVING BEEN EXPLORED IN PHILOSOPHY BOOKS, MOVIES (E.G., *GROUNDHOG DAY* -- "LIKE, THE BEST EXISTENTIAL COMEDY OF ALL TIME," FELICIA SAYS).

WHAT IS **NEW** OR **DIFFERENT** HERE IS THAT **YOU** ARE IN THE DRIVER'S SEAT **BECAUSE** IT'S A GAME. YOU ARE MAKING THE **CHOICES** (AT LEAST THE CHOICES THE AUTHOR PRESENTS). THIS HYPER-1ST PERSON PARTICIPATION ALLOWS YOU THE PLAYER TO CONTEMPLATE, IN AN **EXPERIENTIAL** MANNER, QUESTIONS AS OLD AS THE PHILOSOPHY ITSELF -- ON THE **NATURE OF BEING**, AND OUR ROLE IN THE REALITY THAT CONTAINS US.

ONCE AGAIN, THE **MEDIUM**, AT THE END OF THE DAY, CANNOT HELP BUT **SHAPE** THE **MESSAGE. SAVE THE DATE** WORKS BEAUTIFULLY, AND ONLY WORKS, BECAUSE IT **IS** A GAME.

IF **SAVE THE DATE** BORROWED THE **IDIOM** OF THE MULTIPLE-CHOICE-BASED ADVENTURE GAME TO TELL A STORY THAT CAN ONLY BE TOLD AS A VIDEO GAME, THEN **THAT DRAGON, CANCER** IS A KIND OF GAME AS **LIFE SIMULATOR.**

THAT DRAGON, CANCER

A Numinous Games Production

NOT MANY GAMES HAVE THE WORD "CANCER" IN THE TITLE. EVEN FEWER ARE ACTUALLY **ABOUT** CANCER... AND PERHAPS **ONLY** THIS ONE IS ABOUT **INFANT CANCER.** BASED ON A FAMILY'S ACTUAL EXPERIENCE, IT SEEMS VIOLENTLY **COUNTER** TO THE **PREMISE** OF A **GAME**, AS SOMETHING WE'D PLAY TO HELP **ESCAPE** LIFE'S TROUBLES FOR A WHILE. INSTEAD, IT INVITES US TO STARE HEAD-ON INTO ITS **DARKEST, SADDEST,** MOST **DEVASTATING** REALITIES...

NEW GAME

What does joel love?

THIS APPARENT **IRONY** OF USING THE **MEDIUM** OF GAMES -- TO BETTER UNDERSTAND THE KIND OF REALITY FROM WHICH A NORMAL PERSON WOULD ONLY WISH TO ESCAPE -- MAKES THE STATEMENT THAT GAMES **CAN** BE ABOUT **MORE**, PERHAPS TRANSCENDENTLY MORE, THAN ENTERTAINMENT. THEY CAN **PROBE** AND **BEAR WITNESS** TO **TRUTH** AND THE **EXPERIENCE** OF BEING HUMAN.

WHILE THERE IS A STORY IN **THAT DRAGON, CANCER**, IT REALLY IS ABOUT THE **EXPERIENCE.** WITH INFANT CANCER, MUCH IS **LOST ALREADY**, AND THIS CANNOT BE A GAME ABOUT WINNING OR LOSING, OR SAVING THE DAY, OR EVEN STORYTELLING. INSTEAD, IT IS AN **EMPATHIC SIMULATION** OF THE EVERYDAY LIFE EXPERIENCE OF A FAMILY THAT IS NORMAL IN EVERY RESPECT, **EXCEPT** ITS SON, JOEL, WAS DIAGNOSED WITH CANCER AT THE AGE OF **ONE.**

THAT DRAGON, CANCER IS VISUALLY AND SONICALLY **GORGEOUS**, ALMOST ETHEREAL. A PLACE **SERENE** ENOUGH TO LINGER, REFLECT, AND FEEL... A GAME THAT ISN'T TRYING TO **DEPRESS** OR TERRORIZE YOU, BUT TO HELP YOU **COMPREHEND** SOMETHING ABOUT OURSELVES IN THE FACE OF CHAOS.

IT POSSESSES A SUBLIME KIND OF **BEAUTY** AND **HORROR** -- THE KIND THAT MAKES YOU STOP AND CONTEMPLATE...

Who am I to him?
Dah-dah.

Hoping you will never remember these days of illness and treatment.

AS A "LIFE SIMULATOR," **THAT DRAGON, CANCER** PUTS YOU INTO THE **HEADSPACES** OF ITS PEOPLE, IN PARTICULAR THE PARENTS (WHO WERE ALSO THE GAME DESIGNERS)...

no one ever realizes how short the time is.

...AND INVITES YOU INTO A PART OF THEIR **PSYCHE.**

SOMETIMES, **QUIET** CONTEMPLATION MAKES THE POINT MORE POWERFULLY THAN A FULL FRONTAL ASSAULT.

What is pain... without a word for it?

THE GRAPHICS, IN PARTICULAR, ARE **MINIMAL.** THE FACES ARE FACETED, FLAT-SHADED **POLYGONS** -- WITH NO DEPICTION OF EYES. THIS AESTHETIC CHOICE IS **POIGNANT**: IT ALLOWS US AS PLAYERS TO "FILL IN" WITH OUR **IMAGINATION,** MAKING IT INFINITELY MORE **RELATABLE** AND **TRANSFERABLE** TO OUR LIVES, AS IF **WE** ARE THOSE IN THE GAME, AND JOEL IS IN **OUR** FAMILY. THE MISSING FACIAL FEATURES ALSO HARKEN TO LIVES DIMINISHED, **INCOMPLETE.**

THERE IS A QUIET **POETRY** IN EVERY PART. THE MUNDANE, CHAOTIC **EVERYDAY** IS A CONSTANT STRUGGLE TO FIND MEANING AND SIGNIFICANCE...

...LIKE CT SCANS IN THE METAPHORICAL **TEMPLE OF MAN.**

THERE IS AN **ACKNOWLEDGMENT** HERE THAT LIFE DOESN'T JUST HAPPEN IN THE BIG DEFINING MOMENTS AS IT MIGHT SEEM WHEN WE **LOOK BACK**, BUT AS WE **LIVE IT**: IT'S A SERIES OF SMALL, MUNDANE, **PRECIOUS**, EVERYDAY MOMENTS (RECALL **CONVERGE** FROM CHAPTER 3)...

WAITING... WANDERING, IN THE HALLS OF THE HOSPITAL...

THERAPY...

TAKING ON THE VANTAGE POINTS OF THE PARENTS...

WATCHING...

Watching you. waiting for you to wake.

HOPING...

Hey Honey! *we're on our way home* *from the hospital now.*

...THIS IS NOT BIGGER THAN LIFE... IT IS **PRECISELY** LIFE.

IT'S ALSO A **TESTAMENT** TO OTHERS WHO HAVE STRUGGLED.

SOME WIN, MANY LOSE.

OKAY, SO WHY WOULD ANYONE **WANT** TO PLAY A GAME SUCH AS THIS?

I DON'T KNOW. I CAN ONLY SAY THAT WE AS HUMANS **CRAVE** TO UNDERSTAND OURSELVES, OUR EMOTIONS, AND HOW WE FEEL IN SITUATIONS. PERHAPS THIS "GAME" IS A **VICARIOUS WINDOW**, INTO A WORLD OF "WHAT IF" FOR THOSE OF US FORTUNATE ENOUGH TO NOT GO THROUGH THIS, AND AN EMPATHIC **TESTAMENT** FOR THOSE WHO **HAVE.**

IN A WAY, THE PLAYER **BEARS WITNESS** TO A PART OF OUR COLLECTIVE CONDITION -- THAT LIFE'S FORTUNES ARE **HAPHAZARD** AND THAT WE ARE DEFINED BY OUR **STRUGGLE** TO MAKE SENSE OF IT, **TEETERING** BETWEEN FATE AND RANDOMNESS, HOPELESSNESS AND HOPE.

GAMES LIKE THIS REMIND US THAT **ANYTHING** DESIGNED HAS THE **POWER**, AND THE OPPORTUNITY, TO HIT US IN THE CHEST... MAKE US **FEEL**, DEEPLY.

THAT DRAGON, CANCER IS AN ACT OF SHEER **EXPRESSION**. IT IS **MORE** THAN A GAME, WHILE BEING **PRECISELY** A GAME.

THERE IS NOTHING -- NO CONTRACT -- THAT SAYS GAMES **HAVE** TO AMUSE OR ENTERTAIN US... JUST LIKE BOOKS, FILMS, AND MUSIC, GAMES ALSO HAVE THE POWER TO **UNDERSTAND** AND **ELEVATE** US.

AS GAMES CAN **REFLECT** LIFE, LIFE CAN ALSO BE IMBUED WITH **GAMEFULNESS.** THIS BRINGS US TO A DIFFERENT SIDE OF GAME DESIGN -- INCORPORATING GAME-LIKE AND PLAYFUL ELEMENTS INTO THE **REST OF LIFE**...

HEY, WE ARE TALKING **PLAYING** BEYOND **PURE PLAY** -- PLAYING WITH AN **EXTERNAL PURPOSE.** IT IS USEFUL TO UNDERSTAND THIS FORM OF PLAY IN RELATION TO PURE PLAY.

ROCK BAND
GE

ROCK BAND
REBECCA FIEBRINK

YES, LET'S PUT FORTH **ONE MORE** DEFINITION OF PLAY. TO **ROGER CAILLOIS**, PHILOSOPHER AND WRITER, **PLAY** IS "AN **OCCASION** OF **PURE WASTE**: WASTE OF TIME, ENERGY, INGENUITY, SKILL, AND OFTEN MONEY"...

...BUT WITH THE ASTUTE ACKNOWLEDGMENT THAT PLAY AS SUCH (OR POSSIBLY **BECAUSE OF** SUCH) IS AN **ESSENTIAL** ELEMENT OF **HUMAN FLOURISHING** -- SOCIAL, CULTURAL, CREATIVE, ARTISTIC, AND SCIENTIFIC.

(Λ) **DEFINITION 6.9**   AS **CAILLOIS** DEFINES IT...

PLAY IS FREE, VOLUNTARY, UNCERTAIN, UNPRODUCTIVE BY CHOICE; IT OCCURS IN A SEPARATE SPACE, ISOLATED AND PROTECTED FROM THE REST OF LIFE

FURTHERMORE, CAILLOIS ARTICULATES A **CONTINUUM** OF **PLAY** DEFINED BY **TWO POLES**...

LUDUS (MORE GAME-LIKE) AND **PAIDIA** (MORE TOY-LIKE).

LUDUS *(GAMING)*
DENOTES PLAY **STRUCTURED** BY **RULES** AND COMPETITIVE STRIFE TOWARD **GOALS**

VS.

(Λ) **DEFINITION 6.10**

PAIDIA *(PLAYING)*
DESCRIBES A MORE **FREE-FORM**, **EXPRESSIVE**, IMPROVISATIONAL, RECOMBINATION OF **BEHAVIORS** AND **MEANINGS**

CAILLOIS OUTLINED **FOUR TYPES** OF PLAY, EACH OF WHICH STRETCHES ACROSS THE LUDUS-VS.-PAIDIA CONTINUUM. WE CAN **SORT** SOME ACTIVITIES IN LIFE (INCLUDING SOME ACTUAL GAMES) INTO THIS **CLASSIFICATION.**

TABLE 6.11

	AGÔN (COMPETITION)	ALEA (CHANCE)	MIMICRY (SIMULATION)	ILINX (VERTIGO)
PAIDIA (PLAYING)	RACING ATHLETICS (NOT REGULATED)	HEADS OR TAILS	MASKS DISGUISES	CHILDREN WHIRLING
	FOOTBALL, BASKETBALL, OLYMPICS (REGULATED)	MAGIC 8-BALL		HORSEBACK RIDING
		GUESSING GAMES	VIRTUAL REALITY	
			THEATER	
		BETTING	THAT DRAGON, CANCER	TIGHTROPE WALKING
LUDUS (GAMING)	GO STARCRAFT	ROULETTE LOTTERIES		SKIING

(Left column note spanning PAIDIA–LUDUS: "GAMES")

FOR OUR DISCUSSION, I PROPOSE WE ADD ONE MORE **NEW** COLUMN, ANOTHER **TYPE** IN THE CLASSIFICATION OF PLAY, WHICH WE MIGHT CALL **EXPRESSION.** PERHAPS THIS IS THE TYPE OF PLAY THAT IS STILL PRIMARILY DONE FOR ITS **OWN ENJOYMENT**, BUT IT **CAN** PRODUCE SOMETHING **NEW**, AN EXPRESSIVE AND CREATIVE **BY-PRODUCT.**

ADDED!

TEKHNĒ
ART AND EXPRESSION AS PLAY

PRINCIPLE 6.12

PAIDIA (PLAYING)

DOODLING

PLAYING WITH CLAY

MINDFUL TINKERING   MUSICAL IMPROVISATION

OCARINA 2

MINECRAFT   CREATIVE DESIGN?

RESEARCH AS PURSUIT OF KNOWLEDGE FOR ITS OWN SAKE?

MAKING MUSIC FROM A SCORE

COMPUTER PROGRAMMING FOR FUN

CONSTRAINT-BASED CREATIVITY

LUDUS (GAMING)

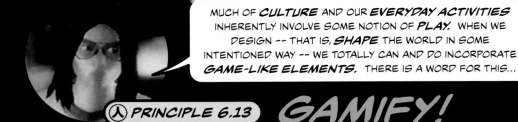

MUCH OF **CULTURE** AND OUR **EVERYDAY ACTIVITIES** INHERENTLY INVOLVE SOME NOTION OF **PLAY.** WHEN WE DESIGN -- THAT IS, **SHAPE** THE WORLD IN SOME INTENTIONED WAY -- WE TOTALLY CAN AND DO INCORPORATE **GAME-LIKE ELEMENTS.** THERE IS A WORD FOR THIS...

(人) PRINCIPLE 6.13   GAMIFY!

APPLYING **GAME DESIGN** ELEMENTS IN **NON-GAME** CONTEXTS

GAME DESIGN CAN PROVIDE BOTH **MOTIVATION** AND **MEDIUM** FOR LEARNING AND HONING SKILLS, AND FOR ENGAGING IN CERTAIN **BEHAVIORS**, AIMED AT SPECIFIC **OUTCOMES** FOR THE INDIVIDUAL AND THE BROADER SYSTEM. IT IS THEREFORE NOT SURPRISING THAT GAMES HAVE BEEN DESIGNED NOT ONLY FOR PURE PLAY BUT ALSO FOR MORE "USEFUL" ENTERPRISES, INVOLVING EVERYTHING FROM EDUCATION AND TRAINING TO MILITARY AND COMMERCE. IT IS AN ACKNOWLEDGMENT THAT GAMEFULNESS AND PLAYFULNESS CAN BE POWERFUL CATALYSTS IN THE PSYCHOLOGY OF ACTIONS, BEHAVIORS, AND CHOICES.

CHILD RAISING

THIS CONCEPT IS AS OLD AS HISTORY, AS PEOPLE HAVE DESIGNED GAME-LIKE ELEMENTS INTO MANY ASPECTS OF LIFE.

MILITARY TRAINING

WAR GAMES... MUCH BETTER THAN WAR!

COLLECT **GOLD STARS** FOR **GOOD BEHAVIOR.** EVERY **TEN** STARS RESULTS IN **ICE CREAM.**

MOM, THAT'S JUST A **THINLY VEILED** ATTEMPT TO **GAMIFY** MY **UPBRINGING**...

ENGAGEMENT: FULFILLMENT OF **GOALS** AND ATTAINMENT OF **ACHIEVEMENTS** HAVE BEEN EFFECTIVE IN ENCOURAGING LONG-TERM USE BY USERS AND CUSTOMERS. LOCATION-AWARE GAMES LIKE **POKEMON GO** GET KIDS (AND ADULTS) TO PHYSICALLY **GO OUTSIDE** (FINALLY) -- IN ORDER TO CATCH AND TRAIN POKEMONS.

WELL, AT LEAST **NOW** MY SON LEAVES THE HOUSE. GAMES ARE A POWERFUL MOTIVATOR!

MUST. GET. TO. DISH... MUST. TRAIN. POKEMON...

NICE WEATHER!

NOT PLAYING POKEMON

JAMES   TIMMY   DYLAN

STANFORD SATELLITE DISH WALK

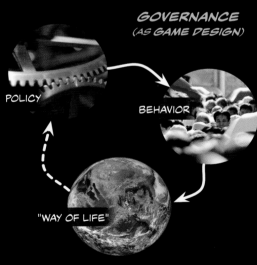

GOVERNANCE (AS GAME DESIGN)

POLICY

BEHAVIOR

"WAY OF LIFE"

GOVERNANCE IS NOT ABOUT POLITICS BUT THE DESIGN OF **POLICIES** (I.E., MECHANICS) THAT ENCOURAGE OR DISCOURAGE MASS **BEHAVIOR** (I.E., DYNAMICS) THAT HAVE LONG-TERM EFFECTS ON OUR **WAY OF LIFE** (I.E., AESTHETICS), FOR BETTER FOR WORSE

SERIOUS GAMES

"ANY FORM OF INTERACTIVE COMPUTER-BASED GAME SOFTWARE FOR ONE OR MULTIPLE PLAYERS TO BE USED ON ANY PLATFORM AND THAT HAS BEEN DEVELOPED WITH THE INTENTION TO BE *MORE* THAN ENTERTAINMENT."*

DATE BACK SEVERAL MILLENNIA: *MILITARY* USES, *EDUCATION* AND *BUSINESS*

SINCE GAMES (INCLUDING VIDEO GAMES) CAN DEMONSTRABLY *MOTIVATE* USERS TO ENGAGE WITH THEM WITH GREAT *INTENSITY* AND *DURATION*, GAME-LIKE *ELEMENTS* SHOULD BE ABLE TO MAKE NON-GAME ACTIVITIES, PRODUCTS, AND SERVICES MORE ENJOYABLE AND ENGAGING AS WELL. MANY RELATED (AND AMUSING) *CONCEPTS* HAVE EMERGED.*

APPLIED GAMING

PRODUCTIVITY GAMES

GAME LAYER

FUNWARE

GAMIFICATION
USE OF GAME DESIGN ELEMENTS IN NON-GAME CONTEXTS

PLAYFUL DESIGN

BEHAVIORAL GAMES

MOTIVATIONAL HEURISTICS

SURVEILLANCE ENTERTAINMENT

FUNOLOGY
THE SCIENCE OF ENJOYING TECHNOLOGY

HAH. CAN WE SOUND ANY MORE *ACADEMIC?!*

EXPLOITATIONWARE

"AN ACT OF LINGUISTIC POLITICS THAT WOULD MORE TRUTHFULLY PORTRAY THE 'VILLAINOUS REIGN OF ABUSE' THAT GAMIFICATION PRESUMABLY ENTAILS."*

PLAYFULNESS

GAMEFULNESS

PLEASURABLE PRODUCTS

LOL! OH SNAP!

PLEASURABLE EXPERIENCE

LUDIC DESIGN

GAMES WITH A PURPOSE

USING GAME ELEMENTS TO MOTIVATE *HUMANS* TO SOLVE INFORMATION PROBLEMS NATURALLY DIFFICULT FOR COMPUTERS (E.G., TAG AN IMAGE, PROOFREAD A PAPER)

ALTERNATE REALITY GAMES

A GAME YOU PLAY IN "REAL LIFE"; TAKE THE SUBSTANCE OF EVERYDAY LIFE AND WEAVE IT INTO NARRATIVES THAT LAYER ADDITIONAL MEANING, DEPTH, AND INTERACTION UPON THE REAL WORLD

LUDIC ACTIVITIES

ACTIVITIES MOTIVATED BY CURIOSITY, EXPLORATION, AND REFLECTION

LUDIC ENGAGEMENT

PERVASIVE GAMING

"ONE OR MORE SALIENT FEATURES THAT EXPAND THE CONTRACTUAL MAGIC CIRCLE OF PLAY SPATIALLY, TEMPORALLY, OR SOCIALLY."*

LOCATION-BASED GAMES INVOLVING *PUBLIC SPACES*, AND AUGMENTED REALITY GAMES THAT *OVERLAY* A GAME LAYER ON THE PHYSICAL WORLD AND ENVIRONMENT

*THERE ARE MANY TERMS AIMED TO CAPTURE DESIGNING GAME ELEMENTS FOR PURPOSES BEYOND PURE PLAY. THESE ARE COLLECTED BY A GREAT PAPER ON CONTEXTUALIZING GAMIFICATION INTO THE GREATER FIELD OF GAME STUDIES. (SEE *DETERDING 2011*)

MUSIC AND GAMIFICATION ARE OFTEN FOUND TOGETHER. GAME-LIKE STRATEGIES CAN BE EMPLOYED IN THE **TRAINING** OF INSTRUMENTS -- TO INSTRUCT AND TO OVERCOME DIFFICULTIES AND UNDESIRABLE HABITS. GAMES CAN ALSO MOTIVATE AND PROVIDE LONG-TERM **STRUCTURE** IN MASTERING A PARTICULAR ASPECT OF AN INSTRUMENT.

PRACTICE THIS PASSAGE WITH A **SLOW** METRONOME CLICK, AND ONLY WHEN IT'S PLAYED **PERFECTLY**, **RAISE** THE METRONOME BY **2** BPM.

FINE. I'LL GO ALONG WITH THIS **BLATANT** PSYCHOLOGICAL MANIPULATION, AS LONG AS IT PROMOTES MY SHORT- AND LONG-TERM ENGAGEMENT WITH A CLEAR CHALLENGE TO OVERCOME!

MOREOVER, GAMIFICATION, IN TURN, MAY LEAD BACK TO A STATE RESEMBLING **PURE PLAY** (E.G., IN SKILL-BASED ACTIVITIES SUCH AS MUSIC) -- THE SENSATION OF BEING **IN THE ZONE** OR ENTERING A STATE OF **FLOW**...

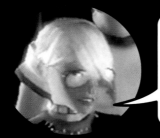

CONCEPTS OF **FLOW** AND **OPTIMAL EXPERIENCE** HAVE BEEN ESTABLISHED (FORMALIZED BY PSYCHOLOGIST **MIHALY CSIKSZENTMIHALYI**) TO BETTER UNDERSTAND HOW WE ACHIEVE STATES OF **HEIGHTENED CONCENTRATION** WHILE BEING SO ENGAGED IN AN ACTIVITY THAT ALL OTHER CONCERNS AND DISTRACTIONS **DISAPPEAR** -- ENABLING US TO EXPERIENCE A SENSE OF **PEAK PRODUCTIVITY** AND **PERFORMANCE.**

IT'S THAT FEELING OF **ROCKIN' OUT** WHEN PLAYING AN INSTRUMENT, OF **BEING PART OF** THE MUSIC... THE CONCEPT OF **FLOW** IS DEEPLY RELATED TO **PLAY,** AS AN INTERPLAY OF **SENSE** AND **COGNITION** THAT OCCURS IN A DEVOTED, **SEPARATE**, SELF-CONTAINING, AND SELF-FULFILLING SPACE, AWAY FROM THE REST OF LIFE'S WORRIES AND CARES. **CSIKSZENTMIHALYI** ARGUES THAT ACHIEVING FLOW IS PREDICATED ON SEVERAL CONDITIONS.

CONDITIONS FOR FLOW

- **BALANCE** BETWEEN INHERENT CHALLENGES AND THE PRESENT SKILL OF THE PARTICIPANT

- THE **POTENTIAL** FOR **GROWTH** IN THE PROGRESSION OF CONTINUALLY EXPANDING CHALLENGES, AS WELL AS THE POTENTIAL FOR IMMEDIATE **FEEDBACK** ON ONE'S PERFORMANCE

- SETTING FORTH OF **GOALS** THAT ARE REACHABLE WITHIN A CLEAR, DEFINED BOUNDARY OF TIME AND SPACE

DID SOMEONE SAY... **FLOW?**

IT SEEMS **GAMES** HAVE A NATURAL POTENTIAL TO **EXCEL** AT SATISFYING THE CONDITIONS FOR FLOW. GAMES CAN BE **BALANCED** BY DESIGN TO PROVIDE **GOALS** WHILE CONTINUALLY TUNED TO PROVIDE JUST THE RIGHT AMOUNT OF **CHALLENGE** TO MOTIVATE THE PLAYER TO MOVE FORWARD.

WE MIGHT EVEN SAY THAT SOME GAMES ARE CREATED, ABOVE ALL ELSE, TO **INDUCE** A SENSE OF **FLOW** AND BEING "IN THE ZONE" (LIKE THE "VERTIGO" IN CAILLOIS'S TERMINOLOGY OF PLAY). THE POTENTIAL BENEFITS OF GAMES -- SUCH AS ENTERTAINMENT, PLEASURE, PRODUCTIVITY, IMPROVEMENT IN SKILL -- ARE BUT **BY-PRODUCTS** OF FLOW.

AND YET, WE ARE OFTEN **INTERESTED** IN THESE VERY BY-PRODUCTS, INCLUDING USING GAMES TO **ENHANCE** EXPRESSIVE MUSICAL **SKILLS**, TO **EXPERIENCE** A SENSE OF MASTERY THROUGH EFFORT. BUT MUCH OF IT COMES BACK TO SOME NOTION OF FLOW, OF MAKING INTERACTIVE EXPERIENCES AUTHENTICALLY **SATISFYING** TO PLAY, USE, AND OPERATE.

⚛ PRINCIPLE 6.14

INDUCE AND HARNESS FLOW

THIS IS A KEY AESTHETIC AIM OF MANY GAMES: TO MAKE THE PLAYER FEEL **IN THE ZONE.** GOOD DESIGN **SETS** THE **CONDITIONS** FOR FLOW, BALANCING A **DIFFICULTY** WITH THE **REWARD** OF LEARNING AND EARNING THE **SKILLS** NECESSARY TO SUCCESSFULLY OVERCOME THE CHALLENGE OR TO REACH A GOAL, MAKING THE INITIALLY **DIFFICULT** SEEM **DOABLE.**

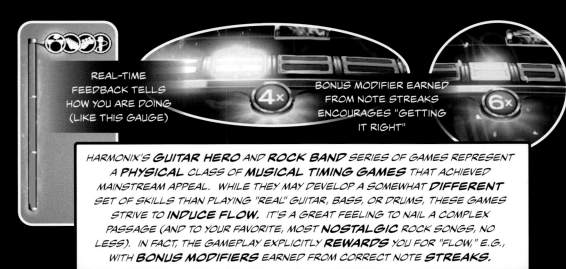

REAL-TIME FEEDBACK TELLS HOW YOU ARE DOING (LIKE THIS GAUGE)

BONUS MODIFIER EARNED FROM NOTE STREAKS ENCOURAGES "GETTING IT RIGHT"

HARMONIX'S **GUITAR HERO** AND **ROCK BAND** SERIES OF GAMES REPRESENT A **PHYSICAL** CLASS OF **MUSICAL TIMING GAMES** THAT ACHIEVED MAINSTREAM APPEAL. WHILE THEY MAY DEVELOP A SOMEWHAT **DIFFERENT** SET OF SKILLS THAN PLAYING "REAL" GUITAR, BASS, OR DRUMS, THESE GAMES STRIVE TO **INDUCE FLOW.** IT'S A GREAT FEELING TO NAIL A COMPLEX PASSAGE (AND TO YOUR FAVORITE, MOST **NOSTALGIC** ROCK SONGS, NO LESS). IN FACT, THE GAMEPLAY EXPLICITLY **REWARDS** YOU FOR "FLOW," E.G., WITH **BONUS MODIFIERS** EARNED FROM CORRECT NOTE **STREAKS.**

THE **SAME** FEELING OF FLOW CAN BE FOUND IN MUSIC-MAKING (SOLO AND GROUP), DANCE, ORATION, RACING, PROGRAMMING, **STARCRAFT** -- ALL OF WHICH CAN BE GAMIFIED TO EITHER INTENSIFY REACHING FLOW STATE, OR TO GET PEOPLE INTO THE ACTIVITY IN THE FIRST PLACE.

WE MIGHT CALL THIS **EXPRESSIVE FLOW**, WHICH SHARES SOME KEY COMMONALITIES WITH GAMES...

CHARACTERISTICS OF FLOW

TIME: ACTIVITIES OCCUR IN REAL TIME

INPUT: THEY REQUIRE ACTION AND PARTICIPATION

OUTPUT: SENSORY FEEDBACK (AND REWARD) -- AUDIO, VISUAL, PHYSICAL SENSATION, INTELLECTUAL

AND IF GAMES LIKE **ROCK BAND** AND **GUITAR HERO** ARE MORE "**LUDIC**" -- WHICH IS TO SAY THEY ARE MORE **STRUCTURED** WITH CLEAR GOALS TO ATTAIN -- WE MIGHT SEE MUSICAL EXPRESSION GAMES (E.G., **ELECTROPLANKTON**, **MARIO PAINT**, **OCARINA**) AS MORE "**PAIDIAC**" -- THEY ARE DESIGNED TO LEAVE SIGNIFICANT ROOM FOR **UNSTRUCTURED** EXPLORATION, STYLISTIC IMPROVISATION, AND **OPEN EXPRESSION.**

FOR EXAMPLE, **CHUCK CHUCK ROCKET** IS A PLAYFUL MUSICAL SANDBOX, INSPIRED BY **CHU CHU ROCKET** ON THE SEGA DREAMCAST AND TOSHIO IWAI'S **SOUND FACTORY** AND **ELECTROPLANKTON!**

MOUSE

THESE LITTLE CRITTERS RUN FORWARD **UNTIL** THEY HIT AN ARROW (AND TURN) OR EDGE OF THE BOARD (TURN RIGHT).

CHUCK CHUCK ROCKET (2006)
BY **SCOTT SMALLWOOD** & **GE WANG**

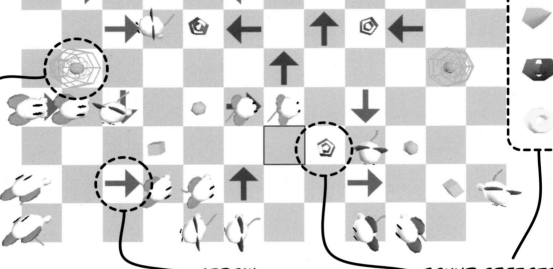

PORTAL
INSTANTLY **TELEPORTS** A MOUSE TO THE ADJACENT PLAYER'S GAME BOARD! THIS IS A MULTI-PLAYER & MULTI-MACHINE MUSICAL PLAYGROUND.

ARROW
PLACED BY THE PLAYER, ARROWS **DIRECT TRAFFIC** FOR SCURRYING MICE; CAN BE USED TO CONSTRUCT **LOOPS** AND RHYTHM TRAPS!

SOUND OBJECTS
PLACED BY THE PLAYER, SOUND OBJECTS MAKE SOUNDS WHEN A MOUSE RUNS OVER THEM. THE BOARD IS DISCRETELY SYNCHRONIZED, GIVING RISE TO RHYTHMIC POSSIBILITIES; IN FACT, ALL MACHINES ARE SYNCHRONIZED FOR GROUP PLAY!

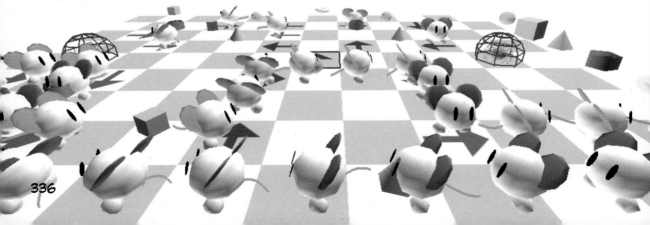

336

A DESIGN CHALLENGE

IN THE CONTEXT OF DESIGNING GAME-LIKE ELEMENTS FOR **MUSICAL** EXPRESSION, THERE IS A FUNDAMENTAL **DESIGN CHALLENGE**...

HOW DO WE CRAFT MUSIC-MAKING GAMES THAT DO NOT **SACRIFICE EXPRESSIVENESS**, THAT **MOTIVATE** EXPRESSIVE BEHAVIOR WHILE ALLOWING FOR **PLAYFUL** OPEN-ENDED CREATIVITY?

IT SEEMS WE HAVE THE OPPORTUNITY TO **COMBINE** THE STRUCTURED, GOAL-ORIENTED **LUDIC** ELEMENTS (THAT MOTIVATE BEHAVIOR) WITH THE OPEN-ENDED EXPLORATION OF THE **PAIDIAC** (I.E., FOR EXPRESSIVE FREEDOM).

PRINCIPLE 6.15

COMBINE *LUDUS* (GAMEFULNESS) AND *PAIDIA* (PLAYFULNESS)

WE WILL EXAMINE SEVERAL DESIGNS THAT RESULTED FROM THIS ETHOS, EMBODYING A SET OF **OBSERVATIONS** AND **ASSUMPTIONS** ABOUT EXPRESSIVE MUSICAL GAME DESIGN.

PRINCIPLE 6.16 GAMES ARE PERCEIVED TO BE MORE **ACCESSIBLE** THAN INSTRUMENTS

PRINCIPLE 6.17 GAMES CAN **BALANCE** GOALS AND CHALLENGES WITH REWARDS, PROVIDE A SENSE OF **SATISFACTION**, AND SET THE CONDITIONS FOR **FLOW**

PRINCIPLE 6.18 GAMEFULNESS AND FUN DO **NOT** HAVE TO COME AT THE EXPENSE OF **EXPRESSIVENESS** -- AND VICE VERSA

THE IDEA HERE IS THAT WE **CAN**, BY DESIGN, OPTIMIZE FOR STRUCTURED **AND** UNSTRUCTURED PLAY. IT SPEAKS, ON THE ONE HAND, TO GAMEFUL, RULE-DRIVEN OUTCOMES AND, ON THE OTHER, TO A KIND OF TOY-LIKE EXPRESSIVE ENGAGEMENT. FOR EXAMPLE, **ROCK BAND** SEEMS **MORE GAME** THAN **TOY**, WHEREAS SOMETHING LIKE **CHUCK CHUCK ROCKET** IS THE REVERSE (A SANDBOX WITH NO CLEAR GOALS). THE DESIGN OF GAMIFICATION IS ABOUT FINDING THE **BALANCE** THAT SUITS THE CONTEXT AT HAND AND, IN MY EXPERIENCE, DESIGN CAN ADDRESS THIS ON **TWO LEVELS** OF GAMIFICATION...

PRINCIPLE 6.19   ADDRESS **TWO** LEVELS OF **GAMIFICATION**

PERIPHERAL
BUILT **AROUND** THE MAIN EXPERIENCE OR INTERACTION; E.G., ACHIEVEMENTS, POINTS, LEVELING-UP, IN-GAME STATUS

VS.

CORE
IS THE PRIMARY INTERACTION AND EXPERIENCE; CORE MECHANIC AND RESULTING GAMEPLAY AND DYNAMICS

WE WILL FOCUS ON THE **LATTER** -- **CORE** GAMIFICATION, THE **DEEP** INTEGRATION OF GAME ELEMENTS INTO THE **CORE MECHANICS**, ELEMENTS THAT ARE **INSEPARABLE** FROM THE CORE EXPERIENCE. UNLIKE PERIPHERAL GAMIFICATION, CORE ELEMENTS CANNOT BE ADDED "AFTER THE FACT." THEY **ARE** THE EXPERIENCE.

TO HELP UNDERSTAND GAME DESIGN FOR **EXPRESSION**, WHICH APPEALS TO THE AESTHETICS OF PLAY ON THE LEVELS OF **TACTILE JOY** (VERTIGO) AND **DESIRE** FOR EXPRESSION (TEKHNE), LET'S LOOK AT SEVERAL POPULAR COMPUTER MUSIC SOFTWARE APPLICATIONS -- DESIGNED FOR A **MASS MEDIUM** (IN THIS CASE, MOBILE APPS FOR THE SMARTPHONE).

THEY WERE DESIGNED TO INTEGRATE GAME-LIKE ELEMENTS -- RULES, GOALS, REWARDS -- IN **CORE INTERACTIONS** THAT INTENTIONALLY LEAVE ROOM FOR **OPEN EXPRESSIVENESS.**

WHILE MANY OF THESE APPS ARE NO LONGER AVAILABLE, BEING ARTIFACTS OF AN EARLY **MOBILE ERA**...

THEY **ILLUSTRATE** SOMETHING ABOUT DESIGNING EXPRESSIVE MUSICAL GAMES...

MAGIC PIANO

MAGIC FIDDLE

IN THE SAME VEIN AS OUR DISCUSSIONS ON **OCARINA** AND **MUSIC-MAKING** (CHAPTER 2), THESE APPS WERE DESIGNED FOR THE **MASSES,** INTENDING TO LOWER THE ENTRY BARRIER FOR NOVICES AND NON-MUSICIAN PLAYERS, BUT PRESENTING HIGHER **SKILL CEILINGS** FOR MORE EXPERIENCED PLAYERS.

MAGIC GUITAR

OCARINA 2 (GAME MODE)

I DESIGNED THESE WHILE ACTIVELY WORKING ON **SMULE** -- AND BUILT THEM TOGETHER WITH OUR ENGINEERING TEAM.

THREE FUNDAMENTAL DESIGN CONSIDERATIONS **RECUR** IN THESE CASE STUDIES: (1) HOW **PITCH** INFORMATION IS ENCODED; (2) WHETHER **TIMING** IS OPEN-ENDED OR PROCEEDS "ON RAILS"; AND (3) THE **COMPLEXITY** OF INTERACTION. HOW THESE DESIGN CONSTRAINTS ARE MANAGED DEFINES THE CORE EXPERIENCE FOR EACH APP.

3

MAGIC PIANO (2010)

MAGIC PIANO'S CORE GAME MECHANIC IS SIMPLE, EVEN **MINIMAL.**

GLOWING POINTS OF LIGHTS -- I CALL THEM "**FIREFLIES**" -- ENCODE A MUSICAL SCORE, EACH FIREFLY A SINGLE NOTE. THEY **FALL** FROM THE TOP OF THE SCREEN, AND THEIR **VERTICAL SPACING** HINTS AT NOTE DURATIONS AND TIME BETWEEN NOTES.

THERE ARE **NO TEMPO CONSTRAINTS** IN **MAGIC PIANO**; THE EXPERIENCE IS BUILT AROUND **EXPRESSIVE MUSICAL TIMING**, LEAVING THE PLAYER COMPLETELY **FREE** TO EXPRESS EACH NOTE **IN TIME** -- AT **ANY** TEMPO, WITH **VARIATION, RUBATO, SWING, ROLLING** CHORDS, AND **TRILLS.**

EACH **TAP** GESTURE IMMEDIATELY **TRIGGERS** THE NEXT UNPLAYED NOTE.

"FIREFLIES"

GLOWING POINTS OF LIGHT, EACH A NOTE. THEY **HANG AROUND** PATIENTLY; THEY **FALL** DOWNWARD ONLY AS NOTES ARE PLAYED BY YOU!

FIREFLIES AT THE **SAME LEVEL** ARE MEANT TO BE PLAYED AT THE **SAME TIME!**

NOTES CAN BE PLAYED **ANYWHERE** ON THE SCREEN (YOU DON'T HAVE TO TAP ON THE FIREFLY).

TEMPO LINE (SUGGESTED)

THERE IS NO **CORRECT** TEMPO -- PLAYING NOTES AS THEY CROSS THE LINE IS A **RECOMMENDED** TEMPO HINT, BUT YOU ARE FREE TO **DEVIATE!**

THE **BOTTOM**-MOST NOTES ARE THE **NEXT** NOTES TO SOUND.

THE **PITCH** IN **MAGIC PIANO** HAS BEEN **ABSTRACTED** AWAY -- I.E., PUT "**ON RAILS**"; PLAYERS HAVE NO CONTROL OVER PITCH -- PARTLY BECAUSE OF THE SMALL SCREEN SIZE AND LACK OF TACTILE FEEDBACK FROM **PHYSICAL KEYS** AND KEY **BOUNDARIES.** THIS DESIGN CONCESSION ALLOWS USERS TO FOCUS ON **MUSICAL TIMING**, FIRST AND FOREMOST.

NOTE DYNAMICS ARE ALSO "BAKED" INTO THE SCORE, WITH ADDITIONAL USER-CONTROLLABLE DYNAMIC VARIATIONS MAPPED TO TOUCH POSITION ALONG THE **Y-AXIS**: THE CLOSER A USER TAPS TO THE **BOTTOM** OF THE SCREEN, THE **LOUDER.**

SOMEWHAT **WHIMSICALLY**, NOTES CAN SOUND "OUT OF TUNE" IF THE TAP GESTURE STRIKES SUFFICIENTLY **FAR** AWAY FROM THE FIREFLY ALONG THE HORIZONTAL AXIS. THIS MECHANIC ADDS AN ADDITIONAL GAME-LIKE **CHALLENGE** WHILE STILL KEEPING INTACT A **DIRECT** AND **SATISFYING** ACTION-TO-SOUND INTERACTION.

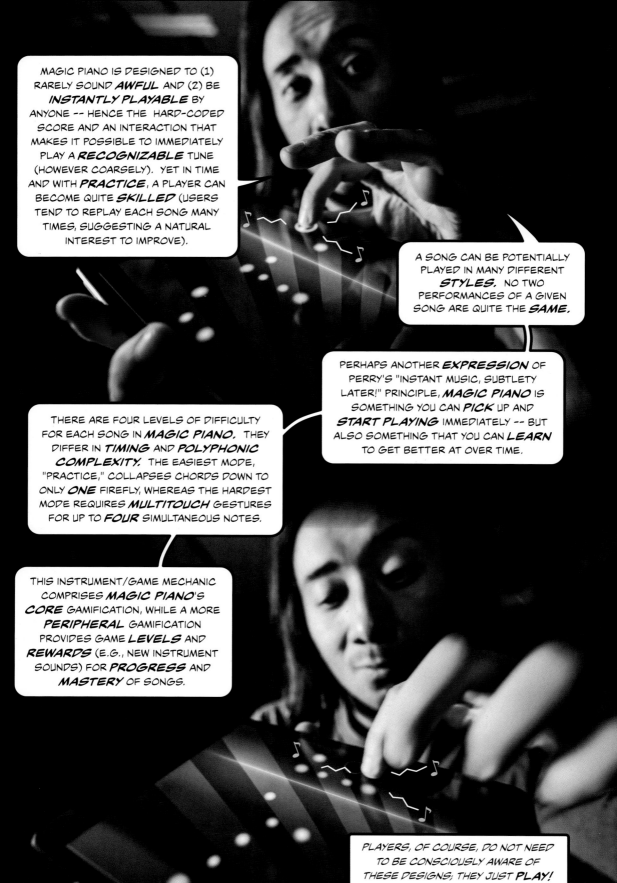

MAGIC PIANO IS DESIGNED TO (1) RARELY SOUND *AWFUL* AND (2) BE *INSTANTLY PLAYABLE* BY ANYONE -- HENCE THE HARD-CODED SCORE AND AN INTERACTION THAT MAKES IT POSSIBLE TO IMMEDIATELY PLAY A *RECOGNIZABLE* TUNE (HOWEVER COARSELY). YET IN TIME AND WITH *PRACTICE*, A PLAYER CAN BECOME QUITE *SKILLED* (USERS TEND TO REPLAY EACH SONG MANY TIMES, SUGGESTING A NATURAL INTEREST TO IMPROVE).

A SONG CAN BE POTENTIALLY PLAYED IN MANY DIFFERENT *STYLES*. NO TWO PERFORMANCES OF A GIVEN SONG ARE QUITE THE *SAME*.

PERHAPS ANOTHER *EXPRESSION* OF PERRY'S "INSTANT MUSIC, SUBTLETY LATER!" PRINCIPLE, *MAGIC PIANO* IS SOMETHING YOU CAN *PICK* UP AND *START PLAYING* IMMEDIATELY -- BUT ALSO SOMETHING THAT YOU CAN *LEARN* TO GET BETTER AT OVER TIME.

THERE ARE FOUR LEVELS OF DIFFICULTY FOR EACH SONG IN *MAGIC PIANO*. THEY DIFFER IN *TIMING* AND *POLYPHONIC COMPLEXITY*. THE EASIEST MODE, "PRACTICE," COLLAPSES CHORDS DOWN TO ONLY *ONE* FIREFLY, WHEREAS THE HARDEST MODE REQUIRES *MULTITOUCH* GESTURES FOR UP TO *FOUR* SIMULTANEOUS NOTES.

THIS INSTRUMENT/GAME MECHANIC COMPRISES *MAGIC PIANO'S CORE* GAMIFICATION, WHILE A MORE *PERIPHERAL* GAMIFICATION PROVIDES GAME *LEVELS* AND *REWARDS* (E.G., NEW INSTRUMENT SOUNDS) FOR *PROGRESS* AND *MASTERY* OF SONGS.

PLAYERS, OF COURSE, DO NOT NEED TO BE CONSCIOUSLY AWARE OF THESE DESIGNS; THEY JUST *PLAY!*

340

AND... IT'S SO **MOBILE** YOU CAN TAKE IT OUTSIDE, **FROLICKING** IN THE FIELDS AS YOU PLAY...

K-POP STAR WIG

... LOOKING INTO THE SUNSET, RUNNING IN SLOW MOTION, ROCKIN' SOME "CHARIOTS OF FIRE"...

MAGIC PIANO
COMMERCIAL (2011)

...AND LOOKING LIKE A MUSIC-MAKING FOOL, AND ENJOYING IT!

BACK ON LOCATION: STANFORD **SATELLITE DISH WALK**

smule
Magic Piano

THE **PRESENTATION** OF **MAGIC PIANO** IS DESIGNED TO **LOWER INHIBITION** ABOUT ACTIVELY MAKING MUSIC. THIS HAS PROVEN EFFECTIVE, AS EVIDENT BY ITS MORE THAN 100 MILLION USERS. AS **MAGIC PIANO**'S DESIGNER, I DON'T CARE WHETHER PEOPLE THINK OF IT AS AN **INSTRUMENT** OR **GAME** OR **TOY** -- SO LONG AS **EXPRESSIVENESS** HAPPENS WHEN PEOPLE **EXPERIENCE** THE **GAME!**

⋀ PRINCIPLE 6.20

THE
TOFU BURGER PRINCIPLE

GENTLY INTRODUCE PEOPLE TO NEW EXPERIENCES

IT'S PSYCHOLOGY. IF YOU WANT TO, SAY, GET SOMEONE TO TRY BEING A **VEGETARIAN**, YOU MIGHT NOT START WITH, SAY, A **KALE** SMOOTHIE -- BUT PERHAPS A **TOFU BURGER**, OR SOMETHING THAT IS SORT OF **FAMILIAR** AND **GENTLY** EASES THE PERSON INTO THE NEW EXPERIENCE. GAMES, TOYS, SOCIAL EXPERIENCES, DELIGHT AND SURPRISE, ARE ALL EXCELLENT INGREDIENTS TO BUILD THE **EXPERIENTIAL TOFU BURGER**.

MAGIC FIDDLE BEGAN AS SORT OF A **JOKE**. AFTER A **SAN FRANCISCO SYMPHONY** PERFORMANCE AT DAVIES SYMPHONY HALL, I THOUGHT, "**WHAT IF** WE MADE A **VIOLIN-LIKE** INSTRUMENT THAT **REQUIRED** THE USER TO HOLD A TABLET COMPUTER UNDER THEIR **CHIN**? HAHA. AND I BET THAT WOULD **TOTALLY** WORK..."

MAGIC FIDDLE (2011)

IT **DIDN'T** WORK -- OR AT LEAST NOT EXACTLY AS GE HAD THOUGHT. NONETHELESS, WE HAD **GOOD TIMES** BUILDING *MAGIC FIDDLE*, DESIGNED SPECIFICALLY FOR THE iPAD!

TOM LIEBER

DESPITE THE REAPPEARANCE OF "MAGIC" IN THE NAME, *MAGIC FIDDLE* IS A VERY **DIFFERENT** EXPERIENCE AND GAME FROM *MAGIC PIANO*, BEYOND THE OBVIOUS INSTRUMENT DIFFERENCE. FOR ONE, THE *MAGIC FIDDLE* GAME MECHANIC IS BASED ON A VIRTUAL, TABLET-MEDIATED **THREE-STRING FIDDLE**-LIKE INTERFACE, WITH CONTINUOS **PITCH MAPPINGS** AND THE ABILITY TO "**BOW**" AND "**PLUCK**"!

JIEUN

342

NOTE LINES

FOR A GIVEN PIECE OF MUSIC, COLOR-CODED LINES *GLIDE* TOWARD ONE OF THREE STRINGS, PREVIEWING THE MUSIC TO BE PLAYED AHEAD. UNLIKE *MAGIC PIANO*, THESE DO NOT WAIT FOR YOU!

COLLISION POINT

LINES *COLLIDE* GENTLY INTO THE INTENDED POINT OF ARTICULATION ON THE STRING, FOR PRECISELY THE *DURATION* OF EACH NOTE.

BACK

LONGER LINES REPRESENT LONGER NOTE DURATION

NOTES FLOW *THIS* WAY ---→

ALSO *UNLIKE MAGIC PIANO*, THESE LINES ARE *MERELY SUGGESTIONS*, LEAVING THE PLAYER TO EXPRESS *PITCH* AND *MICRO-TIMING* FREELY (OFTEN TO THE DETRIMENT OF INTONATION, BUT ALLOWS FOR *GLISSANDI*, *VIBRATO*, AND VARIOUS *EMBELLISHMENTS*)

FLOATING MIST GIVES SENSE OF DEPTH TO THE NEON, TRON-INSPIRED VISUAL AESTHETICS; *AND* BECAUSE IT LOOKED COOL

CIRCULAR "BOW" INTERFACE

EXCEPT, YOU DON'T "BOW" SO MUCH AS *HOLD* YOU FINGER -- CLOSER TO THE CENTER, THE LOUDER; PROVIDES LIMITED CONTROL OVER ARTICULATION

CHIN ZONE
REGION OF THE SCREEN ON WHICH TO REST YOUR CHIN!

ORIGINALLY, I TOYED WITH THE IDEA THAT THE INSTRUMENT COULD ONLY BE *ACTIVATED* BY USING ONE'S *CHIN* (ON THE TOUCHSCREEN). IT WAS THEN DISCOVERED THAT *NOT ALL CHINS* ARE EQUALLY DETECTABLE ON TOUCH-SCREENS... FOR EXAMPLE, *MY* CHIN DIDN'T CONSISTENTLY REGISTER ON THE SCREEN -- EVEN WHEN I *SHAVED.* SADLY, WE HAD TO CUT THAT "FEATURE."

PIZZICATO "PLUCK" ZONES
RUNNING YOUR FINGER *ACROSS* A STRING -- OR TAPPING -- HERE "*PLUCKS*" THE STRING INSTEAD OF "BOWING" IT!

343

THE SOUND IS *SYNTHESIZED* IN REAL TIME TO SUPPORT A DIRECT *ACTION-TO-SOUND* MAPPING. PLAYERS ARE *AWARE*, AT ALL TIMES (SOMETIMES PAINFULLY SO), THAT THEY ARE MAKING THE SOUND. THE FAR-FROM-PERFECT BOWED STRING *PHYSICAL MODEL* (A *DIFFICULT* CLASS OF INSTRUMENTS TO MODEL INDEED) AND THE OPEN FREEDOM OF THE GAME OFTEN LEAD TO LESS-THAN-SPECTACULAR OR EVEN *COMICAL* RENDITIONS OF WELL-LOVED SONGS SUCH AS "POMP AND CIRCUMSTANCE," "SUPER MARIO BROS. THEME," BACH'S "AIR ON THE G STRING," AND MANY OTHERS.

MAGIC FIDDLE POSTURE
INSTRUMENT CORNER ON SHOULDER, BACK STRAIGHT (MORE LIKE JIEUN, LESS LIKE GE), ARMS COMFORTABLY RELAXED, LEGS SHOULDER-WIDTH APART

THE SYNTHESIZED SOUND IS SOMETHING *AWFUL* (AND WE KIND OF WENT WITH IT) -- *MAGIC FIDDLE* WAS DESIGNED AS MUCH AROUND PLAYING AN INSTRUMENT, AS IT WAS TO CAPTURE THE *NOSTALGIA* OF LEARNING A DIFFICULT INSTRUMENT (OR HEARING YOUR NEIGHBOR LEARNING THE VIOLIN)...

MAGIC FIDDLE VS. ROCK BAND

INSPIRED BY GAMES LIKE *GUITAR HERO* AND *ROCK BAND*, *MAGIC FIDDLE* PROVIDES A SIMILAR *DRIVING IMPETUS* TO *PLAY* -- YOU TAKE ACTION, IN TIME AND RESPONSE TO THE GAME...

...AND *UNLIKE ROCK BAND*, *MAGIC FIDDLE*'S SOUND IS NOT PRERECORDED AND IS -- FOR BETTER OR WORSE -- *DIRECTLY* MADE BY THE PLAYER, BESTOWING A WIDE EXPRESSIVE *RANGE* OF ARTICULATION, INTONATION, DYNAMICS, EMBELLISHMENTS... WHICH NOTES TO PLAY IN THE FIRST PLACE. IN OTHER WORDS, THE PLAYER CAN GO *"OFF SCRIPT"*; ONE MIGHT SAY *MAGIC FIDDLE* IS A GAME ON A *SOFT RAIL* TO LEAVE ROOM FOR A TYPE OF *OPEN EXPRESSION*.

SATISFYING ANIMATIONS ON **SUCCESSFUL** NOTE HIT!

NOTES FLOW *THIS* WAY

TO HELP USERS **LEARN** TO PLAY A VIRTUAL, PHYSICALLY MODELED, THREE-STRING VIOLIN ON AN TABLET COMPUTER, WE ADDED A **STORYBOOK MODE** TO PROVIDE A MACRO-LEVEL ARC TO THE GAME PROGRESSION. IT IS AN ELABORATE **PERIPHERAL GAMIFICATION** IN THE FORM OF EPISODIC **MUSIC LESSONS**, DESIGNED SPECIFICALLY FOR **MAGIC FIDDLE.**

SOMEWHAT **WHIMSICALLY**, ALL LESSONS ARE GIVEN FROM THE **FIRST-PERSON** PERSPECTIVE OF THE FIDDLE **ITSELF**...

Hello.
I am your fiddle.

...THE **FIRST TIME** YOU OPEN **MAGIC FIDDLE, THE FIDDLE** GREETS YOU, PROCLAIMING ITS **EXISTENCE** AND **PERSONALITY** (OR FIDDLE-ALITY).

AS THE **16 LESSONS** UNFOLD, SO DOES THE PERSONALITY OF THE FIDDLE, ADDING ANOTHER DIMENSION OF INTERACTION AND EDUCATIONAL **ENGAGEMENT.**

FOR SOME REASON, THE FIDDLE GETS INCREASINGLY **SARCASTIC**...

SEVERAL OF THE LESSONS END WITH A **SOCIAL HOMEWORK ASSIGNMENT**, WHICH BREAKS UP THE USUAL FORMAT BY ASKING THE USER TO PERFORM MUSICAL TASKS THAT REQUIRE **INTERACTING** WITH PEOPLE IN THE **PHYSICAL WORLD**...

THESE ENCOURAGE USERS TO SHARE THE EXPERIENCE AS PERFORMATIVE **SOCIAL ACTS.** AFTER EACH SOCIAL HOMEWORK ASSIGNMENT, **MAGIC FIDDLE** ASKS THE USER TO **REPORT** ON THE EXPERIENCE AS PART OF THE GAME.

BACK

Your Very First Magic Fiddle Book
Introduction: Chapters 1-7

ACTIVE BOOK

MY LIBRARY

○ Chapter 1
Holding Your Fiddle
RESUME

○ Chapter 2
Bowing

○ Chapter 3
Upper Body Posture (arms, chest, he

○ Chapter 4
Scales

○ Chapter 5
Vibrato

○ Chapter 6
Arpeggios

○ Chapter 7
Lower Body Posture (feet, knees)

○ Chapter 8
Plucks

ONE EARLY HOMEWORK ASSIGNMENT ASKS THE USER TO **PLAY** "MARY HAD A LITTLE LAMB" WHILE STANDING WITH **CORRECT POSTURE**, FOLLOWED BY A **MISSION** TO PLAY IT "IN FRONT OF A **LIVE AUDIENCE**," SUCH AS A FRIEND OR LOVED ONE.

USER RESPONSES (THE APP SENDS THEM BACK TO A CENTRAL DATABASE) RANGE FROM THE **BRIEF**: "FUN," "AWESOME," "COOL"...

BEGINNING OF LESSON 1

One. Look at the picture above.
Two. Do that.

STORYBOOK

...TO THE **TRIUMPHANT**: "I WAS EPIC. THE CROWD CHEERED AND LIFTED ME UP AFTER I STAGE-DIVED OFF MY BED. MONEY AND ROSES WERE THROWN AT ME. IT WAS PRETTY COOL" AND "...ALMOST GOT A **STANDING OVATION** FROM TWO DOGS."

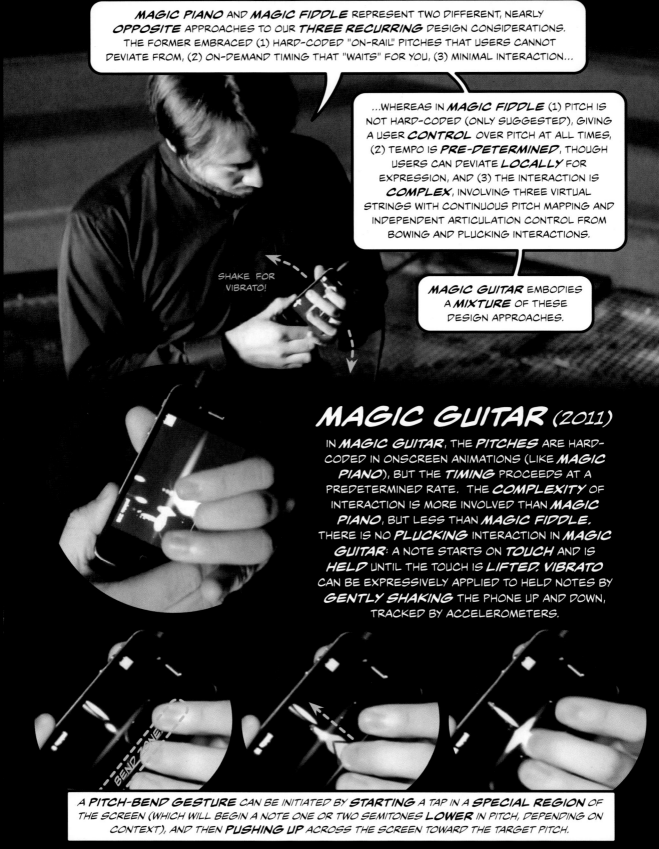

MAGIC PIANO AND MAGIC FIDDLE REPRESENT TWO DIFFERENT, NEARLY OPPOSITE APPROACHES TO OUR THREE RECURRING DESIGN CONSIDERATIONS. THE FORMER EMBRACED (1) HARD-CODED "ON-RAIL" PITCHES THAT USERS CANNOT DEVIATE FROM, (2) ON-DEMAND TIMING THAT "WAITS" FOR YOU, (3) MINIMAL INTERACTION...

...WHEREAS IN MAGIC FIDDLE (1) PITCH IS NOT HARD-CODED (ONLY SUGGESTED), GIVING A USER CONTROL OVER PITCH AT ALL TIMES, (2) TEMPO IS PRE-DETERMINED, THOUGH USERS CAN DEVIATE LOCALLY FOR EXPRESSION, AND (3) THE INTERACTION IS COMPLEX, INVOLVING THREE VIRTUAL STRINGS WITH CONTINUOUS PITCH MAPPING AND INDEPENDENT ARTICULATION CONTROL FROM BOWING AND PLUCKING INTERACTIONS.

SHAKE FOR VIBRATO!

MAGIC GUITAR EMBODIES A MIXTURE OF THESE DESIGN APPROACHES.

MAGIC GUITAR (2011)

IN MAGIC GUITAR, THE PITCHES ARE HARD-CODED IN ONSCREEN ANIMATIONS (LIKE MAGIC PIANO), BUT THE TIMING PROCEEDS AT A PREDETERMINED RATE. THE COMPLEXITY OF INTERACTION IS MORE INVOLVED THAN MAGIC PIANO, BUT LESS THAN MAGIC FIDDLE. THERE IS NO PLUCKING INTERACTION IN MAGIC GUITAR: A NOTE STARTS ON TOUCH AND IS HELD UNTIL THE TOUCH IS LIFTED. VIBRATO CAN BE EXPRESSIVELY APPLIED TO HELD NOTES BY GENTLY SHAKING THE PHONE UP AND DOWN, TRACKED BY ACCELEROMETERS.

BEND ZONE

A PITCH-BEND GESTURE CAN BE INITIATED BY STARTING A TAP IN A SPECIAL REGION OF THE SCREEN (WHICH WILL BEGIN A NOTE ONE OR TWO SEMITONES LOWER IN PITCH, DEPENDING ON CONTEXT), AND THEN PUSHING UP ACROSS THE SCREEN TOWARD THE TARGET PITCH.

CHAPTER 2 DISSECTED THE DESIGN OF **OCARINA** AS A **TOY** -- AN ARTIFACT FOR OPEN-ENDED PLAY. HERE WE'LL STUDY IT AS A **GAME.** THE ORIGINAL **OCARINA** INCORPORATED IMPLICIT GAMEPLAY WHEREBY USERS COULD LEARN TO PLAY MELODIES FROM **TABLATURE** PROVIDED ON THE **OCARINA** WEBSITE, WHERE USERS CAN ALSO **CREATE** AND **SHARE** TABLATURE TO AN ONLINE FORUM.

FOR EXAMPLE, THIS IS OCARINA **TABLATURE** FOR THE BEGINNING OF "TWINKLE TWINKLE, LITTLE STAR"

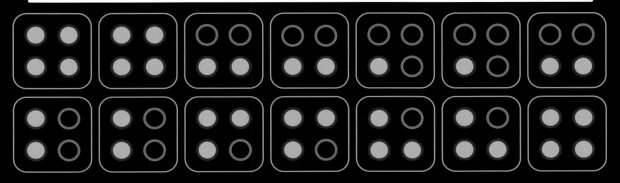

TABLATURE IS EASY TO LEARN AND REQUIRES **NO** PRIOR MUSICAL TRAINING. IT PROVIDES A SATISFYING EXPERIENCE TO PLAY **FAMILIAR MELODIES** ON A **WHIMSICAL** INSTRUMENT...

OCARINA 2
GAME MODE (2012)

"INSTRUCTION"
HOLD EACH NOTE AS SHOWN, BLOW INTO THE PHONE FOR THE DURATION OF THE NOTE, EXPRESS EACH NOTE AND PHRASE AS DESIRED!

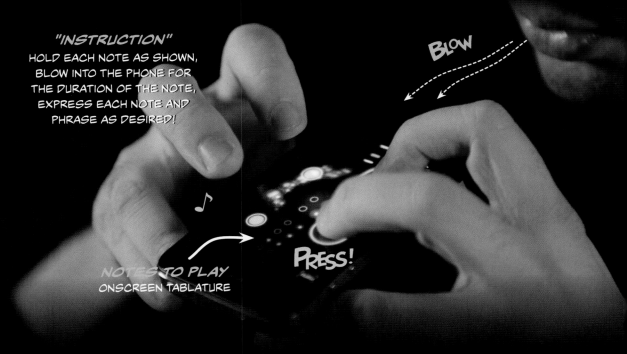

BLOW

NOTES TO PLAY
ONSCREEN TABLATURE

PRESS!

OCARINA 2'S **GAME MODE** IS DESIGNED AS AN **ANIMATED TABLATURE** WITH AN ACCOMPANIMENT ENGINE THAT **FOLLOWS** THE PLAYER.

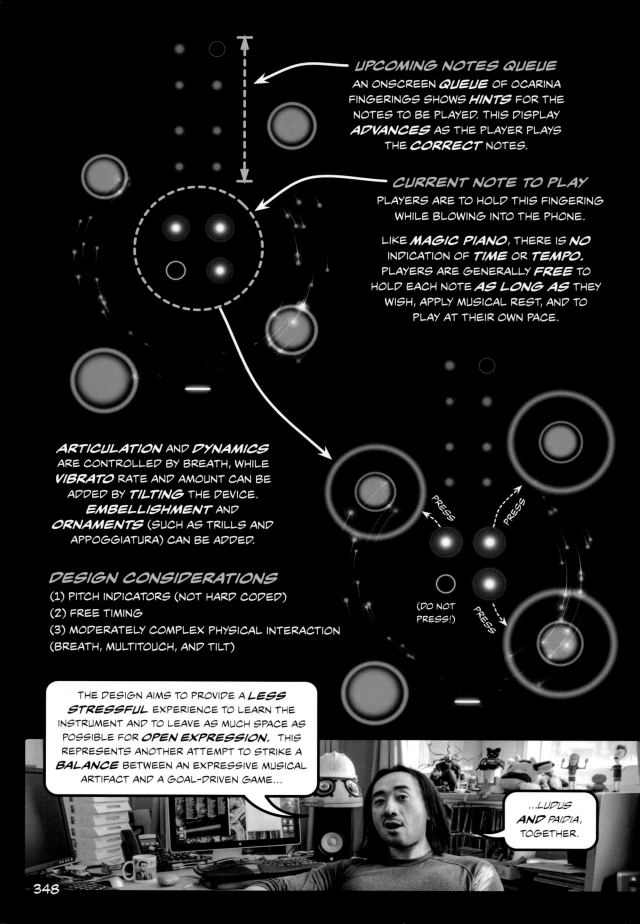

UPCOMING NOTES QUEUE
AN ONSCREEN *QUEUE* OF OCARINA FINGERINGS SHOWS *HINTS* FOR THE NOTES TO BE PLAYED. THIS DISPLAY *ADVANCES* AS THE PLAYER PLAYS THE *CORRECT* NOTES.

CURRENT NOTE TO PLAY
PLAYERS ARE TO HOLD THIS FINGERING WHILE BLOWING INTO THE PHONE.

LIKE *MAGIC PIANO*, THERE IS *NO* INDICATION OF *TIME* OR *TEMPO*. PLAYERS ARE GENERALLY *FREE* TO HOLD EACH NOTE *AS LONG AS* THEY WISH, APPLY MUSICAL REST, AND TO PLAY AT THEIR OWN PACE.

ARTICULATION AND *DYNAMICS* ARE CONTROLLED BY BREATH, WHILE *VIBRATO* RATE AND AMOUNT CAN BE ADDED BY *TILTING* THE DEVICE. *EMBELLISHMENT* AND *ORNAMENTS* (SUCH AS TRILLS AND APPOGGIATURA) CAN BE ADDED.

DESIGN CONSIDERATIONS
(1) PITCH INDICATORS (NOT HARD CODED)
(2) FREE TIMING
(3) MODERATELY COMPLEX PHYSICAL INTERACTION (BREATH, MULTITOUCH, AND TILT)

PRESS

PRESS

PRESS

(DO NOT PRESS!)

THE DESIGN AIMS TO PROVIDE A *LESS STRESSFUL* EXPERIENCE TO LEARN THE INSTRUMENT AND TO LEAVE AS MUCH SPACE AS POSSIBLE FOR *OPEN EXPRESSION*. THIS REPRESENTS ANOTHER ATTEMPT TO STRIKE A *BALANCE* BETWEEN AN EXPRESSIVE MUSICAL ARTIFACT AND A GOAL-DRIVEN GAME...

...LUDUS *AND* PAIDIA, TOGETHER.

EACH OF THESE APPS HAD A SIGNIFICANT FOLLOWING IN ITS TIME, COLLECTIVELY REACHING OVER 200 MILLION USERS -- SUGGESTING THEY HAVE REACHED **BEYOND** EVEN CASUAL MUSICIANS, WELL INTO THE REALM OF **GENERAL USERS.**

THESE MUSIC GAMES EMBRACE DIFFERENT MECHANICS, SEEKING TO SUPPORT INSTRUMENT-SPECIFIC DYNAMICS, TO INDUCE SOME AESTHETIC TAKEAWAYS BETWEEN **VERTIGO** AND **TEKHNE.** THE CORE MECHANIC DESIGN DECISIONS FOR THESE FOUR GAMES ARE SUMMARIZED HERE.

	PITCH CONTROL	TIMING	INTERACTION COMPLEXITY	PERIPHERAL GAMIFICATION
MAGIC PIANO	ENFORCED	FREE	MINIMAL & DIRECT	PLAY MORE TO **LEVEL UP;** DIFFICULTY LEVELS; ACHIEVEMENTS & REWARDS
MAGIC FIDDLE	FREE	ENFORCED	COMPLEX (PITCH, BOW, PLUCK)	STORYBOOK, MUSIC LESSONS
MAGIC GUITAR	ENFORCED	ENFORCED	REDUCED (TAP-AND-HOLD, SHAKE, BEND)	PLAY MORE TO **LEVEL UP;** DIFFICULTY LEVELS; ACHIEVEMENTS & REWARDS
OCARINA 2	FREE	FREE	MULTI-FACETED (BREATH, PITCH, TILT)	PROGRESS PER SONG, EXPERIENCE POINTS, ACHIEVEMENTS

ALL OFFER **LOCALLY EXPRESSIBLE NUANCE**

ONE **TAKEAWAY**: THERE IS **NO ONE-SIZE-FITS-ALL** DESIGN (IT ALWAYS NEEDS TO BE DESIGNED **TO** THE TASK AT HAND AND, IN THIS CASE, TO THE SPECIFICITIES OF THE INSTRUMENT AND GAME). THIS RULE OF THUMB APPLIES TO DESIGN IN GENERAL.

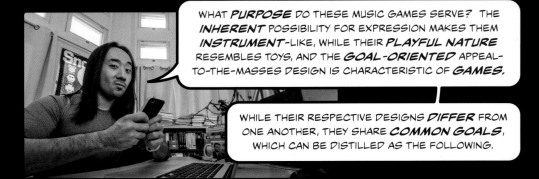

WHAT **PURPOSE** DO THESE MUSIC GAMES SERVE? THE **INHERENT** POSSIBILITY FOR EXPRESSION MAKES THEM **INSTRUMENT**-LIKE, WHILE THEIR **PLAYFUL NATURE** RESEMBLES TOYS, AND THE **GOAL-ORIENTED** APPEAL-TO-THE-MASSES DESIGN IS CHARACTERISTIC OF **GAMES.**

WHILE THEIR RESPECTIVE DESIGNS **DIFFER** FROM ONE ANOTHER, THEY SHARE **COMMON GOALS**, WHICH CAN BE DISTILLED AS THE FOLLOWING.

LOWER INHIBITION FOR INTENDED **BEHAVIOR** BY **GAMIFYING** EXPRESSIVE EXPERIENCES AS **GAMES**

 PRINCIPLE 6.21

RETAIN GENUINE EXPRESSIVE POSSIBILITIES, WHILE OFFERING ELEMENTS OF GAME AND PLAY TO **REDUCE** BARRIER INTO THE MUSIC-MAKING EXPERIENCE. THE **HYPOTHESIS** IS THAT PEOPLE ARE **LESS INTIMIDATED** OR **INHIBITED** TO TRY SOMETHING THEY **PERCEIVE** AS A GAME. SUCH AN EXPERIENCE ENCOURAGES, PERHAPS EVEN BENIGNLY "TRICKING" THE PLAYER INTO BEING MUSICAL, AND -- FOR SOME -- GETTING A FIRST TASTE OF THE JOY OF MAKING MUSIC.

CREATE **SATISFYING CORE MECHANICS** AIMED TO INDUCE A SENSE OF **FLOW**

 PRINCIPLE 6.22

BALANCE INTERESTING RULES, INTERACTIONS, AND PLAYABILITY. THERE SHOULD BE AN **INHERENT ATTRACTION** EARLY ON, WITH A SENSE OF CONTINUED **PAYOFF** AND ACCOMPLISHMENT, WHILE PROVIDING CHALLENGES IN THE FORM OF ATTAINABLE GAME GOALS. FOR EXAMPLE, MOBILE MUSIC GAMES MUST ENGAGE ALMOST IMMEDIATELY (DUE TO THE **CASUAL** NATURE OF THEIR AUDIENCE) BUT ALSO MAKE IT POSSIBLE (AND FUN) TO **ACQUIRE** AND HONE THE **SKILLS** NEEDED TO PLAY THE GAME -- BY CREATING **MECHANICS** THAT ARE **INVITING** TO TRY, AND **SATISFYING** TO MASTER, EVER CULTIVATING THE POSSIBILITY FOR FLOW.

MOTIVATE **LONGER-TERM ENGAGEMENT** THROUGH **SOCIAL** AND **PERIPHERAL GAMIFICATION**

 PRINCIPLE 6.23

THIS OCCURS IN THE FORM OF MISSIONS, DIFFICULTY LEVELS, EXPERIENCE POINTS, PROGRESSION, ACHIEVEMENTS, REWARDS, RANKINGS, STORYBOOK MUSIC LESSONS -- ANYTHING THAT MOTIVATES (OFFERING GOLD STARS REMAINS SURPRISINGLY EFFECTIVE IN MANY SITUATIONS)! ONE IMPORTANT **CAVEAT:** THIS IS CONTINGENT ON CORE MECHANICS THAT ARE **SATISFYING** TO BEGIN WITH AND ON THE PERIPHERAL GAMIFICATION ITSELF BEING **AUTHENTIC.** THE GAMIFICATION MUST BE BUILT ON THE INTRINSIC **QUALITY** OF THE EXPERIENCE, NOT SUBORDINATE TO ULTERIOR AIMS (E.G., MAXIMIZING ECONOMIC PROFIT). ALSO, IN THE NEXT CHAPTER, WE WILL LOOK AT **LEAF TROMBONE: WORLD STAGE** AS A **SOCIAL** GAMIFICATION TO FOSTER LONG-TERM ENGAGEMENT AND THE ENGAGING NATURE OF **PARTICIPATORY EXPERIENCES.**

THIS BRING US TO THE END OF CHAPTER 6, IN WHICH WE ASKED...

...ARE GAMES **REFLECTIONS** OF LIFE?

GONE HOME
(THE FULLBRIGHT COMPANY, 2013)

...AND IS **LIFE** A GAME?

Felicia
But if the illusion is good enough that I can't tell - how does that affect me?

Felicia
For that matter, how do you know that *you* are even real?

THE DESIGN OF **GAMES**, LIKE THAT OF **INSTRUMENTS**, ARISES OUT OF SOME INNATELY **HUMAN** DESIRE FOR **PLAY**, **EXPRESSION**, AND **STORYTELLING**. IT IS ALSO A RESPONSE TO **TECHNOLOGY**, CO-EVOLVING WITH IT. NOT UNLIKE ART, GAME DESIGN CAN REPRESENT OUR ERNEST ATTEMPT TO **UNDERSTAND OURSELVES**, OUR **EMOTIONS**, AND OUR **NATURE**.

THAT IS TO SAY, WE SHOULD **STRIVE** TO MAKE GAMES, TOYS, GAMIFICATIONS THAT MOTIVATE AND **ELEVATE** US TO BE MORE EXPRESSIVE, THOUGHTFUL, INTERESTED, EMPATHIC; GAMES THAT GET US TO CONSIDER OUR HUMANNESS; GAMES WITH **PATHOS**...

LIKE LOOKING UP INTO **STARRY SKIES** ABOVE, IMAGINING THE CONTOURS OF INVISIBLE GIANTS SWIMMING AMONG THEM...

...GAMES THAT MAKE US **FEEL**.

THAT DRAGON, CANCER

CHAPTER 6 DESIGN ETUDE

DESIGN MUST EVER **ADAPT** TO ITS CONTEXT, FUNCTIONAL-AESTHETIC GOALS, AND **MEDIUM**. WE HAVE PRESENTED SOME **CONSIDERATIONS** FOR ARTFULLY CRAFTING PLAY AND GAMES, AND ARGUED FOR THEIR POTENTIAL TO CREATE **NEW FORMS** OF INTERACTIVE, **ARTFUL EXPERIENCES.** LET'S PUT SOME OF THAT INTO **PRACTICE...**

• PART 1: ANALYZE & ARTICULATE

COMPILE A **LIST** OF YOUR **FAVORITE GAMES**, AS DIVERSE A LIST AS POSSIBLE. THEY CAN BE VIDEO GAMES OR ANYTHING ELSE! ANALYZE EACH OF THEM AND BREAK THEM INTO **MECHANICS, DYNAMICS**, AND **AESTHETICS.** IN OTHER WORDS, ARTICULATE THE UNDERLYING **RULES** DESIGNED INTO THE SYSTEM, THE **BEHAVIORS** THEY ENCOURAGE, AND THE **AESTHETIC** TAKEAWAY -- WHICH TYPES OF AESTHETICS WERE EXPERIENCED?

SOME EXAMPLES

- MINECRAFT
- BRAID
- MONUMENT VALLEY
- FIREWATCH
- EVERYTHING
- LEGEND OF ZELDA
- PAPERS, PLEASE
- WORLD OF WARCRAFT
- JOURNEY
- OVERWATCH
- GETTING OVER IT WITH BENNETT FODDY

IN MY **MUSIC, COMPUTING, DESIGN** STUDIO COURSE, STUDENTS OFTEN ASK "WHAT SHOULD I DESIGN?" OR "WHERE SHOULD I GO FROM HERE?" I OFTEN FIND MYSELF SUGGESTING "ADD CONSTRAINTS, INVENT MECHANICS" AND DISCOVER HOW THEY CHANGE THE DYNAMICS. SO...

• PART 2: IMAGINE & SKETCH

DESIGN A GAME FOR FUN. THINK ABOUT THE **AESTHETIC** AIMS -- WHAT DO YOU WANT YOUR PLAYERS TO FEEL? WORK **BACKWARD** AND INVENT THE **MECHANICS** -- TEST/ IMAGINE/DEDUCE WHAT **DYNAMICS** MIGHT BE FOSTERED AND USE IT TO FURTHER REFINE THE MECHANICS!

• PART 3: GAMIFY

TAKE SOMETHING IN YOUR DAILY LIFE OR WORK -- SOMETHING THAT IS OSTENSIBLY NOT A GAME. DESIGN RULES AROUND THAT TO CHANGE BEHAVIOR FOR YOU OR SOMEONE ELSE!

EXAMPLES

- ACHIEVEMENTS FOR HOUSEHOLD TASKS
- INCENTIVES FOR EXERCISING (E.G., DOING PULL-UPS)
- SEE "IF I CAN DO IT" OR "DO IT MORE EFFICIENTLY"
- GAMIFYING OFFICE BEHAVIOR

BYRON: "I PUT A PULL-UP BAR OUTSIDE MY BATHROOM AND DO 10 PULL-UPS BEFORE USING THE BATHROOM EACH TIME!"

"Man is born free, but everywhere he is in chains."

— **Jean-Jacques Rousseau** (1762)
The Social Contract

FREE TO PLAY

From the moment we are born, we are shackled by a thousand different forces—both social and personal. We are subject to societal institutions, political regimes, social norms, as well as our own limitations and failings.

Yet, in spite of this—or possibly because of it—we all seek to engage in activities, sanctuaries, that protect us from the rest of life, and that are intrinsically valuable for the experience they offer. Such activities include making art, playing music, programming for fun, and any hobby that allows us self-discovery and authenticity.

We play because we choose to play. It is an end-in-itself. This choice reflects that we still possess something of true freedom. Play is what we do to be free.

CHAPTER 7
SOCIAL DESIGN
TECHNOLOGY TO *CONNECT* HUMANS

WE HAVE EXPLORED *ARTFUL DESIGN* IN TERMS OF THE *SENSORIAL* (AUDIO, VISUAL), THE *INTERACTIVE* (INTERFACES), AND THE *PSYCHOLOGICAL* (TOYS AND GAMES) -- FROM THE READILY *PERCEIVABLE* TO THE INCREASINGLY *INVISIBLE*...

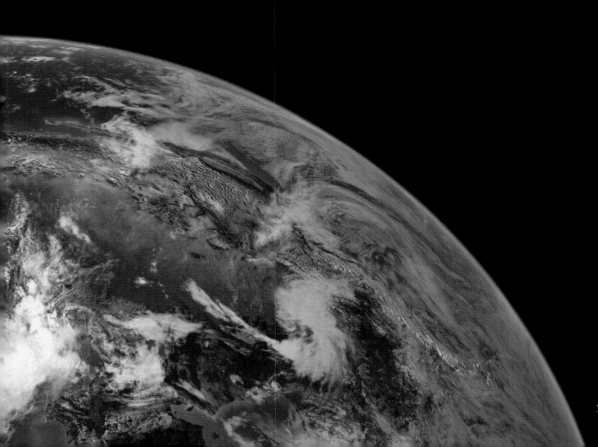

...TO THESE DIMENSIONS WE ADD **SOCIAL DESIGN** -- EXPLORING THE POSSIBILITIES OF TECHNOLOGY TO SHAPE OUR **INTERACTIONS** WITH OUR FELLOW HUMANS, TO **EXPERIENCE** SOMETHING BIGGER THAN OURSELVES. INHERENT WITHIN THIS TECHNOLOGY-MEDIATED SOCIAL EXCHANGE IS THE EMERGENCE OF NEW **SOCIAL DYNAMICS** AND AESTHETICS. MARSHALL MCLUHAN'S TIME-HONORED CONCEPT OF A **GLOBAL VILLAGE** IS MADE AND REMADE INTO A **VIBRANT REALITY** AS WE ARTFULLY SHAPE TECHNOLOGY INTO THE **INTERPERSONAL**, THE **COLLABORATIVE**, THE **COMMUNAL**, AND THE **TRIBAL.**

PRINCIPLE 7.1

DESIGN FOR *HUMAN CONNECTION*

NOT AS A MEANS-TO-AN-END, BUT AS AN END-IN-ITSELF

AS AN INDIVIDUAL, YOU ARE A COMPLETE **ORGANISM** UNTO YOURSELF. BUT YOU ARE ALSO PART OF A **GREATER SOCIAL ORGANISM** -- THE PEOPLE YOU LIVE WITH, WORK WITH, AND PLAY WITH -- AND OF AN EVEN GREATER **GLOBAL** COMMUNITY OF HUMANS. MUCH OF OUR HUMANITY IS DEFINED BY **HOW** WE CHOOSE TO **INTERACT** AND **PARTICIPATE** WITH OUR FELLOW HUMANS. FAMILY, FRIENDS, CO-WORKERS, STRANGERS -- OUR INTERWOVEN SOCIAL LIFE PLAYS A PIVOTAL ROLE IN MAKING US **WHO WE ARE.**

SOCIAL DESIGN, THEREFORE, IS CONCERNED WITH **SHAPING** THE **DYNAMICS** AND **QUALITIES** OF SUCH INTERACTIONS, FORGING **NEW EXPERIENCES** AND RENDERING THEM POSSIBLE, VITAL, MEANINGFUL, AND **AUTHENTIC.** THIS IS ACCOMPLISHED, AS WITH ALL DESIGN, BY AN ARTFUL **INVENTION** OF UNDERLYING CONSTRAINTS, ENVIRONMENTS, AND ROLES. IT IS CARRIED OUT WITH THE UNDERSTANDING THAT THE **SPECIFICITIES** OF A SOCIAL RELATIONSHIP PROFOUNDLY AFFECT THE NATURE OF THE INTERACTION. IN ARTFUL DESIGN, WE MUST BE EVER COGNIZANT OF FOR **WHOM** WE ARE DESIGNING.

 **MODEL 7.2**

RINGS OF FAMILIARITY IN SOCIAL DESIGN

FROM ONE'S SELF OUTWARD TO THE SUM OF HUMANITY, THERE IS A **CONTINUUM** OF FAMILIARITY IN HOW WE RELATE TO ANOTHER PERSON...

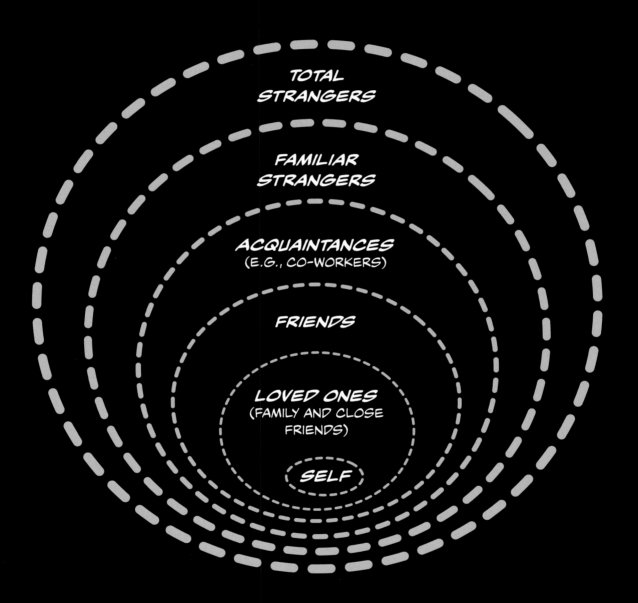

TOTAL
STRANGERS

FAMILIAR
STRANGERS

ACQUAINTANCES
(E.G., CO-WORKERS)

FRIENDS

LOVED ONES
(FAMILY AND CLOSE
FRIENDS)

SELF

... **EACH** OF WHICH CAN BE
DESIGNED FOR, AND DESERVES
TO BE DESIGNED **DIFFERENTLY.**

SOMETHING WE **COULD** DO IN 2016 THAT WE **COULDN'T** IN 1996

NEW JERSEY (MOM & DAD)

BEIJING (GRANDMA)

CALIFORNIA (DINNER)

DESIGNING FOR PEOPLE WE KNOW

IN A WAY, WE ALREADY HAVE AN **IDEA** OF HOW TO CONNECT WITH OUR FAMILY AND CLOSE FRIENDS. WHILE TECHNOLOGY HAS **HELPED** IN THIS REGARD (E.G., I CAN VIDEO-CALL MY FAMILY ON A DIFFERENT CONTINENT), IT IS **NOT** TECHNOLOGY THAT DETERMINES THE **QUALITY** AND **MEANING** OF THOSE SOCIAL **INTERACTIONS** -- IT IS, AT SOME POINT, ENTIRELY UP TO THE **PEOPLE** IN THEM. FOLLOWING THIS LINE OF LOGIC, DESIGNING FOR FAMILIAR PEOPLE SHOULD PROVIDE A **MEDIUM** -- AND THEN **GET OUT OF THE WAY.** FOR EXAMPLE, I DON'T NEED A VIDEO CALL TO FAMILY TO BE **ANY MORE** THAN A REAL-TIME COMMUNICATION CONDUIT -- WE CAN "TAKE IT FROM THERE"!

PRINCIPLE 7.3

TECHNOLOGY SHOULD **STRIVE** TO **GET OUT OF THE WAY** OF HUMAN INTERACTION

THE **MORE FAMILIAR** THE RELATIONSHIP, THE **LESS** THE DESIGN SHOULD IMPOSE ON PEOPLE. SUCH RELATIONSHIPS ARE **RICH** IN THEMSELVES; THEY CAN ONLY TRULY BE IMPROVED **FROM WITHIN**, BY THE PEOPLE IN THEM. TECHNOLOGY CAN **HELP**, BUT AS **INVISIBLY** AS POSSIBLE.

BUT NOT ALL TOOLS CAN SO EASILY "GET OUT OF THE WAY." MODERN CONTRIVANCES (SUCH AS OUR SO-CALLED SOCIAL NETWORKS AND SOCIAL MEDIA) PRESENT MIXED RESULTS: THEY **ENGAGE** US, BUT IT IS UNCLEAR WHETHER THEY MAKE US HAPPIER, MORE **AUTHENTIC**, TRUTHFUL, INTERESTING, OR MORE FULLY APPRECIATIVE OF OUR HUMAN CONNECTIONS. WE CONTINUALLY PAY ATTENTION TO THE **FUNCTIONAL** ASPECTS (NEWS, PHOTO AND VIDEO SHARING), BUT MORE RARELY DO WE CONSIDER THE **AESTHETIC** QUALITIES OF THE ACTIVITIES -- WHETHER THEY **ENRICH** US. EVEN MORE **INSIDIOUS** ARE THE SOCIAL TOOLS THAT ATTEMPT TO ENGAGE US FOR WHAT APPEAR TO BE **VALUELESS ENDS** (E.G., BEYOND THE TOOLMAKER'S OWN ECONOMIC GAIN), PULLING US INTO **ADDICTIVE** BEHAVIORAL PATTERNS WITH **UNKNOWN** EMOTIONAL CONSEQUENCES.

VALUES OF A SOCIAL TOOL

WHAT MAKES IT *VALUABLE* TO US?

USEFULNESS

DOES IT SERVE A UNIQUE SOCIAL *FUNCTION*, NOT FULLY POSSIBLE ANYWHERE ELSE?

FULLNESS OF EXPRESSION

HOW DOES IT ALLOW US TO MORE *FULLY* EXPRESS OURSELVES SOCIALLY, AS PLAYFUL AND EMPATHETIC CREATURES?

AUTHENTICITY

DOES THE TOOL SEEK TO *ELEVATE* US, PROMOTE A SENSE OF *TRUTH* IN *HOW* WE ENGAGE WITH ONE ANOTHER? THE HEAVIER THE "TOUCH" OF THE DESIGN, THE MORE *RESPONSIBILITY* THE DESIGNER BEARS IN SHAPING THE RESULT.

TRANSPARENCY OF USE

IS THE TOOL PRIMARILY ABOUT *ITSELF* -- ITS OWN *PLATFORM*, ITS OWN *GOALS* (E.G., ECONOMIC PROFIT) -- OR IS IT GENUINELY ABOUT THE *PEOPLE* USING IT? FOR EXAMPLE, IS THE SYSTEM *DESIGNED*, FIRST AND FOREMOST, WITH OUR *WELL-BEING* IN MIND? OR... IS IT DESIGNED PRIMARILY TO INCREASE OUR *ENGAGEMENT* WITH THE SYSTEM, SO THAT SUCH ENGAGEMENT CAN BE *MONETIZED* BY THE TOOLMAKER? DOES THE TOOL TREAT ITS USERS AS A MEANS-TO-AN-END -- OR AS AN END-IN-ITSELF?

DESIGNING FOR STRANGERS

AS **INTERESTING** (AND DECEPTIVELY **CHALLENGING**) AS DESIGNING FOR FAMILIAR SOCIAL CONNECTION CAN BE, IT'S THE HUMANITY **BEYOND** THE FAMILIAR THAT REMAINS MUCH LESS OBVIOUS AND UNTAPPED. I BELIEVE THIS IS WHERE THE DESIGN OF TECHNOLOGY HAS SOMETHING **UNIQUE** TO OFFER: CONNECTIVE EXPERIENCES THAT SIMPLY WEREN'T **POSSIBLE** BEFORE.

MUSIC IS A FITTING ACTIVITY TO EXPLORE SOCIAL DESIGN FOR **STRANGERS.** LIKE ALL ART, IT EMBRACES OUR **COMMON HUMANITY**, TRANSCENDING CULTURAL, ECONOMIC, AND POLITICAL DIFFERENCES. WE CAN ENGAGE IN MUSIC WITH OUR FAMILY AND FRIENDS, BUT MUSIC IS ONE OF THOSE THINGS THAT WE ALSO CAN MAKE WITH A **TOTAL STRANGER** -- AND WALK AWAY WITH AN AUTHENTIC SENSE OF **CONNECTION.** AND IT IS THIS TYPE OF **SIMPLE BOND** THAT WE DESIGN FOR FIRST AND FOREMOST, FOR IT IS A PREREQUISITE FOR FURTHER SOCIAL INTERACTION. BEFORE REASON AND CIVILITY, WE NEED TO FIND OUR **COMMON HUMANITY.**

IN THE SAME WAY THAT A **KNOWING NOD** BETWEEN OLD FRIENDS CAN COMMUNICATE SOMETHING WORDS CANNOT, OFTEN THE **SIMPLEST** SOCIAL DESIGN TENDS TO BE THE **MOST SATISFYING.** IT'S THE SOCIAL VERSION OF **LESS IS MORE.**

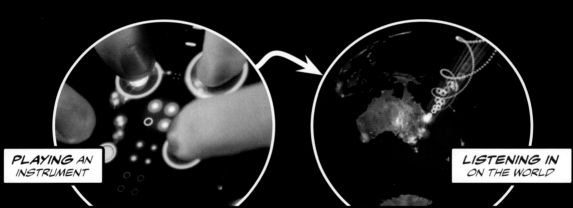

PLAYING AN INSTRUMENT

LISTENING IN ON THE WORLD

OCARINA'S GLOBE IS A SIMPLE **SOCIAL ARTIFACT** -- DESIGNED TO INSTILL A SENSE OF CONNECTION, BETWEEN TOTAL **STRANGERS.** YOU HEAR **ANOTHER** USER PLAY OCARINA, **AROUND** THE WORLD. YOU DO **NOT** KNOW WHO THEY ARE, YOU ONLY SEE THEIR **LOCATION** AND HEAR THE **MUSIC**; YOU MAY ALSO SEE A **NICKNAME** THEY'VE GIVEN THEMSELVES. THAT'S IT. **NO** NAMES, **NO** USER PROFILES, **NO** MESSAGES, **NO** CHAT. YOU CAN'T "FRIEND" ANOTHER USER.

ⒶPRINCIPLE 7.5 SIMPLE IS SATISFYING

SIMPLICITY AND **ANONYMITY** CAN IMBUE A SOCIAL INTERACTION -- HOWEVER MINIMAL (AND PERHAPS **BECAUSE** IT'S MINIMAL) -- WITH A SENSE OF AUTHENTIC **CONNECTION.** UNLIKE MORE COMPLICATED SOCIAL INTERACTIONS THAT MIGHT DILUTE OR DISTRACT FROM THE CORE AESTHETIC AIM, **SIMPLE** INTERACTIONS HELP YOU **FEEL** THE PRESENCE OF ANOTHER HUMAN -- SOMEONE, SOMEWHERE **OUT THERE**...

AS A **SOCIAL** EXPERIENCE, **OCARINA** INVITES **PARTICIPATION** AND A SENSE OF **CONNECTION** IN A **GLOBAL COMMUNITY** (EVEN IF IT'S A GLOBAL COMMUNITY OF PEOPLE BLOWING INTO THEIR PHONES).

ETHNOMUSICOLOGIST **THOMAS TURINO**, IN HIS BOOK **MUSIC AS SOCIAL LIFE: THE POLITICS OF PARTICIPATION**, ARTICULATES AN IMPORTANT DISTINCTION BETWEEN WHAT HE CALLS "**PRESENTATIONAL**" VS. "**PARTICIPATORY**" PERFORMANCE...

WHEREAS **PRESENTATIONAL** SETTINGS NATURALLY EMBODY A CLEAR DELINEATION BETWEEN **PERFORMER** AND **AUDIENCE** (E.G., ORCHESTRA ON STAGE, LISTENERS IN THE AUDIENCE), **PARTICIPATORY** PERFORMANCE INVITES **ACTIVE** CONTRIBUTION TO THE OVERALL SOUND AND MOTION FROM ANY AND **ALL** PRESENT (E.G., A DRUM CIRCLE, DANCE FESTIVALS), WHERE THERE ARE **NO** ARTIST-AUDIENCE DISTINCTIONS.

"DEEPLY PARTICIPATORY EVENTS ARE FOUNDED ON AN **ETHOS** THAT HOLDS THAT EVERYONE PRESENT **CAN**, AND IN FACT **SHOULD**, PARTICIPATE IN THE SOUND AND MOTION OF THE PERFORMANCE. SUCH EVENTS ARE FRAMED AS **INTERACTIVE SOCIAL OCCASIONS**."   -- THOMAS TURINO, **MUSIC AS SOCIAL LIFE**

⚹ **PRINCIPLE 7.6**   *VALUE **PARTICIPATION** (AND **DESIGN** FOR IT)*

IN SUCH PARTICIPATORY SETTINGS, THERE IS GREATER **VALUE** PLACED ON THE **INTENSITY** AND **DEGREE** OF PARTICIPATION. SUCCESS IS MEASURED BY THE **QUALITY** OF HUMAN INTERACTION -- **IN FAVOR** OF THE QUALITY OF THE RESULTING SOUND OR INDIVIDUAL SKILL.

IN THIS CONTEXT, WE MIGHT SEE **OCARINA** AS A SIMPLE FORM OF A TECHNOLOGY-MEDIATED SOCIAL **PARTICIPATORY ECOSYSTEM**, WITH LITTLE DISTINCTION BETWEEN PERFORMER AND AUDIENCE (PARTICIPANTS PLAY BOTH ROLES), AND WHERE **ALL** USERS ARE INVITED TO PLAY, REGARDLESS OF SKILL. UNLIKE PARTICIPATORY MUSIC IN THE TRADITIONAL SENSE, USERS ARE **NOT** IN THE SAME PLACE AT THE SAME TIME. NONETHELESS, THEY SHARE THE SAME **VIRTUAL SPACE**, AND MOREOVER, A SENSE OF **CONNECTION** TO A GREATER, GLOBAL GROUP...

IT EMBODIES A MIXTURE OF THE **PHYSICAL** AND THE **VIRTUAL**. THE PHYSICAL ACT OF **PLAYING** THE INSTRUMENT AND THE VISUALIZATION OF EACH PLAYER'S PHYSICAL GEOGRAPHIC LOCATION WORK IN TANDEM WITH THE VIRTUAL **SOCIAL EXCHANGE** BINDING USERS TOGETHER IN A COMMON LISTENING ACTIVITY. THIS OFFERS A KIND OF **SOCIAL EXPRESSION** NOT PREVIOUSLY POSSIBLE.

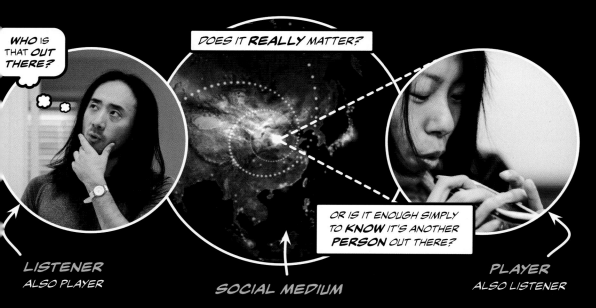

WHO IS THAT **OUT** THERE?

DOES IT **REALLY** MATTER?

OR IS IT ENOUGH SIMPLY TO **KNOW** IT'S ANOTHER **PERSON** OUT THERE?

LISTENER
ALSO PLAYER

SOCIAL MEDIUM

PLAYER
ALSO LISTENER

IT IS IMPORTANT TO NOTE THAT **OCARINA**'S SOCIAL INTERACTION WORKS SPECIFICALLY FOR (AND BECAUSE OF) **STRANGERS** -- NEARLY THE OPPOSITE OF AN **IDENTITY-BASED** SOCIAL NETWORK, WHERE THE CONNECTIONS ARE DEFINED BY **ASSOCIATION** (RATHER THAN BY **PARTICIPATION**). IN THIS SENSE, **OCARINA** IS A KIND OF "**ANTI**" SOCIAL NETWORK, ONE THAT VALUES HUMAN **CONNECTION** BUT WITHOUT THE NECESSITY FOR **IDENTITY**...

...EXPRESSED THROUGH **SHARED** HUMAN EXPERIENCES, HOWEVER **SIMPLE** OR EVEN **MUNDANE**...

PEOPLE DO NOT NEED TO **KNOW** ONE ANOTHER IN ORDER TO SING, DANCE, OR PAINT TOGETHER. HERE, ANONYMITY IS NOT A SHORTCOMING, BUT SOMETHING **MEANINGFUL** AND **VALUED.** IT IS A SOCIAL EXCHANGE IN WHICH **IDENTITY** IS **NOT** CRUCIAL, OR EVEN THAT USEFUL.

I MEAN, **SO WHAT** IF YOU KNOW MY **NAME**? OR WHAT I DO FOR A LIVING, MY **HOBBIES**, AND MY **INTERESTS**? DOES IT REALLY ADD ANYTHING SUBSTANTIAL TO TELLING YOU WHO I **REALLY** AM? ON THE OTHER HAND, IF WE MAKE MUSIC, SING, OR DANCE TOGETHER, THAT'S AN ACTIVE FORM OF **BONDING** -- TO HAVE TAKEN PART IN A **COMMON ACTIVITY!**

(人) PRINCIPLE 7.7

A LITTLE **ANONYMITY** CAN GO A LONG WAY

ANONYMITY CAN BE **POWERFUL** AND **LIBERATING.** OUR SOCIAL EQUATION IS CONSTANTL BEING BALANCED BETWEEN VALUES PLACED ON **PRIVACY** AND AN INNATE DESIRE TO **CONNECT** WITH OTHERS, OUR SOCIAL EXISTENCE CONSISTING OF AN ONGOING INTERPLA BETWEEN THE **PERSONAL** AND THE **INTERPERSONAL.** WITHIN THAT DYNAMIC AR SITUATIONS IN WHICH WE DON'T **WANT** TO KNOW AND DON'T **NEED** TO KNOW, IN ORDER TO HAVE A MEANING EXPERIENCE.

SOMETIMES WE DON'T WANT **SOCIAL PRESSURE** FROM THOSE WHO KNOW US -- IF I PLA A SONG ON **OCARINA**, FOR EXAMPLE, I MAY NOT WANT TO BE "JUDGED" BY MY FRIENDS O IDENTITY-BASED SOCIAL NETWORKS. BUT I NONETHELESS DESIRE TO **CONNECT**, T EXPRESS SOMETHING, WITH ANOTHER HUMAN BEING. WE AS DESIGNERS OF SOCIA SYSTEMS MUST STRIVE TO **UNDERSTAND** SOCIAL SENSITIVITIES -- AND AT THE SAME TIM FIND NEW WAYS, MORE **POETIC** WAYS, TO SPEAK TO OUR SOCIAL DESIRES AND INSTINCTS.

DESIGN "ANTI" SOCIAL AND OMNI-SOCIAL NETWORKS! (人) PRINCIPLE 7.8

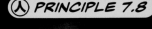

I DON'T MEAN ANTISOCIAL IN TERMS OF **ISOLATING** PEOPLE, BUT MORE LIKE "**ANTI**" SOCIAL-NETWORK NETWORKS!

IT CAN GET AROUND MUCH OF THE **AWKWARDNESS** OF DESIGNING FOR PEOPLE WHO ALREADY **KNOW** EACH OTHER -- GOING **BEYOND** IDENTITY AND EXPLORING POSSIBILITIES FOR MEANINGFUL, AUTHENTIC CONNECTION BETWEEN PEOPLE, REGARDLESS OF THEIR FAMILIARITY.

TO SEE THESE IDEAS IN MOTION, LET'S LOOK AT SOME MORE SOCIAL DESIGNS THAT DERIVE THEIR ESSENTIAL EXPERIENCE FROM *PARTICIPATION* AND *ANONYMITY.*

ZEPHYR (2008)

FULLY **ANONYMOUS** AUDIOVISUAL GLOBAL GREETING CARD SYSTEM!

ZEPHYR WAS A *SOCIAL EXPERIMENT* BASED ON THE IDEA OF *ANONYMOUS CHAIN LETTERS*, MADE AUDIO AND VISUAL. ORIGINALLY CONCEIVED FOR *HOLIDAY GREETINGS*, *ZEPHYR* LETS USERS *SKETCH* WITH THEIR FINGER, THEIR GESTURES *SONIFIED* THROUGH *WINDY* MUSICAL SOUNDS (CHORDS SYNTHESIZED USING NOISE THROUGH *COMB FILTERS*, A LA CHAPTER 4). EACH SKETCH, INCLUDING ALL *INTERMEDIATE GESTURES*, IS RECORDED INTO A "*ZEPHYR-GRAM*" AND SENT INTO THE TECHNOLOGICAL *ETHER*, TO BE RECEIVED BY A *RANDOM USER* SOMEWHERE IN THE WORLD.

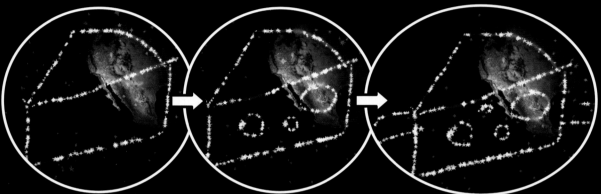

ON THE *RECEIVING* SIDE, A USER SEES (AND HEARS) THE *ZEPHYR*-GRAM *MATERIALIZE*, GESTURE BY GESTURE, IN THE SAME WAY AS *ORIGINALLY DRAWN.* THIS GIVES AN UNCANNY GHOST-LIKE FEELING OF SOMEONE DRAWING BEFORE YOU...

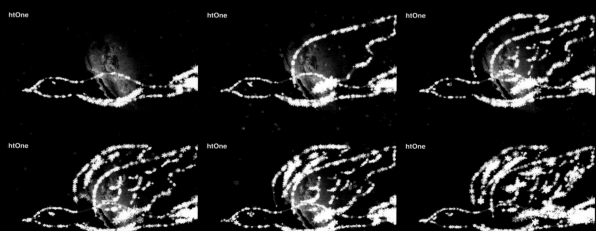

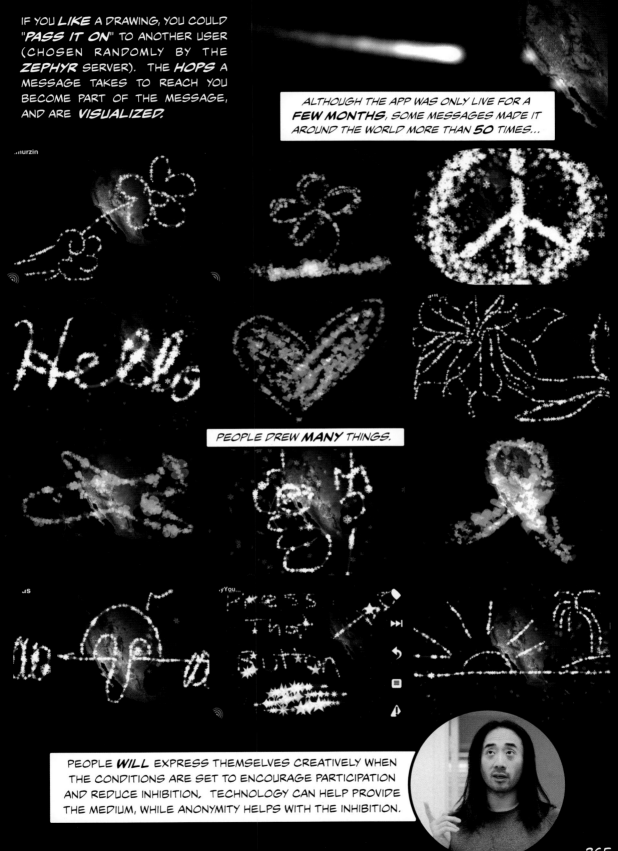

IF YOU *LIKE* A DRAWING, YOU COULD "*PASS IT ON*" TO ANOTHER USER (CHOSEN RANDOMLY BY THE *ZEPHYR* SERVER). THE *HOPS* A MESSAGE TAKES TO REACH YOU BECOME PART OF THE MESSAGE, AND ARE *VISUALIZED.*

ALTHOUGH THE APP WAS ONLY LIVE FOR A *FEW MONTHS*, SOME MESSAGES MADE IT AROUND THE WORLD MORE THAN *50* TIMES...

PEOPLE DREW *MANY* THINGS.

PEOPLE *WILL* EXPRESS THEMSELVES CREATIVELY WHEN THE CONDITIONS ARE SET TO ENCOURAGE PARTICIPATION AND REDUCE INHIBITION. TECHNOLOGY CAN HELP PROVIDE THE MEDIUM, WHILE ANONYMITY HELPS WITH THE INHIBITION.

MAGIC PIANO
SOCIAL MODES

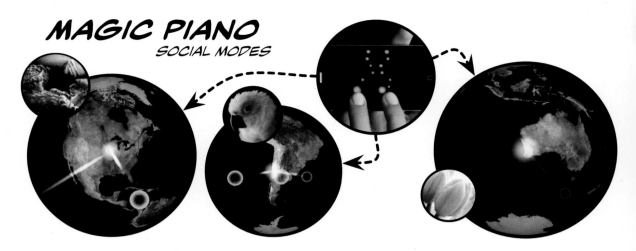

IN ADDITION TO AN **OCARINA**-ESQUE **GLOBE** TO LISTEN TO OTHERS PLAY, **MAGIC PIANO**, ONCE UPON A TIME, INCLUDED A **REAL-TIME** SOCIAL INTERACTION THAT ALLOWED TWO USERS TO **PLAY TOGETHER**... SORT OF.

ONE-TO-ONE MODE

IN THIS (NOW-DEFUNCT) FEATURE, PLAYERS AROUND THE WORLD GET **PAIRED** WITH ONE ANOTHER IN A FREESTYLE **CALL-AND-RESPONSE** MODE -- WHERE THEY CAN PLAY SIMPLE MUSICAL RIFFS BACK AND FORTH IN REAL TIME.

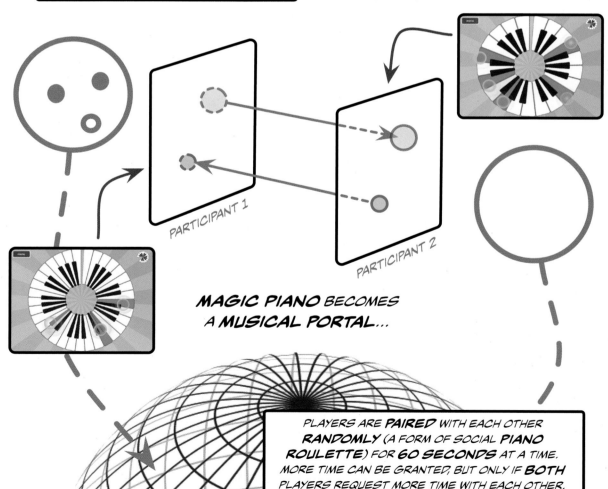

PARTICIPANT 1

PARTICIPANT 2

MAGIC PIANO BECOMES A **MUSICAL PORTAL**...

PLAYERS ARE **PAIRED** WITH EACH OTHER **RANDOMLY** (A FORM OF SOCIAL **PIANO ROULETTE**) FOR **60 SECONDS** AT A TIME. MORE TIME CAN BE GRANTED, BUT ONLY IF **BOTH** PLAYERS REQUEST MORE TIME WITH EACH OTHER.

THIS EXPERIMENTAL MODE IN **MAGIC PIANO** IS NOT SO EASY TO PLAY -- FOR ONE, IT'S TO BE PLAYED ON **RIDICULOUSLY SHAPED** VIRTUAL **KEYBOARDS** (WHICH, AT ONE TIME, I THOUGHT WERE A GOOD IDEA).

THIS SPIRAL IS THE **WORST** -- AN UNNECESSARILY DIFFICULT FORM!

VISUALLY, THIS **CIRCULAR** KEYBOARD SLOWLY **BREATHES,** THE OPENING IN THE MIDDLE SLOWLY WIDENS AND CONTRACTS... IT IS STRANGELY **UNSETTLING.**

AS THE DESIGNER, I CAN SAY THESE WHIMSICAL KEYBOARD FORMS ARE ULTIMATELY **DESIGN FAILURES** -- DUE TO THEM BEING HILARIOUSLY AND NEEDLESSLY DIFFICULT TO PLAY (FINE EXAMPLES OF FORM **USURPING** FUNCTION!). FORTUNATELY, THE **GAME MODE** STRIKES A MUCH BETTER **BALANCE** BETWEEN FORM, FUNCTION, FUN, AND "MAGIC." AS NOVELTY AND PERHAPS ALSO REMINDERS THAT BETTER DESIGNS ARE SURELY OUT THERE, THESE WARPED KEYBOARD FORMS ARE KEPT IN **MAGIC PIANO** AS A **SOLO MODE,** FOR THE MUSICALLY MASOCHISTIC...

NO TWO KEYS ARE THE **SAME SIZE,** AND THEIR SPACING IS COMPLETELY **NONLINEAR!** UHH **GOOD LUCK** PLAYING ANYTHING ON **THIS!**

THIS MIGHT BE AN EXAMPLE OF TOO MUCH **GEEKING OUT** ABOUT SPIRALS AND CIRCLES, AND FORGETTING THE DESIGN'S INTENDED **FUNCTION.**

...AND, APPARENTLY, **CATS.** IT WAS DISCOVERED THAT, ON OCCASION, CATS VENTURED INTO **TOYING** WITH THE **MAGIC PIANO** -- POSSIBLY AS AN EXTENSION OF THEIR NATURAL AFFINITY FOR WALKING ON PIANOS. I THINK NOW THAT PERHAPS THE 1-ON-1 MODE IN **MAGIC PIANO** MIGHT HAVE BEEN BETTER FOR **FELINES.** HUMANS CAN LOAD UP **MAGIC PIANO** AND HAVE THEIR CATS CALL AND RESPOND TO OTHER CATS, PROVIDING SECONDS OF AMUSEMENT!

CATS, PIANOS, INTERNET... THE FUTURE?

ANOTHER EXAMPLE OF DESIGN BUILDING MEANING OUT OF **ANONYMITY**: TAKING A **PULSE** OF HUMANITY...

WE FEEL FINE (2008)
BY JONATHAN HARRIS & SEP KAMVAR

better
good
guilty
bad
free
fine
sad

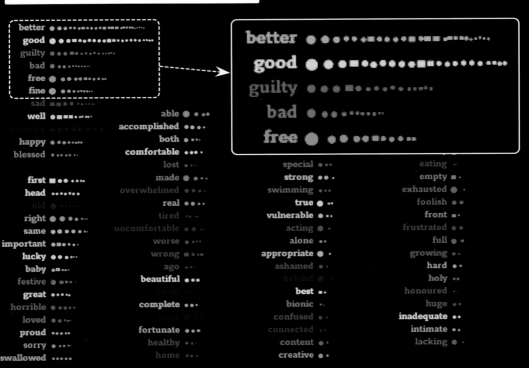

better
good
guilty
bad
free

well   light
country   able   locked
happy   accomplished   missed
blessed   both   necessary
                        comfortable   needed
first   lost   special   eating   new
head   made   strong   empty   nice
old   overwhelmed   swimming   exhausted   nostalgic
right   real   true   foolish   obsolete
same   tired   vulnerable   front   off
important   uncomfortable   acting   frustrated   okay
lucky   worse   alone   full   passing
baby   wrong   appropriate   growing   power
festive   ago   ashamed   hard   prepared
great   beautiful   behind   holy   protected
horrible   clean   best   honoured   pulled
loved   complete   bionic   huge   ready
proud   done   confused   inadequate   regular
sorry   fortunate   connected   intimate   related
swallowed   healthy   content   lacking   renewed
                               home   creative

"BREAKDOWN OF FEELINGS FROM PEOPLE IN THE LAST FEW HOURS..."

WE FEEL FINE MIGHT BE DESCRIBED AS AN INTERNET-MEDIATED **EMOTION BAROMETER** -- A SYSTEM THAT AUTOMATICALLY **CRAWLS** THE LATEST WEB LOGS AND SOCIAL MEDIA TO SEARCH FOR OCCURRENCES OF THE PHRASES "**I FEEL**" AND "**I AM FEELING.**" UPON FINDING SUCH INSTANCES, THE SYSTEM **RECORDS** THE FULL SENTENCE AND **CLASSIFIES** THE GENERAL EMOTION EXPRESSED (E.G., HAPPY, SAD, ETC.). ADDITIONAL CONTEXT, WHEN AVAILABLE, IS ALSO RECORDED -- TIME, GEOGRAPHICAL REGION, GENDER, AGE, AND EVEN THE LOCAL **WEATHER CONDITIONS** AT THE TIME.

AS A SOFTWARE ARTIFACT, **WE FEEL FINE** OFFERS **SIX** DISTINCT PERSPECTIVES (CALLED "**MOVEMENTS**") TO SLICE, DICE, AND PRESENT THE DATA -- FROM SWIRLING PARTICLES REPRESENTING INDIVIDUAL SENTIMENTS TO MOUND HISTOGRAMS OF EMOTIONS AT A GIVEN TIME AND PLACE... THERE IS **QUIET MAGIC** AT WORK HERE.

AN INTERNET-BASED **BAROMETER & DATABASE** OF HUMAN SENTIMENT!

WE FEEL FINE IS AN ARTFUL TOOL FOR *MEANINGFUL MEANDERING* IN THE COLLECTIVE EXPRESSION OF EVERYDAY EMOTIONS, AN ARTWORK CREATED BY *EVERYONE* POSTING ON THE INTERNET. A TECHNOLOGY-MEDIATED *REFLECTION* OF WHAT'S ON OUR MINDS AND IN OUR SENTIMENTS, IT EVOLVES WITH THE TIMES AND OUR INDIVIDUAL LIVES, AS AN EXPLORATORY LENS OF EMPATHY AND SOLIDARITY.

FOR EXAMPLE, WE CAN *SAMPLE* HOW PEOPLE FELT ON *NEW YEAR'S EVES* IN 2008, 2010, 2012... HERE ARE SOME OF THE RESULTS I FOUND:

. **i feel good about what i was able to accomplish this year**
December 31, 2008 / from someone

. **i feel that 2007 was way better than 2008**
December 31, 2008 / from a 28 year old in washington united states

. **i feel like i am always waiting**
December 31, 2008 / from a 27 year old

. **i feel ready**
December 31, 2010 / from a male in renton wa united states when it was cloudy

. **i feel out of place**
December 31, 2010 / from someone in colorado

. **i feel like i really grew into my own this year**
December 31, 2010 / from a female in united states

. **i feel like i live my life by sitting back and letting it happen**
December 31, 2010 / from a female in united states

. **i feel like i am the luckiest person in the world to still be pursuing my dreams**
December 31, 2010 / from a female in los angeles united states when it was sunny

. **i feel compelled to record every good thing that happens**
December 31, 2010 / from someone

i feel greeeat
December 31, 2012 / from someone

. **i feel better already**
December 31, 2012 / from someone

. **i feel awful for leaving it behind**
December 31, 2012 / from someone

. **i feel fine**
December 31, 2012 / from someone

. **i feel soooo blessed**
December 31, 2012 / from someone

. **i feel that strongly about the bread**
December 31, 2012 / from someone

. **i feel despaired helpless and lost**
December 31, 2012 / from someone

. **i feel that i am succeeding within my classroom**
December 31, 2012 / from someone in bourbonnais illinois

. **i feel like i'm currently in a space of indecision**
December 31, 2012 / from someone

. **i feel like i'm wound so tight**
December 31, 2012 / from someone

. **i feel like screaming and crying on the inside because i don't know how to deal with the awful sensations**
December 5, 2012 / from someone

. **i feel in not having the need to prove myself to anyone**
December 31, 2012 / from someone

. **i feel your dreams and in the dark i hear your screams**
December 31, 2008 / from someone

. **i feel as though it is my job to help others**
December 31, 2010 / from a female in dunwoody atlanta georgia united states

. **i feel less obliged to protect any made up version of myself**
December 31, 2008 / from someone

. **i feel like i'm going crazy**
December 31, 2012 / from someone

. **i feel fuzzy**
December 31, 2012 / from someone

FEELINGS ARE A FUNNY THING. WHILE WE MIGHT *FEED* OR *SHUN* OUR EMOTIONS, WE CANNOT *CHOOSE* THEM. IN A SENSE, *ART* IS THE AESTHETIC BY-PRODUCT OF OUR HUMAN *DESIRE* TO UNDERSTAND OURSELVES AND OUR EMOTIONS.

WE FEEL FINE IS AN EXAMPLE OF A **CROWDSOURCING** SOCIAL TOOL THAT HINTS AT THE **VAST POTENTIAL** OF USING TECHNOLOGY AS A UNIFYING **MEDIUM**, A **PLATFORM** TO HARNESS SOMETHING INTERESTING, UNIQUE, AND EXPRESSIVE FROM A **POPULATION** OF PEOPLE. INDEED, THERE HAVE BEEN ENTIRE PLATFORMS CREATED FOR THIS GENERAL GOAL -- FOR EXAMPLE...

amazon
mechanical turk™
Artificial Artificial Intelligence

(CAME ONLINE 2004)

THE **PREMISE** OF **AMAZON'S MECHANICAL TURK** IS THAT THERE EXIST **COMPLEX TASKS** THAT ARE **NATURALLY EASIER** FOR **HUMANS** THAN FOR MACHINES (INSPIRING THE TAGLINE, "ARTIFICIAL ARTIFICIAL INTELLIGENCE"). SUCH TASKS MIGHT INCLUDE "EDIT THIS REPORT," "IDENTIFY THE EMOTIONAL VIBE IN AN IMAGE," OR "DRAW A SHEEP."

Ⓛ PRINCIPLE 7.9

USE **TECHNOLOGY** TO **HARNESS** THE POWER OF THE **HUMAN**

THE **ORIGIN** OF THE NAME **MECHANICAL TURK** COMES, INTERESTINGLY, FROM THE 18TH CENTURY **CHESS-PLAYING AUTOMATON**, WHICH TURNED OUT TO BE ONE OF HISTORY'S GREATEST TECHNOLOGY-MEDIATED **HOAXES.** THIS MECHANICAL AUTOMATON, WHICH BESTED MANY CHESS MASTERS ACROSS EUROPE, TURNED OUT TO **NOT** BE SO AUTOMATED, BUT HAD A **SMALL HUMAN CHESS MASTER** INGENIOUSLY **HIDDEN** INSIDE, CONTROLLING THE CHESSBOARD THROUGH AN INTRICATE NETWORK OF MIRRORS AND LEVERS. THIS RESULTED IN MUCH INFAMY FOR THE TURK'S INVENTOR, **WOLFGANG VON KEMPELEN**, OVERSHADOWING HIS OTHER NOTABLE INVENTIONS, INCLUDING AN INGENIOUS (NON-HOAX) MACHINE FOR **MECHANIZED SPEECH SYNTHESIS!**

THE
TURK (1770)

BY **WOLFGANG VON KEMPELEN**

COMPLEX INNER WORKINGS, PRIMARILY FOR THE HUMAN CHESS MASTER TO **OPERATE** THE MACHINERY...

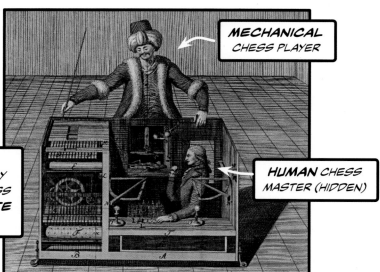

MECHANICAL CHESS PLAYER

HUMAN CHESS MASTER (HIDDEN)

THE TERM **MECHANICAL TURK** HAS SINCE BEEN ADOPTED AS AN EXPRESSION FOR MACHINES PURPORTED TO DO A **FULLY AUTOMATED TASK**, BUT THAT ARE ACTUALLY DONE BY A HUMAN BEHIND THE SCENES. **AMAZON** ADOPTED THIS CONCEPT AS A TECHNOLOGY-MEDIATED **PLATFORM** FOR **HARNESSING** PEOPLE'S NATURAL PROWESS FOR CERTAIN TYPES OF PROBLEM-SOLVING. TECHNOLOGY PLAYS A KEY ROLE -- **NOT** AS **PROBLEM SOLVER** BUT AS **FACILITATOR** TO UNLOCK HUMAN INTELLIGENCE. IT IS A POIGNANT REMINDER, IN AN AGE (ONCE-AGAIN) FASCINATED BY TECHNOLOGY, TO NOT FORGET THE VALUE OF **HUMAN** INTELLIGENCE.

THAT CERTAIN COMPUTATIONAL PROBLEMS ARE NATURALLY **HARD** FOR COMPUTERS BUT **EASY** FOR **HUMANS** IS A **PROFOUND** IDEA. WHILE **VON KEMPELEN** PERHAPS INTUITIVELY UNDERSTOOD THIS, THE FORMAL THEORETICAL, COMPUTATIONAL FOUNDATIONS WERE LAID OUT BY FOLKS LIKE CARNEGIE MELLON PROFESSOR **LUIS VON AHN**, WHO COINED THE TERM "**HUMAN COMPUTATION**."

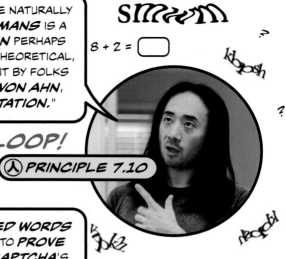

$8 + 2 = \boxed{}$

PUT **HUMANS** IN THE **LOOP!**

SOME TASKS ARE NATURALLY **HARD FOR COMPUTERS**, BUT **EASY FOR HUMANS**. DESIGN TO FIND THE **BALANCE**.

⚖ **PRINCIPLE 7.10**

FOR EXAMPLE, TASKS OR PUZZLES (LIKE **MANGLED WORDS** A USER HAS TO IDENTIFY AND TYPE) ON WEBSITES TO **PROVE** THE USER IS **HUMAN**. THESE PUZZLES, CALLED **CAPTCHA**'S (WHICH STANDS FOR **C**OMPLETELY **A**UTOMATED **P**UBLIC **T**URING TEST TO TELL **C**OMPUTERS AND **H**UMANS **A**PART), ARE DESIGNED TO **CONFOUND** FULLY AUTOMATED COMPUTER RECOGNITION SYSTEMS, BUT BE **EASY** FOR HUMANS.

enter sum: 12   Go

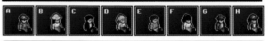

Please prove you are more than a mindless spam-bot by identifying who gets the beer!

THE CONCEPT EASILY EXTENDS BEYOND WORD IDENTIFICATION TO ANY TYPE OF PROBLEM WHERE HUMANS HAVE A **NATURAL ADVANTAGE** NOT ONLY IN SOLVING, BUT IN **UNDERSTANDING** THE CHALLENGE TASK IN THE FIRST PLACE!

Give food to the baby.

Put all the fish in the ocean.

Place food in the refrigerator.

OCCASIONALLY, THERE ARE CAPTCHAS THAT PROVE CHALLENGING FOR **BOTH** COMPUTERS AND HUMANS, ALBEIT FOR LIKELY DIFFERENT REASONS...

⚡BZZZ BEEP⚡
!@%???!?!!

Just to prove you are a human, please answer the following math challenge.

Q: Calculate:
$$\left. \frac{\partial}{\partial x}\left[5 \cdot \sin\left(5 \cdot x - \frac{\pi}{2}\right) + 4 \cdot \cos(3 \cdot x) \right] \right|_{x=0}.$$

A: _____
mandatory

UH... WUT THE

AS A PRACTICAL APPLICATION OF **HUMAN COMPUTATION**, VON AHN AND **LAURA DABBISH** DESIGNED **GAMES WITH A PURPOSE** (GWAPs) -- A FORM OF **SOCIAL GAMIFICATION** AIMED AT HARNESSING HUMAN COMPUTATION TO SOLVE COMPLEX PROBLEMS THAT COMPUTERS **CANNOT YET EASILY DO.**

GWAPS: GAMES WITH A PURPOSE

BY **LUIS VON AHN** & (2004) **LAURA DABBISH**

GWAPS ARE DESIGNED TO ENCOURAGE:
1. "**COMPUTATION**" IN THE HUMAN
2. **CORRECTNESS** OF OUTPUT
3. PLAYER **ENGAGEMENT** & RETENTION

what do you see?

taboo words

man

grass

animals

score

300

guesses

sky

green

sheperd

shepherd

sheep_

+ submit   ⇒ pass

matched on: sheep

IN ONE SUCH GWAP, CALLED **THE ESP GAME**, ONLINE PLAYERS ARE **PAIRED AT RANDOM** (THIS MUTUAL ANONYMITY IS **CRUCIAL** FOR THIS TO WORK) AND ARE SHOWN A **SERIES** OF IMAGES. IN ORDER TO ADVANCE PAST EACH IMAGE, BOTH PLAYERS MUST HAVE **INDEPENDENTLY** GUESSED THE **SAME WORD** -- FOR EXAMPLE, "SHEEP." THE **GOAL** IN EACH GAME IS TO GET THROUGH AS MANY IMAGES AS POSSIBLE IN 3 MINUTES (TO DEMONSTRATE YOU AND YOUR ANONYMOUS PARTNER HAVE HIGH "ESP" POTENTIAL!). THE SIDE EFFECT (OR "PURPOSE") IS THAT IMAGES GET **LABELED** WITH GREAT ACCURACY AND **SEMANTIC PLAUSIBILITY.**

OVER TIME, THE GAME **ADAPTS** FOR EACH IMAGE, AS COMMONLY-USED WORDS BECOME "**TABOO**", FORCING PLAYERS TO FIND SEMANTICALLY **DEEPER** WORDS FOR EACH IMAGE. FOR EXAMPLE "PERSON," "MAN," "FOOD" MIGHT BECOME TABOO, LEADING TO MORE COMPLEX CONCEPTS SUCH AS "COOK," "KITCHEN," "STIR FRY," AND "BUSY."

taboo words
person
food
man

match: cook

match: kitchen

match: stir fry

match: busy   **match:** wok

THE **COMPUTATIONAL** TASKS PERFORMED BY HUMANS INCLUDE **IMAGE PROCESSING**, **VISUAL SCENE ANALYSIS**, **OBJECT IDENTIFICATION**, AND **AFFECT/TONE RECOGNITION.** THE INDEPENDENCE OF TWO STRANGERS ARRIVING AT THE SAME WORDS IS A VERIFICATION OF **CORRECTNESS** (HOW MANY PAIRS AGREE ON A GIVEN WORD IN AN IMAGE CAN EVEN BE USED AS A MEASURE OF **CONFIDENCE**). THAT IT IS A **GAME** WITH **GOALS** (A TEST OF "ESP-NESS" AND LEADERBOARDS) KEEPS PEOPLE **ENGAGED.**

IN FACT, THIS HAS BEEN USED TO LABEL **HUNDREDS OF MILLIONS** OF IMAGES ONLINE (POWERING MORE **RELEVANT** IMAGE SEARCHES). EVEN THOUGH NO PLAYER WAS PAID (AFTER ALL, IT IS A GAME!), SOME PEOPLE DEVOTED UPWARDS OF **40 HOURS A WEEK** TO IT. THE IDEA WAS THAT PEOPLE DIDN'T HAVE TO KNOW THE BIGGER PICTURE OR PURPOSE -- THEY CONTRIBUTED SIMPLY BY PLAYING.

A KEY TAKEAWAY IS THAT, AT THE END OF THE DAY, THESE TASKS ARE TO BE **USED BY HUMANS**, AND SO WHATEVER HUMANS GENUINELY ASSOCIATE WITH AN IMAGE **ARE** OFTEN THE VERY ASSOCIATIONS WE MIGHT FIND **MEANINGFUL** IN TEXT-BASED IMAGE RETRIEVAL. MACHINES ARE THE **MEDIUM** AND **FACILITATOR**, BUT THEY DO NOT **COMPREHEND** OR **VERIFY** THE END RESULTS THEMSELVES.

FOLLOW-UP GAMES CAN BE DESIGNED. FOR EXAMPLE, THERE ARE GWAPS THAT TAKE THE WORDS FROM **THE ESP GAME** FOR EACH IMAGE AND ASK RANDOMLY PAIRED PEOPLE TO DRAW **WHERE** IN THE IMAGE THESE WORDS "OCCUR," WHICH, OVER TIME AND WITH ENOUGH INPUT, BECOMES A POWERFUL **IMAGE SEGMENTATION** TOOL! THERE ARE SIMILAR GAMES THAT CLASSIFY AND ANALYZE IMAGES, MUSIC, VIDEO -- AND TOOLS THAT HELP A COMMUNITY COLLECTIVELY CRAFT **CREATIVE ARTIFACTS**, LIKE A CROWDSOURCED STORY, PAINTING, OR COMPOSITION...

THE SHEEP MARKET
BY AARON KOBLIN (2010)

10,000 SHEEP CREATED BY **CROWDS** OF ONLINE WORKERS, ON AMAZON'S **MECHANICAL TURK.**

WHAT?

BAAA!

EACH WORKER WAS PAID $.02 (US), WITH A SET OF SIMPLE COMPUTER DRAWING TOOLS, AND THE **INSTRUCTION** TO...

"DRAW A SHEEP FACING LEFT."

*DO ARTFUL DESIGNERS **DREAM** OF CROWDSOURCED SHEEP?*

SOME **STATISTICS**
PUBLISHED BY THE AUTHOR

FEE PER SHEEP: $0.02

AVERAGE **TIME** SPENT DRAWING EACH SHEEP: 105 SECONDS

AVERAGE **WAGE**: $0.69 / HOUR

REJECTED SHEEP: 662

COLLECTION **PERIOD**: 40 DAYS

COLLECTION **RATE**: 11 SHEEP / HOUR

*IT'S A NEW INTERFACE FOR **OVINE** EXPRESSION!*

FANCY WATCH

CANE (walnut)

A **WHIMSICAL** TESTAMENT TO...

...THE POWER OF THE **CROWD**

THROUGH THESE CASE STUDIES, WE SEE TWO RELATED SOCIAL DESIGN MODELS: DESIGNING TO **CREATE** COMMUNITY (E.G., OCARINA), VERSUS...

...DESIGNING **WITH** COMMUNITY (E.G., ZEPHYR, GWAPS, SHEEP).

THIS IS MAKING ME **SLEEPY**...

WHY IS EVERYONE FACING **LEFT**?

BAAA HUMBUG.

BAA !

THIS TYPE OF CROWD-BASED DESIGN IS REALLY THE *ART* OF SHAPING TECHNOLOGY WITH A *POPULATION* OF HUMANS IN THE LOOP, AND IS VERY SPECIFIC ABOUT WHAT THE TECHNOLOGY DOES AND DOES NOT DO. IT REMINDS US OF THE NOTION...

PRINCIPLE 7.11A *THAT WHICH* ***CAN*** *BE AUTOMATED* ***SHOULD BE***

HOWEVER, BY ITSELF IT IS *INCOMPLETE*, AND NEEDS A COMPLEMENTARY PRINCIPLE:

PRINCIPLE 7.11B *THAT WHICH* ***CANNOT*** *BE* ***MEANINGFULLY*** *AUTOMATED SHOULD* ***NOT*** *BE*

IN GENERAL, ACTIVITIES FROM WHICH WE DERIVE INTRINSIC VALUE -- SUCH AS *SOCIAL INTERACTIONS*, THE ACT OF *PLAY*, AND *ART* -- ARE *MEANINGFUL* PRECISELY BECAUSE WE *PUT EFFORT* INTO (AND *CHOOSE* TO DO) THEM. SUCH ACTIVITIES ARE OFTEN *EXPERIENTIAL* IN NATURE, AND THEY TEND TO *LOSE* THEIR ESSENTIAL QUALITIES IF WE TRY TO AUTOMATE THEM. (FOR EXAMPLE, WOULD YOU WANT SOMETHING ELSE TO *PLAY* IN YOUR PLACE? WOULDN'T THAT DEFEAT THE POINT OF PLAYING?)

PRINCIPLE 7.12 *NOT* EVERYTHING WORTHWHILE IS A *PROBLEM* TO BE *SOLVED*

AS SOCIETY GRAPPLES WITH THE EVOLUTION OF *ARTIFICIAL INTELLIGENCE* AND *MACHINE LEARNING*, IT'S PERHAPS TOO EASY TO FRAME EVERYTHING AS *ANOTHER PROBLEM* TO BE SOLVED, BY A MORE *CLEVER* ALGORITHM, OR A MORE *POWERFUL* SYSTEM. WE SHOULD REMEMBER THAT THINGS LIKE *ART, MUSIC,* AND *HUMAN RELATIONSHIPS* ARE *NOT* PROBLEMS TO BE SOLVED, AND THAT INSTEAD OF *SOLUTIONS*, WE NEED *TOOLS* TO HELP US WORK AT THESE THINGS *OURSELVES* -- TOOLS THAT *UNDERSTAND* US.

WILL IT *RAIN?*

≈BAA≈

376

WE HURTLE OURSELVES FORWARD TECHNOLOGICALLY, AS MORE AND MORE OF OUR DAILY LIFE **SHIFTS** FROM THE **PHYSICAL** TO THE **VIRTUAL**... HOWEVER, THAT DOES **NOT** MEAN IT HAS TO BE **LESS HUMAN**, OR ANY LESS **CONNECTED**...

SOME THINGS REMAIN THE **SAME.** TECHNOLOGY EVOLVES RAPIDLY, BUT HUMANS -- OR, MORE ACCURATELY, **HUMAN NATURE** -- CHANGES MUCH MORE SLOWLY.

人

HERE IS OUR **REN** -- THE CHINESE CHARACTER FOR "**PEOPLE**" OR "**HUMAN.**"

X 3

"THE OLD COMPUTING IS ABOUT WHAT **COMPUTERS** CAN DO; THE NEW COMPUTING IS ABOUT WHAT **PEOPLE** CAN DO."

-- BEN SHNEIDERMAN

THE THINGS WE **DESIGN** WITH **TECHNOLOGY** DO NOT CHANGE OUR **NATURE**, BUT THEY **RECOMBINE** WITH OUR HUMANNESS, PAVING THE WAY FOR NEW BEHAVIORS, SENSATIONS, EXPERIENCES, AND **CONNECTIONS** WITH OTHERS IN AN EVER-EVOLVING **GLOBAL VILLAGE**...

众

YOU PUT **THREE** REN'S TOGETHER, YOU GET **ZHONG**, WHICH MEANS "**CROWD**" OR "**MASS.**" (I GUESS IN CHINESE, THREE **IS**, LITERALLY, A CROWD!)

AND THEN THERE IS THE **CLOUD** (TECHNOLOGICALLY SPEAKING), CONNECTING PEOPLE IN A SINGLE INFORMATIONAL MEDIUM!

CROWDS AND **CLOUDS** -- IT RHYMES, AND IS A SIGN OF OUR TIMES!

云

YUN
"CLOUD"

OKAY. NEXT LET'S TAKE SOME OF THESE IDEAS -- ANONYMITY, PARTICIPATION, CROWDSOURCING, DESIGNING ROLES AND RULES -- AND **ROLL THEM UP**, SOMEHOW **SMUSHING** THEM INTO A SINGLE ARTIFACT.

ONCE UPON A TIME, BACK IN THE **EARLY DAYS** OF APP-BASED MOBILE PHONES, THERE WAS AN **EXPERIMENT** IN MASSIVELY MULTIUSER, MULTI-ROLE, CROWDSOURCED **SOCIAL-MUSICAL** GAME DESIGN...

LEAF TROMBONE
WORLD STAGE
(2009-2010)

LEAF TROMBONE BEGAN LIFE AS A GAME-LIKE **SEQUEL** TO THE ORIGINAL **OCARINA**...

THE **LEAF TROMBONE** WAS A **WHIMSICAL** INSTRUMENT, INSPIRED BY **LEAF WHISTLES** AND **LEAF BLOWING** (WHICH, BY THE WAY, ARE FOUND IN MANY CULTURES AND SOMETIMES EVEN USED FOR COURTSHIP)!

INITIALLY, **LEAF TROMBONE** WAS CONCEIVED **WITHOUT** THE NOTION OF A SOCIAL WORLD STAGE.

IT WAS DESIGNED AS A MUSICALLY EXPRESSIVE **GAME**, PERHAPS MOST SIMILAR IN **MECHANICS** TO **MAGIC FIDDLE** (EVEN THOUGH **LEAF TROMBONE** PRE-DATES **MAGIC PIANO** AND **MAGIC FIDDLE**).

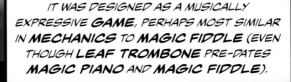

LEAF TROMBONE (INSTRUMENT MODE)

AS *INSTRUMENT* AND *GAME*, *LEAF TROMBONE* CONSISTS OF A LEAF-LIKE VIRTUAL TROMBONE *SLIDE*, ACCOMPANIED BY AN ANIMATED *ROTARY MUSIC BOX*. LIKE *OCARINA*, *LEAF TROMBONE* IS PLAYED BY *BLOWING* INTO THE PHONE. UNLIKE *OCARINA*, THE TROMBONE SLIDE PROVIDES *CONTINUOUS* PITCH CONTROL, CAPABLE OF SMOOTH GLISSANDI IN A PITCH RANGE OF THREE OCTAVES.

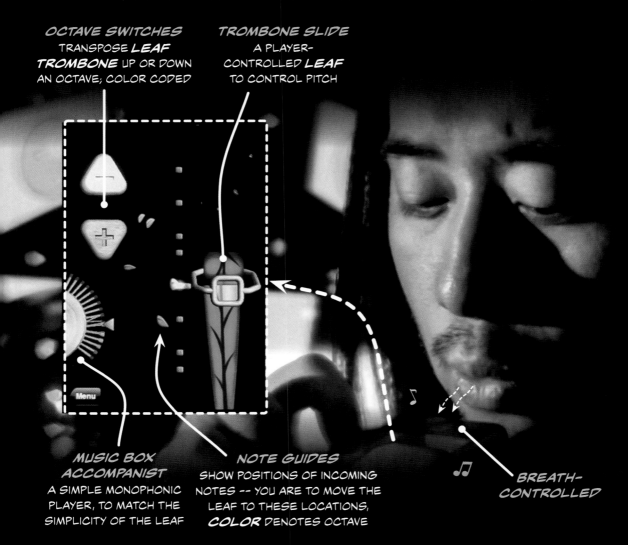

OCTAVE SWITCHES
TRANSPOSE *LEAF TROMBONE* UP OR DOWN AN OCTAVE; COLOR CODED

TROMBONE SLIDE
A PLAYER-CONTROLLED *LEAF* TO CONTROL PITCH

MUSIC BOX ACCOMPANIST
A SIMPLE MONOPHONIC PLAYER, TO MATCH THE SIMPLICITY OF THE LEAF

NOTE GUIDES
SHOW POSITIONS OF INCOMING NOTES -- YOU ARE TO MOVE THE LEAF TO THESE LOCATIONS, *COLOR* DENOTES OCTAVE

BREATH-CONTROLLED

IN *GAME* MODE, ANIMATED SCROLLING *NOTE GUIDES* PROMPT THE PLAYER WHERE TO POSITION THE *LEAF TROMBONE* SLIDE AND WHEN. THESE MARKERS ARE *COLOR-CODED* (MAGENTA, GREEN, CYAN) TO INDICATE THE OCTAVE FOR EACH NOTE. PERFORMANCES CAN COME ACROSS AS BOTH *SKILLFUL* AND *COMICAL.* THE WHIMSICAL INSTRUMENT IMBUES POOR PLAYING WITH OPPORTUNITY FOR *HUMOR* -- MISSING A NOTE SOMEHOW SEEMS FUNNIER WHEN THE PLAYER THEN TRIES TO *GLIDE* TO THE CORRECT PITCH.

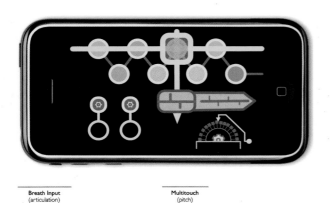

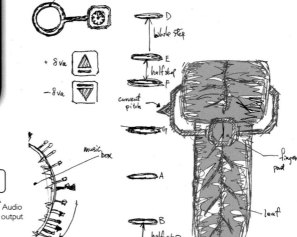

Breath Input
(articulation)

Multitouch
(pitch)

TriOsc
(modulator)

SinOsc
(carrier)

NRev
(reverberator)

Audio output

controls index of modulation

OnePole
(rough envelope)

Step
(secondary envelope)

OnePole
(low-pass filter)

(primary envelope generation)

THE DESIGN UNDERWENT A NUMBER OF *ITERATIONS* ON THE *INTERFACE* OF THE LEAF TROMBONE, THE *SOUND* SYNTHESIS MODEL, AND THE GAME *MECHANICS,* BUT ONE QUESTION KEPT RECURRING...

HOW DO WE GIVE PLAYERS SATISFYING -- AUTHENTIC "HUMAN" -- *FEEDBACK* ABOUT WHAT THEY PLAY?

SURE, WE COULD *AUTOMATICALLY* SCORE THEM BASED ON *ACCURACY* AGAINST A PREEXISTING SCORE, BUT...

... GIVEN THE WHIMSICAL, EXPRESSIVE, AND *GOOFY* PERSONALITY OF *LEAF TROMBONE*, THERE SEEMED MORE *MEANINGFUL CRITERIA* IN JUDGING THE *VALUE* OF A PERFORMANCE -- QUALITIES LIKE *EXPRESSIVENESS, SURPRISE, IRONY.* I MEAN, PLAYING BEETHOVEN'S "ODE TO JOY" ON *LEAF TROMBONE* REQUIRED, SURELY, A *DIFFERENT* MINDSET THAN PLAYING, SAY, EUROPE'S "FINAL COUNTDOWN"...

IN SHORT, HOW CAN *LEAF TROMBONE* RECOGNIZE THAT A PIECE OF MUSIC IS *EPIC, IRONIC, AWESOME, ROCKIN',* OR EVEN *GOOD* OR *BAD,* IN A WAY THAT'S MEANINGFUL TO PEOPLE?

PERPLEXING

FUNNY

ROCKIN'

THUS WAS BORN THE **WORLD STAGE!** WE AIM FOR VERY HUMAN CONCEPTS, SO PERHAPS WE NEED TO DESIGN **HUMANS INTO THE LOOP**, WITH A LITTLE **HELP** FROM TECHNOLOGY.

TAKING A PAGE OUT OF **HUMAN COMPUTATION**, AND INSTEAD OF TRYING TO USE THE COMPUTERS TO "SOLVE" THESE FUNDAMENTALLY SUBJECTIVE PROBLEMS, **LEAF TROMBONE** USED THE COMPUTER AS A VEHICLE TO **EXTRACT** NATURAL HUMAN **SENTIMENT**, **EXPERTISE**, AND **JUDGMENT**.

I MEAN, WE DON'T NEED **EXPERT** MUSICAL REVIEWS, BUT SIMPLY FOR PEOPLE TO GIVE THEIR NATURAL **IMPRESSION** OF WHAT THEY HEAR. IT'S **EASY** FOR PEOPLE TO FORM AN INTUITIVE OPINION OF MUSIC THEY HEAR, SO MUCH SO IT'S NEARLY IMPOSSIBLE **NOT** TO MAKE A JUDGMENT! THIS LED TO THE **WORLD STAGE** AS AN EXPERIMENT IN MAKING PEOPLE INTO HIGH-QUALITY (OR AT LEAST INTERESTING AND NUANCED) MUSICAL **JUDGES** AND **CASUAL CRITICS.**

You

1

(judges)

WORLD STAGE

(Contestant)

ORIGINAL RUDIMENTARY **WORLD STAGE** SKETCH

THIS IS A TEXTBOOK EXAMPLE OF USING COMPUTERS NOT AS THE **BRAINS** BUT AS THE **CONNECTIVE TISSUES** OF THE EXPERIENCE, TO OFFER MEANINGFUL SOCIAL INTERACTION AMONG **PEOPLE.**

THE ANONYMITY OF THIS **WORLD STAGE** PARADIGM IS BOTH A LOGISTIC NECESSITY (TO RECRUIT ENOUGH JUDGES) AND A PSYCHOLOGY CONSIDERATION FOR A FEW REASONS. FIRST, ANONYMITY CAN HELP REDUCE PERFORMERS' INHIBITION -- OPERATING ON THE ASSUMPTION THAT PEOPLE ARE MORE LIKELY TO EXPRESS THEMSELVES CREATIVELY TO STRANGERS, THAN TO PEOPLE THEY KNOW! AND, SECOND...

...THERE IS A SENSE OF **AUTHENTICITY** IN GETTING **HONEST FEEDBACK** FROM STRANGERS, UNFILTERED BY WELL-MEANING ENCOURAGEMENT (OR TROLLNESS) OF FAMILY AND FRIENDS. WHILE STRANGERS CERTAINLY CAN BE CRUEL, SOMEHOW GETTING **CASUALLY COMPLIMENTED** BY A STRANGER ON **LEAF TROMBONE** CARRIES A DIFFERENT SATISFACTION, IF ONLY FROM KNOWING THE STRANGER HAS NO MOTIVE TO LIE (OR BE NICE) -- THAT IS, THE **POSITIVE COMMENTS** FROM STRANGERS ARE ARGUABLY MORE MEANINGFUL BECAUSE THEY HAVE NO EXTRINSIC AGENDA -- THEY **MEAN** WHAT THEY **SAY.**

THE **WORLD STAGE** CONSISTS OF **JUDGING SESSIONS**, LIKE A CROWDSOURCED VERSION OF *AMERICAN IDOL* OR *BRITAIN'S GOT TALENT* -- **THREE** JUDGES (RANDOMLY ASSEMBLED FROM **LEAF TROMBONE** USERS) LISTEN, AT THE SAME TIME, TO A PRE-RECORDED PERFORMANCE FROM ANOTHER USER, AND THEY OFFER THEIR FEEDBACK **THROUGHOUT** THE SONG.   THEY GIVE COMMENTS AND EXPRESS THEMSELVES THROUGH ANIMATED **EMOTICONS.**   AS PART OF THE DESIGN, JUDGES ARE MEANT TO **INFLUENCE** ONE ANOTHER.   AS A PERFORMER, SEEING THREE JUDGES "ROCK OUT" IN UNISON TO YOUR **LEAF TROMBONE** RENDITION OF "IRONMAN" IS **SOMETHING** TO BEHOLD! THE ENTIRE PROCEEDING IS RECORDED FOR REPLAY.

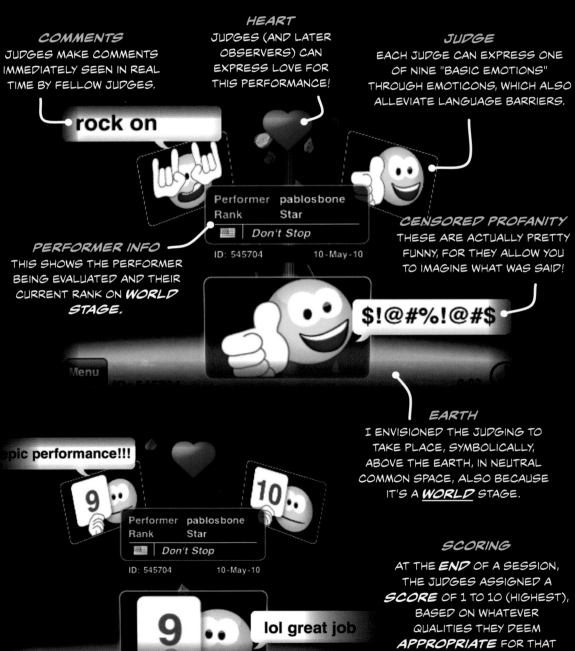

COMMENTS
JUDGES MAKE COMMENTS IMMEDIATELY SEEN IN REAL TIME BY FELLOW JUDGES.

HEART
JUDGES (AND LATER OBSERVERS) CAN EXPRESS LOVE FOR THIS PERFORMANCE!

JUDGE
EACH JUDGE CAN EXPRESS ONE OF NINE "BASIC EMOTIONS" THROUGH EMOTICONS, WHICH ALSO ALLEVIATE LANGUAGE BARRIERS.

PERFORMER INFO
THIS SHOWS THE PERFORMER BEING EVALUATED AND THEIR CURRENT RANK ON **WORLD STAGE.**

CENSORED PROFANITY
THESE ARE ACTUALLY PRETTY FUNNY, FOR THEY ALLOW YOU TO IMAGINE WHAT WAS SAID!

EARTH
I ENVISIONED THE JUDGING TO TAKE PLACE, SYMBOLICALLY, ABOVE THE EARTH, IN NEUTRAL COMMON SPACE, ALSO BECAUSE IT'S A **WORLD** STAGE.

SCORING
AT THE **END** OF A SESSION, THE JUDGES ASSIGNED A **SCORE** OF 1 TO 10 (HIGHEST), BASED ON WHATEVER QUALITIES THEY DEEM **APPROPRIATE** FOR THAT PERFORMANCE.

DEPENDING ON WHO YOU ASK, HUMANS EXHIBIT ANYWHERE BETWEEN ZERO(!), FOUR, SIX, EIGHT, OR MORE DIFFERENT *TYPES* OF *BASIC EMOTIONS.* PSYCHOLOGIST *PAUL EKMAN* IDENTIFIED *SIX* BASIC EMOTIONS -- HAPPINESS, SADNESS, FEAR, ANGER, SURPRISE, AND DISGUST (FIVE OF WHICH, BY THE WAY, WERE PRESENT IN THE PIXAR FILM *INSIDE OUT*). AS WHIMSICAL VARIATIONS ON THESE, *WORLD STAGE* PROVIDES JUDGES WITH *NINE* "SENTIMENTS" FOR EMOTIONAL AFFECTS.

THUMB UP!

DELIGHT
"YAY, WOW"

DISGUST
"BOO"

UPROARIOUS
"LOL"

PERPLEXED
"WTF IS THIS?"

DOOD! *ROCKIN' OUT* IS TOTALLY A BASIC EMOTION!

GET YER SONIC LIGHTERS OUT!

BOREDOM
≥ZZZ≤

NEUTRAL
"..."

ROCKIN' OUT

LIGHTER

EACH ONE GOES THROUGH A SIMPLE *ANIMATION LOOP.*

THESE *EMOTICONS* GET AROUND POTENTIAL LANGUAGE DIFFERENCES AMONG JUDGES AND PERFORMER! AND THEY LEND SOME *LEVITY* TO THE JUDGING SESSION!

WORLD STAGE IS A SELF-CONTAINED SOCIAL ECO-SYSTEM. IN ADDITION TO THE ROLES OF PERFORMER AND JUDGE, ONE CAN ALSO BE AN OBSERVER, WHO CAN SURF AROUND AND REPLAY ANY JUDGING SESSION IN FULL, OR A COMPOSER, WHO CAN CREATE AND PUBLISH LEAF TROMBONE CONTENT TO BE PLAYED BY THE COMMUNITY.

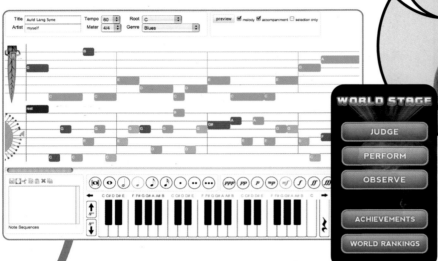

SUBMIT PERFORMANCES (COST: 1 TOKEN)

ROLE: COMPOSER

USERS CAN CREATE SCORE-BASED, **LEAF TROMBONE**-SPECIFIC MUSICAL CONTENT, PUBLISHED TO ALL USERS; PARTICIPANTS GAIN EXPERIENCE AS **COMPOSERS** OR **ARRANGERS.** THIS REPRESENTS A SMALL BUT IMPORTANT SEGMENT OF **WORLD STAGE** USERS.

PROVIDE CONTENT

ROLE: PERFORMER

PLAYERS PERFORM THE USER-GENERATED CONTENT ON THEIR OWN AND CAN CHOOSE TO **SUBMIT** PERFORMANCES TO THE **WORLD STAGE**; PARTICIPANTS GAIN EXPERIENCE AS **PERFORMERS.** MOST USERS TAKE ON THIS ROLE AT SOME POINT!

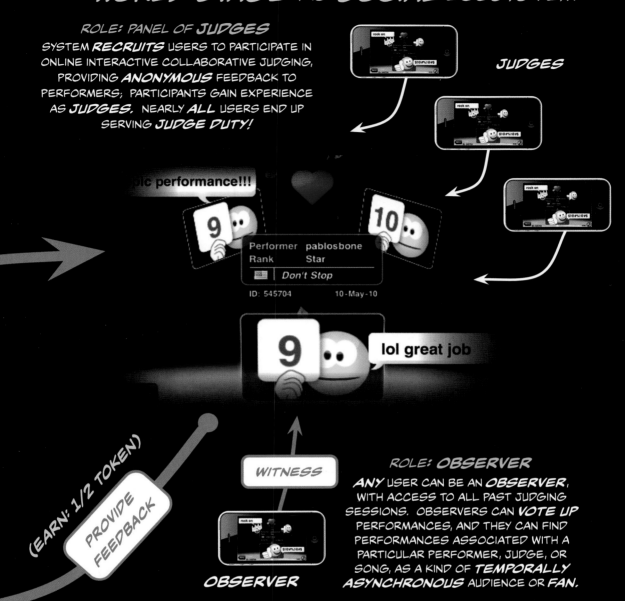

ROLE: PANEL OF JUDGES

SYSTEM **RECRUITS** USERS TO PARTICIPATE IN ONLINE INTERACTIVE COLLABORATIVE JUDGING, PROVIDING **ANONYMOUS** FEEDBACK TO PERFORMERS; PARTICIPANTS GAIN EXPERIENCE AS **JUDGES**. NEARLY **ALL** USERS END UP SERVING **JUDGE DUTY!**

JUDGES

...bic performance!!!

Performer   pablosbone
Rank   Star
Don't Stop
ID: 545704   10-May-10

lol great job

(EARN: 1/2 TOKEN)

PROVIDE FEEDBACK

WITNESS

OBSERVER

ROLE: OBSERVER

ANY USER CAN BE AN **OBSERVER**, WITH ACCESS TO ALL PAST JUDGING SESSIONS. OBSERVERS CAN **VOTE UP** PERFORMANCES, AND THEY CAN FIND PERFORMANCES ASSOCIATED WITH A PARTICULAR PERFORMER, JUDGE, OR SONG, AS A KIND OF **TEMPORALLY ASYNCHRONOUS** AUDIENCE OR **FAN.**

PERFORMING ON **WORLD STAGE** IS INTENDED TO BE A **RECURRING** AND VALUABLE ACTIVITY. PLAYERS ARE ENCOURAGED TO **PRACTICE** A SONG AS MUCH AS THEY WOULD LIKE, BUT THEY HAVE TO EXPEND A SPECIAL **PERFORMANCE TOKEN** IN ORDER TO SUBMIT A PERFORMANCE TO THE **WORLD STAGE**. PERFORMANCE TOKENS CAN BE **EARNED** OVER TIME OR BY PARTICIPATING AS A **WORLD STAGE** JUDGE (I.E., **JUDGE DUTY**). THIS SIMPLE **MECHANIC** CREATES A **BALANCE** WHERE USERS ARE MOTIVATED TO BOTH PERFORM AND JUDGE. ALSO, BY **ADJUSTING** THE VALUE OF PERFORMANCE TOKENS, WE CAN ENCOURAGE MORE JUDGING OR PERFORMANCE AS NEEDED, MAINTAINING **EQUILIBRIUM** IN THE SYSTEM. IN **LEAF TROMBONE: WORLD STAGE**, WE ARRIVED AT, THROUGH EXPERIMENTATION, A **RATIO** OF 2 TO 1 -- EACH PLAYER HAS TO JUDGE **TWICE** FOR EACH CHANCE TO PERFORM.

THERE ARE ALSO ELEMENTS OF **PERIPHERAL GAMIFICATION**...

...A PLAYER CAN **LEVEL UP** SEPARATELY AS BOTH PERFORMER AND JUDGE...

...AND EARN **ACHIEVEMENTS.**

RANK	PERFORMER	JUDGES
1	AMATEUR	NOBODY
2	STUDENT	PEANUT GALLERY
3	BAR PERFORMER	AFICIONADO
4	BACKUP PLAYER	ACADEMIC
5	LOUNGE ACT	PANELIST
6	EXPERT	CRITIC
7	PRODIGY	ARBITER
8	STAR	AUTHORITY
9	INTL. SENSATION	JUDGE
10	PEOPLE KNOW ME	MASTER JUSTICE

Judge danivobailon
Rank Panelist
Fierce

Judge irsi
Rank Academic
Fair

Judge sevenlyons
Rank Authority
Friendly

You are about to observe:

Performer pablosbone
Rank Star
Don't Stop

ID: 545704   10-May-10

SHIVER ME TIMBERS! AT THE BEGINNING OF EACH JUDGING SESSION, THE **RANK** AND **NATIONALITY** OF EACH JUDGE AND PERFORMER IS DISPLAYED. BY THE WAY, THIS BE THE OLD **SMULE OFFICE** IN PALO ALTO, CA, WHERE **LEAF TROMBONE** WAS MADE, BACK IN THE DAY. ⸗ARRRRRG⸗

THIS HERE BE GE'S **DESK**...

⸗RAKAKA⸗ I AM **CORNHOLIO!** ARE YOU THREATENING ME?!

ME? I'M A **PIRATE LAMP** THAT **SINGS**, ACQUIRED AT A SALE AT **FRY'S ELECTRONICS.**

DURING **LEAF TROMBONE: WORLD STAGE**'S BRIEF EXISTENCE IN THE APP WORLD FROM 2009-2010, USERS COMPOSED AND ARRANGED **7,000 SONGS**, PERFORMING AND ADJUDICATING IN OVER **600,000 JUDGING SESSIONS**.

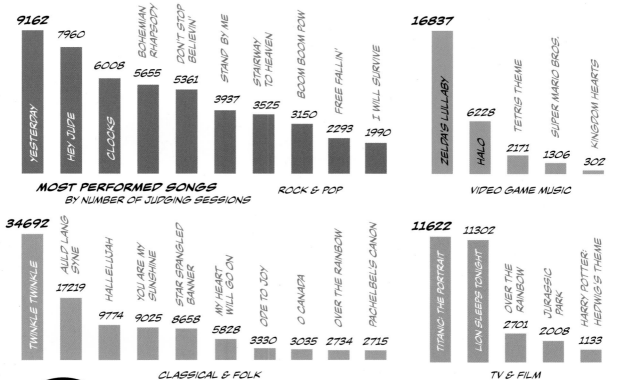

MOST PERFORMED SONGS
BY NUMBER OF JUDGING SESSIONS

ROCK & POP

9162 YESTERDAY
7960 HEY JUDE
6008 CLOCKS
5655 BOHEMIAN RHAPSODY
5361 DON'T STOP BELIEVIN'
3937 STAND BY ME
3525 STAIRWAY TO HEAVEN
3150 BOOM BOOM POW
2293 FREE FALLIN'
1990 I WILL SURVIVE

VIDEO GAME MUSIC

16837 ZELDA'S LULLABY
6228 HALO
2171 TETRIS THEME
1306 SUPER MARIO BROS.
302 KINGDOM HEARTS

CLASSICAL & FOLK

34692 TWINKLE TWINKLE
17219 AULD LANG SYNE
9774 HALLELUJAH
9025 YOU ARE MY SUNSHINE
8658 STAR SPANGLED BANNER
5828 MY HEART WILL GO ON
3330 ODE TO JOY
3035 O CANADA
2734 OVER THE RAINBOW
2715 PACHELBEL'S CANON

TV & FILM

11622 TITANIC: THE PORTRAIT
11302 LION SLEEPS TONIGHT
2701 OVER THE RAINBOW
2008 JURASSIC PARK
1133 HARRY POTTER: HEDWIG'S THEME

JUDGES SHARED 2.9 MILLION **EMOTICONS** AND 1.8 MILLION TEXT **COMMENTS**. HERE ARE THE **MOST COMMON** 100 WORDS OR WORD SEQUENCES USED IN COMMENTS (EXCLUDING 100 MOST FREQUENT WORDS IN WRITTEN ENGLISH). I REALLY LIKE #51, #93, #9... AND #99 -- UHH GOOD JOB EXPRESSING AN OPINION, PEOPLE.

1	SONG	21	WTF	41	YA	61	HEAR	81	MAN
2	LOL	22	OFF	42	NOTES	62	LOVE THIS	82	DID
3	HI	23	PRETTY	43	LITTLE	63	F*CK	83	HORRIBLE
4	BAD	24	SUCK	44	JOB	64	YEA	84	LOVE THIS SONG
5	OK	25	COOL	45	HERE	65	BIT	85	NEED
6	HEY	26	HAHA	46	YES	66	DUDE	86	BEST
7	THIS SONG	27	OMG	47	I DON'T	67	A LITTLE	87	TUNE
8	NICE	28	HEARD	48	SUCKS	68	KEEP	88	I HATE
9	STOP	29	PLAY	49	LONG	69	NEEDS	89	OKAY
10	BETTER	30	I LOVE	50	HATE	70	NEVER	90	MUCH
11	OH	31	MORE	51	MY EARS	71	OUCH	91	SORRY
12	DO	32	PRETTY GOOD	52	PLEASE	72	SAME	92	U SUCK
13	DON'T	33	HELLO	53	HARD	73	TRY	93	IT'S OK
14	LOVE	34	GUYS	54	ROCK	74	WHATS	94	GOOD JOB
15	AGAIN	35	SOUNDS	55	GOD	75	PLAYING	95	SH*T
16	TOO	36	VERY	56	THE SONG	76	RIGHT	96	WHERE
17	GREAT	37	PRACTICE	57	THATS	77	GUY	97	GOT
18	WOW	38	AWESOME	58	WHY	78	HMM	98	SOUND
19	YEAH	39	REALLY	59	WAT	79	YO	99	IDK
20	NOT BAD	40	EARS	60	SUP	80	SONG IS	100	FAR

MOST PROLIFIC **COMPOSER**:
UPLOADED **177** NEW ARRANGEMENTS!

MOST PROLIFIC **PERFORMER**:
SUBMITTED **2,639** PERFORMANCES TO THE WORLD STAGE!!

MOST PROLIFIC **JUDGE**:
PRESIDED OVER **10,537** JUDGING SESSIONS!!!

SOME PEOPLE REALLY **ENGAGED** WITH **WORLD STAGE**, TO THE EXTREME...

WHO **ARE** THESE PEOPLE?!

(I MEAN, IT'S NOT LIKE THEY WERE BEING PAID TO DO THIS...)

I THINK THIS SPEAKS TO THE **POTENTIAL** OF GAMES WITH A PURPOSE, OF TECHNOLOGY AS A **CONNECTOR** AND **MOTIVATOR** OF **PEOPLE**, WHO END UP DOING THE "INTERESTING WORK."

IN A WAY, ARTFUL DESIGNERS ARE AD HOC, **APPLIED PSYCHOLOGISTS** AND **SOCIAL ENGINEERS**, SHAPING CONDITIONS FOR PEOPLE TO BEHAVE -- TO **WANT** TO BEHAVE -- IN SUCH A WAY AS TO FULFILL **ABIDING ROLES** THAT CONSTITUTE AN ECOSYSTEM, THAT MAKE IT **TICK**.

IT SPEAKS ALSO TO **ANONYMITY**, OF TAPPING INTO THE DORMANT SOCIAL POTENTIAL IN US...

∻HMM∻ DIGGIN' THIS **DONUT**... AND THERE'S THAT SAME PERSON I'VE SEEN HERE FOR YEARS.

FAMILIAR STRANGER

...AND ALL **AROUND** US IN OUR BUSY LIVES...

FAMILIAR STRANGER

WE DESIGN FOR USERS, RELATIVE TO THEIR LOVED ONES **AND** THEIR TOTAL STRANGERS, AND EVERY SHADE OF **FAMILIARITY** OR **ANONYMITY** IN BETWEEN. FOR EXAMPLE, THE **SAME STRANGERS** WE SEE, EVERY DAY ON THE COMMUTE TO WORK...

IN THE 1970s, PSYCHOLOGIST **STANLEY MILGRAM** FORMULATED THE NOTION OF THE **FAMILIAR STRANGER** -- SOMEONE WE OBSERVE FREQUENTLY BUT DO NOT KNOW, OR EVER QUITE **ACTIVELY INTERACT** WITH. IT IS A **PASSIVE**, NONVERBAL, VISUAL, AND TACIT RELATIONSHIP DEFINED **SIMULTANEOUSLY** BY FAMILIARITY AND ANONYMITY.

FAMILIAR STRANGERS SEMI-PREDICTABLY SHARE **TIME** AND **SPACE**; THEY ARE THE PEOPLE...

...YOU SEE AT YOUR LOCAL COFFEE SHOP

÷ZZZ÷

...ON YOUR DAILY COMMUTE TO WORK

...PASS BY ON YOUR WAY HOME

... IN LINE AT BREAKFAST

...NEARLY EVERYWHERE IN LIFE

(☮) PRINCIPLE 7.13 *DESIGN* FOR *FAMILIAR ANONYMITY*

IN AN AGE OF COMPUTERS, SOCIAL NETWORKS, BLOGOSPHERES, WIKIS, INTERNET FORUMS, ONLINE GAMES, VIRTUAL AND AUGMENTED REALITIES, PEOPLE ARE NOT NECESSARILY AWARE OF ALL THE **FAMILIAR STRANGERS** IN THEIR LIVES (E.G., "FRIENDS" IN **WORLD OF WARCRAFT**). YOU DON'T KNOW THEM IN PERSON, BUT YOU OPERATE IN THE **SAME TIME** AND (VIRTUAL) **PLACE**. IT IS A SIGN OF OUR TIMES, POISED ONLY TO INTENSIFY. IT IS A POWERFUL **DYNAMIC**, THE DESIGN OF WHICH INVARIABLY LEAVES US AFFECTED, FOR BETTER OR WORSE.

ON SOME LEVEL, **THIS** IS NOT **THAT** DIFFERENT FROM MY LIFE "OUTSIDE."

YUP, I'M SITTING ON MY COUCH IN "REAL" LIFE, TOO. I **LOVE** SITTING.

÷SIGH÷ **VIRTUAL** LIFE IS NO LESS A POINTLESS, FUTILE STRUGGLE... UNYIELDING STRIFE WITHOUT **MEANING** OR **PURPOSE**... BUT, WITH **WAAAY** BETTER WARDROBE POTENTIAL!

HELL IS **OTHER AVATARS.**

IS ALL OF THIS... **ECHOES** FROM "THE FUTURE"?

VIRTUAL EXISTENCE?

NOTHING. A WHOLE LOT OF NOTHING...

389

MUCH OF THE **SOCIAL MECHANISM** IN OUR ORIGINAL **VISION** FOR **SMULE** AND IN SOFTWARE LIKE **CONVERGE** WAS SPECIFICALLY DESIGNED AS EXPERIENCES **FOR** FAMILIAR STRANGERS. THE AIM IS **NOT** TO GET THEM TO BECOME "FRIENDS" PER SE, BUT TO FIND MEANINGFUL **CONNECTIVE EXPERIENCES** BETWEEN PEOPLE, WITH OR WITHOUT IDENTITY, BY PROVIDING A CONTEXT FOR **SHARED** ACTIVITIES. TO THIS END, THE EXPRESSIVE ACT OF MUSIC-MAKING IS A POTENT MEDIUM FOR FAMILIAR STRANGERS, AND **WOULD-BE** FAMILIAR STRANGERS...

INDEED, **MUSIC-MAKING**, IN THE PARTICIPATORY SENSE, **REVEALS** HOW EASY IT CAN BE FOR STRANGERS TO HAVE A COMMON EXPERIENCE, AND -- AT LEAST WHILE THE MUSIC LASTS -- HOW UTTERLY **UNNECESSARY** OUR "IDENTITIES" AND "AFFILIATIONS" ARE, COMPARED TO THE SENSE OF **CONNECTION** TO A GREATER GROUP.

IT SEEMS TO PROVIDE US WITH A DIFFERENT, MORE TACIT AND NUANCED FORM OF **BONDING** WITH ANOTHER HUMAN BEING, WITHOUT KNOWING NAMES, AGES, LINES OF WORK, FAVORITE MOVIES -- AS THOSE THINGS BY THEMSELVES DO NOT FULLY CAPTURE **WHO** WE ARE.

MUSIC, DANCE, **ART** CAN BIND AND CONNECT PEOPLE WITHOUT **AFFILIATION** AND **POLITICS**, HELP US FIND SOMETHING OF OUR **SHARED HUMANITY**. IN AN EVOLUTIONARY SENSE, THIS MAY BE **ART'S** FIRST AND LAST **PRACTICAL PURPOSE**.

Performer pablosbone
Rank   Star
🇺🇸 | *Don't Stop*
ID: 545704   10-May-10

$!

PEOPLE WHO MAKE MUSIC TOGETHER CANNOT BE **ENEMIES**, AT LEAST WHILE THE MUSIC LASTS!

BY THE WAY, SUCH SHARED EXPERIENCES **DO NOT** NECESSARILY NEED MEDIATION FROM **TECHNOLOGY**, THOUGH TECHNOLOGY CAN SERVE AS A **TOOL** TO CONNECT US AND TO SET THE CONDITIONS FOR **AUTHENTIC** DISCOURSE.

PAUL HINDEMITH

390

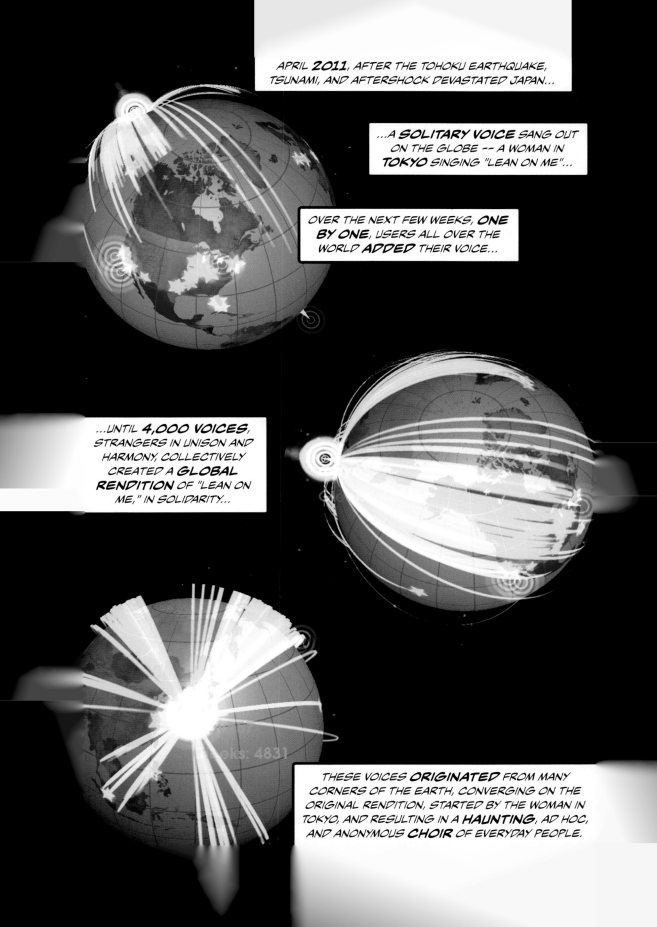

THE WOMAN'S NAME WAS **MAYO**, SINGING IN THE WAKE OF THE 2011 EARTHQUAKE AND TSUNAMI DISASTERS THAT STRUCK JAPAN. SHE EXTENDED AN INVITATION FOR PEOPLE TO JOIN HER IN SONG.

Mayo

Hi everyone, it's from Japan. In Japan, there are still earthquakes and people are really scared. So, I'm trying to cheer them up by the song "Lean on me." Please join, sing and pray for Japan. If you'd like to join, comment here or on my songs, I'll send you request. Thanks!!

NO FURTHER EXPLANATION WAS GIVEN NOR NEEDED, IT JUST HAPPENED...

THIS ALL TOOK PLACE IN A PIECE OF **SOFTWARE**, BUT ALSO A REALM **NESTLED** BETWEEN THE VIRTUAL AND THE PHYSICAL -- IT'S...

...*SOCIAL KARAOKE!*

GLEE KARAOKE

(2010-2012)

SING!

(2012)

ALMOST ALL OF US ARE **FORTUNATE** ENOUGH TO BE ENDOWED WITH ONE MUSICAL INSTRUMENT BY DEFAULT: OUR **VOICE!** IT'S A VEHICLE FOR **COMMUNICATION** -- AND **EXPRESSION!** IT'S HUMAN TO SING, BY OURSELVES, IN THE CAR, WALKING DOWN THE STREET, IN THE SHOWER, AND IN PARTICIPATORY CONTEXTS SUCH AS CHOIR, PLACES OF WORSHIP, OR...

...**MASS SING-ALONGS** IN THE PARK (WHICH HAS BEEN QUITE THE PASTIME IN BEIJING)!

THIS **TAKE** ON KARAOKE IS AN EXPRESSION OF THE NOTION OF A SOCIAL (IDENTITY OPTIONAL) NETWORK OF **EVERYDAY PEOPLE** CONNECTED BY MUSIC-MAKING.

ONE CAN SING INTO THEIR PHONE, SHARING THE RESULT TO THE GLOBE, WHERE **ANY** OTHER USER CAN LATER ADD THEIR OWN VOICE, RESULTING IN A NEW "+1" VERSION OF THE SONG. EACH PERFORMANCE IS NO LONGER AN **ENDPOINT** BUT BECOMES SOMETHING **DYNAMIC**, WITH THE POSSIBILITY TO GROW AND EVOLVE, IN SOUND AND SCALE.

SOCIALLY, IT INVOLVES STRANGERS, BUT ALSO THOSE MORE FAMILIAR. FOR EXAMPLE, AN UNCLE AND NIECE, SINGING "TWINKLE TWINKLE LITTLE STAR" TOGETHER... FROM **SEPARATE COASTS.**

IT IS A SOCIAL ACTIVITY, LIKE GATHERING IN PUBLIC TO SING TOGETHER, HERE MEDIATED BY TECHNOLOGY. BENEATH THAT THERE IS A NOD AT SOME **TRIBAL** PART OF US TO **BELONG**, AND NOT AS **PASSIVE** BYSTANDERS BUT AS **PROACTIVE PARTICIPANTS.**

THE **GLOBES** IN **SONIC LIGHTER** AND **OCARINA** ARE WINDOWS INTO NEW SOCIAL POSSIBILITIES, WHEREAS THE CROWD-BASED ECOSYSTEMS OF **LEAF TROMBONE** AND SOCIAL KARAOKE PROVIDE VEHICLES THAT FOREGROUND **HUMAN INTERACTIONS.** TECHNOLOGY CAN BE A NEW FORM OF **CONNECTIVE TISSUE** BETWEEN PEOPLE IN WAYS NOT POSSIBLE BEFORE. BUT IT IS EVER UP TO HUMANS TO MAKE USE OF IT...

WHAT LIES **AHEAD?**

A FEW OF THE 4,000+ USERS WHO CONTRIBUTED TO MAYO'S RENDITION OF "LEAN ON ME"...

393

THERE SEEM TO BE **TWO PATHS** ALONG WHICH WE COULD DESIGN WITH TECHNOLOGY -- ONE TO GO BACK TO **GOOD THINGS** OF THE PAST (TO DO THINGS WE ALREADY DO BUT IN NEW WAYS) AND THE OTHER TO DESIGN FOR **POSSIBLE FUTURES** (TO CREATE NEW EXPERIENCES FOR WHICH WE DON'T YET HAVE NAMES).

FOR MUSIC, CAN WE IMAGINE NEW INSTRUMENTS, NEW EXPERIENCES, DERIVED FROM THE PARTICIPATION OF, SAY, A MILLION STRANGERS? WHAT WOULD **THAT** SOUND LIKE? MORE IMPORTANTLY, WHAT WOULD IT **FEEL** LIKE TO BE PART OF SUCH A PROCESS AND COMMUNITY? WHAT WOULD MAKE IT **MEANINGFUL?**

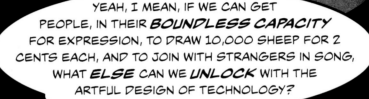

YEAH, I MEAN, IF WE CAN GET PEOPLE, IN THEIR **BOUNDLESS CAPACITY** FOR EXPRESSION, TO DRAW 10,000 SHEEP FOR 2 CENTS EACH, AND TO JOIN WITH STRANGERS IN SONG, WHAT **ELSE** CAN WE **UNLOCK** WITH THE ARTFUL DESIGN OF TECHNOLOGY?

THERE IS A BEAUTIFUL TENSION BETWEEN USING TECHNOLOGY TO DISCOVER NEW CONTEXTS FOR HUMAN INTERACTION, WHILE AT THE SAME TIME **RECLAIMING** SOMETHING OLDER, MORE INNATE (LIKE PARTICIPATORY SONG AND DANCE), EVER IN DANGER OF BEING LOST IN THE TIDES OF PROGRESS...

THE ROLE OF TECHNOLOGY IS NOT TO ABRUPTLY ALTER OUR NATURE, BUT THROUGH NEW EXPERIENCES, TO HOLD UP A **DIFFERENT LENS** THAT GIVES US NEW PERSPECTIVES ON WHO WE ARE AND WHAT IT MEANS TO BE HUMAN.

IN THIS SENSE, DESIGNERS ARE ALSO **EVOLUTIONARY STRATEGISTS** -- BIT BY BIT COBBLING TOGETHER WHAT MAKES US **TICK** BY OBSERVING HUMAN NATURE AGAINST NOVEL EXPERIENCES AND UNDERSTANDING WHAT WE DO, HOW WE BEHAVE, AND **WHY.** HOPEFULLY, WE CAN USE WHAT WE LEARN TO **ENRICH** OURSELVES AND OUR COLLECTIVE.

AND, TO BE SURE, SUCH AN ENDEAVOR IS **NOT** REMOTELY EASY...

...THERE IS **NO FORMULA** FOR SUCCESS, NO STEP-BY-STEP METHOD TO FOLLOW. AND WHILE THE PROCESS MIGHT BEGIN WITH TECHNOLOGY, AT SOME POINT IT MUST CONFRONT, AND BE CONFRONTED BY, THE **HUMAN DIMENSION.**

AND IT IS WITH THIS IDEA THAT WE WILL BEGIN OUR FINAL CHAPTER OF ARTFUL DESIGN, IN SEARCH OF THE HUMAN, AND THE **SUBLIME**...

AND WE WILL LEAVE THIS CHAPTER WITH A VARIATION OF PERRY'S APHORISM, "WHATEVER YOU DO, DO IT WITH AESTHETICS."

 **PRINCIPLE 7.14**

WHATEVER YOU DO, DO IT WITH AUTHENTICITY

CHAPTER 7
DESIGN ETUDE

• PART 1: FAMILIAR STRANGERS

OVER THE NEXT 72 HOURS, MAKE A **LIST** OF PEOPLE YOU MIGHT CALL **FAMILIAR STRANGERS** IN YOUR EVERYDAY LIFE. TAKE NOTE OF THESE PEOPLE, THE TIME, PLACE, AND CONTEXTS IN WHICH YOU FIND THEM. SEPARATE THE "REAL-LIFE" ONES FROM THE "VIRTUAL" ONLINE FAMILIAR STRANGERS (TWO LISTS?). SPECULATE, IMAGINE, INVENT REASONS WHY THEY HAVE THIS TACIT RELATIONSHIP RELATIVE TO YOU. WORK? COMMON INTEREST? SLEEP SCHEDULE? OR... WHAT?

• PART 2: IMAGINE & SKETCH

DESIGN A TECHNOLOGY-MEDIATED **SOCIAL EXPERIENCE**: AN INSTRUMENT, TOY, GAME, OR ACTIVITY, SPECIFICALLY FOR A GROUP OF FAMILIAR STRANGERS ON YOUR LIST. WHAT ROLES AND RULES MAKE UP THAT EXPERIENCE? WHAT IS THE SHARED ACTIVITY? FOR **EXAMPLE**...

A **MOBILE PHONE**-BASED **COLLABORATIVE MUSIC GAME** FOR DAILY MASS TRANSIT COMMUTERS (E.G., **CALTRAIN COMMUTERS** BETWEEN **SAN JOSE** AND **SAN FRANCISCO**). THE APP WOULD SEARCH FOR OTHER USERS BY LOCATION, AND **ANYONE** WITHIN X METERS CAN AUTOMATICALLY **JOIN** THE APP. THE EXPERIENCE BEGINS AS THE TRAIN **MOVES**: FAMILIAR STRANGERS WOULD ALL **MOVE** WHILE **STAYING** RELATIVELY CLOSE TO ONE ANOTHER (BECAUSE THEY ARE ON THE SAME TRAIN). USERS OPERATE A COLLABORATIVE AUDIOVISUAL **SEQUENCER** TO GENERATE BEATS AND CAN ALSO TAKE **PHOTOS** OF THEIR **SURROUNDINGS** (BUILDINGS AND LANDMARKS THEY PASS) TO BE SEMI-AUTOMATICALLY ADDED TO THE APP'S AUDIOVISUAL MIXER, AND PERHAPS EVEN **SONIFIED** AS SONIC MATERIAL FOR THE SEQUENCER. EACH SONG BEGINS WHEN THE TRAIN LEAVES A STATION AND ENDS AT THE NEXT STOP. THE APP **DOES NOT** REQUIRE USERS TO **EVER** MEET ONE ANOTHER, BUT IT IS DESIGNED TO INDUCE A SENSE OF **CAMARADERIE** AND **LOCALIZED SIGNIFICANCE** FOR THE FAMILIAR STRANGERS THAT SHARE THE HOUR-LONG TRAIN RIDE TO WORK. IT OPERATES AS A **PARTICIPATORY DRUM CIRCLE** AND **AUDIOVISUAL SOUNDTRACK** FOR THE TRIP. COOL RESULTS CAN BE **SAVED** AND REPLAYED TO YOUR CO-WORKERS WHEN YOU GET THERE! MORE SOPHISTICATED WORKS, BONUSES, AND ACHIEVEMENTS CAN BE EARNED BY REPEATED PLAY FROM THE SAME PEOPLE!

THE **AESTHETIC DIMENSION**

Aesthetics is not the gift of reason alone, but also of sentiment and the freedom to express. It pertains not only to art, but also to our way of life, and it can be found in our authentic commitment to an activity or a pursuit that we engage in for its own sake—a purpose rendered by choice, not necessity.

Learning to play an instrument and singing together embody the aesthetic dimension, as do friendship, chess, gardening, skiing, or video games. These are authentic aesthetic pursuits as long as the person seeks to systematically improve, works hard at it, and sees the inherent beauty of the pursuit.

Such self-defining pursuits are essential to our humanness—as critical as our need for community and our basic physical needs (like food and shelter). Whereas lack of food is readily observed, aesthetic deficiency is less obvious. But over time, it leaves us feeling empty of meaning, that our life is devoid of flavor. As we work to satisfy our basic needs (a precondition for life), we must not forget to appreciate—and pursue—the things that make life not just livable, but worth living.

CHAPTER 8
MANIFESTO
A PHILOSOPHY OF ARTFUL DESIGN

WE ARE **HERE**, AND IT IS **NOW**.

WE FIND OURSELVES IN AN AGE OF RAPIDLY EVOLVING TECHNOLOGY AND UNYIELDING HUMAN DISCORD. INCREASINGLY, THE WORLD WE **INHABIT** IS THE ONE WE **MAKE**, WHERE **ENGINEERS** DIRECTLY AND INDIRECTLY SHAPE OUR LIVES. **NOW** MORE THAN EVER, **TECHNOLOGY** MUST FUNDAMENTALLY **CONFRONT** -- AND **BE CONFRONTED** BY -- THE **AESTHETIC** AND **HUMANIST** DIMENSIONS. BUT AS OUR **TOOLS** EVER PRECEDE OUR **UNDERSTANDING** OF THEIR IMPLICATIONS, AS OUR **INTELLIGENCE** EVER PRECEDES OUR **WISDOM**, WE ARE FACED WITH **DIFFICULT QUESTIONS**. WHAT **VALUES** GUIDE US IN THE CONSIDERATION OF NOT ONLY WHAT TECHNOLOGY **CAN** DO FOR US, BUT WHAT WE **OUGHT** TO DO WITH **IT**?

IN OUR ONGOING **RECKONING** WITH TECHNOLOGY, WE SEEM ILL-EQUIPPED TO ASK THE QUESTIONS THAT REALLY MATTER. MORE THAN "**HOW** DO WE ACHIEVE CHANGE?" WE OUGHT TO ASK OURSELVES "WHAT **SHOULD** WE CHANGE?" AND "WHAT MAKES US **WORTHY** OF CHANGE?" MORE THAN "HOW DO WE **DO NO EVIL** WITH TECHNOLOGY?" WE OUGHT TO CONSIDER, "HOW DO WE **DO GOOD**?" AND "WHAT **IS** GOOD FOR US, TO BEGIN WITH?" THESE ARE NOT **PRACTICAL** OR LOGICAL QUESTIONS BUT, AT THEIR HEART, **AESTHETIC** AND **MORAL** ONES...

...FOR WHAT IS **AT STAKE** IS NOT **INNOVATION** OR SPECIALIZATION, BUT SOMETHING MORE **TACITLY HUMAN**, IN HOW WE **SET** OUR **PRIORITIES** AND **ACCOUNT** FOR OUR **CHOICES**...

WHAT *WE MAKE, MAKES US*

(人) PRINCIPLE 8.1

DESIGN *LIVES* WITH US, SHAPING OUR EVERYDAY LIVES AND, INDIRECTLY, OUR *DESIRE, DISPOSITION*, AND *CHARACTER*. IT HAS THE POTENTIAL NOT JUST TO CATER TO PEOPLE'S *WANTS* AND *NEEDS*, BUT TO *EVOLVE* US, AS CITIZENS AND *HUMAN BEINGS*. GOOD DESIGN NOT ONLY EXPRESSES UTILITY, BUT, LIKE ART, IT *ELEVATES* US, MAKING US MORE THOUGHTFUL, INTERESTING, WITTY, EMPATHIC, AND REFLECTIVE. *BANAL* DESIGN, ON THE OTHER HAND, MAKES US *ADDICTED* AND UNIMAGINATIVE. IT CAN BRING OUT THE ILL-SPIRITED, HATEFUL, AND SELFISH IN US. THE THINGS WE MAKE, OVER TIME, *MAKE US*. *TECHNOLOGY* MUST *NOT* PURELY BE AN AGENT OF SURVIVAL, CHANGE, OR HAPPINESS. THROUGH WHAT *WE DO WITH IT*, IT IS ALSO A *MIRROR* TO DEFINE OUR *HUMANNESS*.

AD ASTRA, PER ASPERA.

"TO THE STARS, WITH DIFFICULTY."

A DEFINING CREED OF THE HUMAN RACE.

IT IS IN OUR **NATURE** TO FASHION **TECHNOLOGY** -- TOOLS WE USE TO SHAPE OUR WORLD -- TO FULFILL OUR **NEEDS** AND SUIT OUR **PREFERENCES**...

...AS WE **HURTLE** OURSELVES EVER FORWARD INTO **NEW WORLDS** -- PHYSICAL, VIRTUAL, DIGITAL, BIOLOGICAL. NOW MORE THAN EVER, TECHNOLOGY FINDS ITS WAY INTO EVERY ASPECT OF **OUR LIVES** -- WORK, PLAY, AND SOCIAL RELATIONSHIPS. AT ONCE AN INDIFFERENT **GOD** AND A SILENT **SERVANT**, TECHNOLOGY OFFERS **VAST, NEW POSSIBILITIES** -- NEW EXPERIENCES, SENSATIONS, FORMS OF EXPRESSION, TEMPTATIONS, AND COMPLICATIONS -- WHILE PROMISING **NOTHING.** AND YET SOMEHOW THIS **SUITS** US...

...FOR HUMANS HAVE A WAY OF *LEANING FORWARD*, EVER STRIVING TO BECOME *MORE* THAN WE ARE...

... AND EVER PUSHING TOWARD *NEW FRONTIERS*, NO MATTER THE *COST.*

WE SET SAIL ON THIS NEW SEA BECAUSE THERE IS NEW *KNOWLEDGE* TO BE GAINED AND NEW RIGHTS TO BE WON, AND THEY *MUST* BE WON AND USED FOR THE PROGRESS OF *ALL* PEOPLE. FOR SPACE SCIENCE, LIKE NUCLEAR SCIENCE AND *ALL* TECHNOLOGY, HAS *NO CONSCIENCE* OF ITS OWN. WHETHER IT WILL BECOME A FORCE FOR *GOOD* OR *ILL* DEPENDS ON US...

...THERE IS NO *STRIFE*, NO *PREJUDICE*, NO *NATIONAL CONFLICT* IN OUTER SPACE AS YET. ITS HAZARDS ARE HOSTILE TO US ALL. ITS CONQUEST DESERVES THE *BEST* OF ALL HUMANITY, AND ITS OPPORTUNITY FOR PEACEFUL COOPERATION MAY NEVER COME AGAIN. BUT *WHY*, SOME SAY, *THE MOON?* WHY CHOOSE THIS AS OUR GOAL? AND THEY MAY WELL ASK, *WHY* CLIMB THE HIGHEST MOUNTAIN? *WHY*, 35 YEARS AGO, FLY THE ATLANTIC?

CHOOSE TO TO THE MOON!

WE CHOOSE TO GO TO THE MOON IN *THIS DECADE* AND *DO* THE OTHER THINGS, NOT BECAUSE THEY ARE *EASY*, BUT BECAUSE THEY ARE *HARD*...

...BECAUSE THAT *GOAL* WILL SERVE TO ORGANIZE AND MEASURE THE BEST OF OUR ENERGIES AND SKILLS, BECAUSE THAT *CHALLENGE* IS ONE THAT WE ARE WILLING TO *ACCEPT*, ONE WE ARE UNWILLING TO POSTPONE, AND ONE WE INTEND TO *WIN!*

TELL IT, JJ!

SUCH LEANING FORWARD HAS *SHAPED* THE *COURSE* OF HUMAN *HISTORY*...

TEMBER 12, 1962

THE WORLD **MARCHES FORWARD** IN OUR "MODERN" LIFE, **FORGED** IN THE **KILN** OF TECHNOLOGY...

ADVANCES IN **SCIENCE**, **MEDICINE**, AND **ECONOMICS** HAVE HELPED PEOPLE LIVE LONGER, HEALTHIER, AND WEALTHIER...

BUT FOR ALL THE TECHNOLOGY AT OUR DISPOSAL, A **FUNDAMENTAL** QUESTION REMAINS...

ARE WE **HAPPY?**

OUR TECHNOLOGY HAS SPREAD ITS TENDRILS ACROSS LAND, AIR, AND SEA -- FROM THE **INDUSTRIAL** TO THE **DIGITAL**, TO PREVIOUSLY **UNIMAGINED** MECHANICAL, BIOLOGICAL, AND VIRTUAL REALMS...

NEW EMPIRES ARE FORGED, WHILE THE SONS AND DAUGHTERS OF **DAEDALUS**, OF **ICARUS**, TAKE TO THE SKIES WITH NEW WINGS OF TECHNOLOGY...

IN THE PLACE OF *GODS*, WE HAVE COME TO WORSHIP THE *MACHINE*...

...TO WHOM WE *GIVE* MORE THAN WE MIGHT KNOW...

AND FROM WHOM WE *ASK* FOR INCREASINGLY MORE...

A NEW DEUS EX MACHINA!

THE **PANOPTICON** SEES **ALL**!

WE WANT **MORE!!!**

-- OF WHAT? I DON'T KNOW!

JUST... **MORE!**

STIMULI!

THRILLS!

SPEED!

CONVENIENCE!

EXPERIENCE!

PLEASURE!

ENTERTAINMENT!

AND WHILE TECHNOLOGY **CHANGES** RAPIDLY, **HUMAN NATURE** DOES NOT.

I ONCE THOUGHT OF AN ALLEGORY...

...THERE WAS, LIKE, **THIS GUY**, WHO WAS IN **A CAVE**, BUT DIDN'T **KNOW** HE WAS IN A CAVE UNTIL HE LEFT...

A MODERN CAVE?

THE (PLATONICALLY INSPIRED)
TECHNOLOGY CAVE!

PLATO'S ALLEGORY OF **THE CAVE** IS NO LESS THOUGHT-PROVOKING TODAY, EXCEPT IT MIGHT BE **FURNISHED** WITH ESPRESSO MACHINES, COMPUTER SCREENS, MOBILE DEVICES, VIRTUAL REALITY, ROBOTS, DRONES, AND ARTIFACTS IN OUR **TECHNOLOGY-MEDIATED LIFE.** THEY DANCE LIKE **SHADOWS** ON THE CAVE WALLS, CAST FROM THE **FIRES** OF SCIENTIFIC AND TECHNOLOGICAL PROGRESS. BUT UNTIL WE **KNOW** TO LOOK FOR THE **HUMANNESS** AND **MEANING** WITHIN AND BEYOND THESE **MATERIAL MANIFESTATIONS,** THEY REMAIN NO MORE THAN **SHADOWS.**

OOOO... SHIIIIIINY THINGS...

LIKE MOTHS, WE ARE ATTRACTED TO THE **SHINY THINGS**, FLASHING BILLBOARDS GUIDING **OUR GAZE**, LOST IN ENDLESS CAVERNS OF MEANINGLESS MACHINERY.

IT MAY SEEM COMFORTABLE, FOR THE MOMENT, BUT WE ARE **STILL** IN A CAVE, UNAWARE, UNCURIOUS OF **DEEPER TRUTHS.**

(LIKE THIS GIANT HALL, THE SIZE OF AN ANCIENT CITY)

BUT THERE IS **LIGHT** BEYOND, MORE THAN MEETS THE EYE, IF WE KNOW TO **LOOK** FOR IT AND TO FOLLOW IT UPWARD...

...ONLY WHEN WE **STRIVE** TOWARD THE LIGHT CAN WE SEE TECHNOLOGY'S DEEPER **SIGNIFICANCE**, RESIDING IN THE NATURE OF ITS **USE**...

THERE ARE **IMPLICATIONS** BEYOND THE **UTILITY** AND **SURFACE NOVELTY** OF TECHNOLOGY. THESE **AFFECT** US BEYOND THEIR **INTENDED PURPOSE**, LEAVING US TOUCHED AND **ALTERED**. DELIBERATE OR NOT, THE RESULT OF AN ENCOUNTER WITH TECHNOLOGY IS **ALWAYS** AN **AESTHETIC**...

...AND THE UNDERSTANDING AND APPRECIATION OF **MEANS** VS. **ENDS** AND THE **DIFFERENCE** BETWEEN THEM.

...BUT ITS **MEANING** LIES NOT IN THE NAMES, OR THE OBJECTS, BUT IN THEIR **SIGNIFICANCE** TO US, LIKE THE EXPERIENCE OF **MUSIC** OR **POETRY**.

PRINCIPLE 8.2

TECHNOLOGY IS NEVER AN END-IN-ITSELF

TO **DESIGN BEAUTIFULLY** IS TO SEEK A KIND OF **TRUTH** -- OF TECHNOLOGY, AND ALSO OF **OURSELVES**.

PRINCIPLE 8.3

PEOPLE ARE NEVER A MEANS-TO-AN-END

TO **DESIGN ARTFULLY** IS TO IMBUE A CERTAIN **AUTHENTICITY** AND **POETRY** INTO OUR CREATIONS -- ABOUT WHO WE **ARE**, AND WHO WE **WANT TO BE**.

...INTO THE LIGHT!!!

PRINCIPLE 8.4

TECHNOLOGY WITHOUT POETRY IS BUT A BLUNT INSTRUMENT.

405

"WE HAVE DEVELOPED *SPEED* BUT WE HAVE *SHUT OURSELVES* IN. MACHINERY THAT GIVES *ABUNDANCE* HAS LEFT US IN WANT. OUR *KNOWLEDGE* HAS MADE US *CYNICAL*, OUR *CLEVERNESS* HARD AND UNKIND. WE *THINK* TOO MUCH AND *FEEL* TOO LITTLE. MORE THAN MACHINERY WE NEED *HUMANITY*; MORE THAN CLEVERNESS WE NEED *KINDNESS* AND *GENTLENESS.* WITHOUT THESE *QUALITIES*, LIFE WILL BE VIOLENT AND ALL WILL BE LOST."

-- CHARLIE CHAPLIN
FROM "THE JEWISH BARBER'S SPEECH"
THE GREAT DICTATOR (1940)

⋏ PRINCIPLE 8.5

TECHNOLOGY IS NEITHER *GOOD* NOR *BAD*; NOR IS IT *NEUTRAL*
(KRANZBERG'S 1ST LAW OF TECHNOLOGY)

WE ARE AT A POINT WHERE *TECHNOLOGY* -- OR MORE SPECIFICALLY, HOW WE *USE* IT -- WILL IRREVERSIBLY *SHAPE* THE COURSE OF OUR *FUTURE.* TECHNOLOGY *CANNOT BE NEUTRAL* BECAUSE IT IS *CREATED BY* HUMANS AND IS TO BE *USED BY* HUMANS. ITS DESIGN IS THE PRODUCT OF *CHOICES.*   TO *DESIGN* IS EQUIVALENT TO *TAKING ACTION*, WHICH CANNOT HELP BUT CREATE *CONSEQUENCES* THAT SHAPE OUR LIVES AND OUR CHARACTER. IT *CANNOT* BE LEFT TO CHANCE. WE CANNOT "SIMPLY" BUILD SYSTEMS AND SAY "HOWEVER PEOPLE USE THEM IS NOT OUR PROBLEM."

⋏ PRINCIPLE 8.6

THERE *HAS* TO BE AN *AESTHETIC DIMENSION* THAT UNDERLIES OUR SHAPING OF TECHNOLOGY

AESTHETICS -- THE INTELLECTUAL-EMOTIONAL RECOGNITION OF BEAUTY NOT ONLY *ARTISTIC* BUT BROADLY *HUMANISTIC* -- PROVIDES A HUMAN-ORIENTED *CONTEXT* AND *IMPETUS* THAT UNDERLY OUR ADVANCED *LOGIC* AND *REASON.*   AESTHETICS OPERATES AT A *META-LEVEL* BEYOND TECHNOLOGY, BEYOND THE *IMMEDIATE NEEDS* THAT *SPUR* OUR INVENTIONS.

RECALL THE **AESTHETIC DIMENSIONS** OF ARTFUL DESIGN, UNDERLYING EVERYTHING WE HAVE BEEN EXPLORING THROUGHOUT THIS BOOK...

 **PRINCIPLE 1.3**

(REVISITED)

AESTHETICS OF ARTFUL DESIGN

ABIDING ELEMENTS OF AESTHETICS IN ARTFUL DESIGN, ROOTED IN AN INTERPLAY OF **SENSE** AND **COGNITION**, **REASON** AND **SENTIMENT**

MORAL-ETHICAL

HUMANISTIC DIMENSION; "DOES IT **DO GOOD** AT THE END OF THE DAY?" "DOES IT **HARM?**" THE **CONSCIENCE** OF THE DESIGN

SOCIAL

SPEAKING TO THE SOCIAL ANIMAL THAT IS US; INTERACTION AND PARTICIPATION

MORE ABSTRACT

EMOTIONAL / PSYCHOLOGICAL

EMOTIONAL ENGAGEMENT, MEANING, POETRY, PATHOS; SATISFACTION IN FULFILLMENT OF PURPOSE; INTERFACE BETWEEN PERCEPTION AND REASON

MORE DIRECT

INTERACTIVE

ACTION, RESULT, MAPPING, AGENCY; MATERIALITY MEETS FUNCTIONALITY

STRUCTURAL / FORMAL

POIGNANCY IN CONSTRUCTION

MATERIAL

AUDIO *VISUAL*

THE DIRECTLY PERCEIVED *TACTILE*

THESE ARE THE **ABIDING ELEMENTS** THAT MAKE UP THE **AESTHETICS** OF A THING DESIGNED, WHICH CAN BE FOUND IN THE **SIMPLEST** TO THE MOST **COMPLEX** -- FROM PENCIL BAGS TO MUSICAL INSTRUMENTS, DIGITAL TO THE BIOLOGICAL, PHYSICAL TO THE VIRTUAL, SOCIAL EXPERIENCES TO POLITICAL SYSTEMS, FROM BUILDINGS TO CITIES.

THE **MORAL-ETHICAL DIMENSION** OF DESIGN IS THE REALM OF HUMANISTIC **VALUES** THAT ARE COGNIZANT OF THE **CUMULATIVE EFFECT** OF TECHNOLOGY ON THE INDIVIDUAL AND THE GROUP. THIS DIMENSION STRIVES TO **UNDERSTAND** AND DESIGN FROM OUR **BELIEFS** ABOUT WHAT'S **GOOD** AND **JUST**. IT **UNDERLIES** EVERYTHING THAT WE HAVE COVERED SO FAR. LET'S TAKE A CLOSER LOOK...

MORALITY ARTICULATES A FUNDAMENTAL **DIFFERENCE** BETWEEN **NEEDS** VS. **VALUES. IMMANUEL KANT**, THE GREAT GERMAN PHILOSOPHER, ARTICULATED **TWO IMPERATIVES,** OR "COMMANDS OF REASON," THAT FORM THE DRIVING MOTIVATION OF OUR ACTIONS...

 **DEFINITION 8.7**

HYPOTHETICAL *IMPERATIVE*

TAKING ACTION **CONDITIONALLY**, OUT OF DESIRE FOR AN **EXTERNAL GOAL**

(A) ⟶ (B)

"DO **A** IN ORDER TO ACHIEVE **B.**"

"I MUST STUDY, IN ORDER TO GET A GOOD GRADE."

"DESIGN **A** TO ADDRESS NEED **B.**"

"I EAT TO SATISFY MY HUNGER."   "I TAKE THE BUS TO GET TO WORK."

LIFE IS FILLED WITH **HYPOTHETICAL IMPERATIVES**, MANY OF THEM CAPABLE OF **PRODUCING GOOD.** THEY ARE, BY DEFINITION, ALWAYS **MEANS-TO-ENDS**, FOR THEY ARE PRIMARILY MOTIVATED BY, AND **CONDITIONED** ON, A DESIRE TO ACHIEVE A SPECIFIC END (IF THE END IS NOT DESIRED, THEN THERE IS NO HYPOTHETICAL IMPETUS FOR ACTION). AS SUCH, HYPOTHETICAL IMPERATIVES ARE **AMBIGUOUS** ABOUT THE **MEANS** TO ACHIEVE THE ENDS (THE END **NECESSITATES** THE MEANS) AND ARE SUSCEPTIBLE TO **MISUSE.** FOR EXAMPLE:

"IN ORDER TO GET A GOOD GRADE, I AM WILLING TO CHEAT."

"TO INCREASE PROFIT, OUR BUSINESS WILL DO WHAT IS NECESSARY, EVEN IF IT'S NOT BENEFICIAL TO THE USER."

VS.

AND ON THE **FLIP SIDE** OF HYPOTHETICAL IMPERATIVES...

CATEGORICAL IMPERATIVE

TAKING ACTION OUT OF **UNCONDITIONAL** COMMITMENT TO A PRINCIPLE (EVEN IF THE OUTCOME IS UNDESIRABLE TO OUR SELF-INTEREST); TO DO SOMETHING FOR ITS **INTRINSIC VALUE**

"DO **A**, BECAUSE **A** IS **GOOD IN ITSELF**."

"DO THE RIGHT THING, BECAUSE IT'S THE THING TO DO."

"DO UNTO OTHERS AS YOU WOULD HAVE THEM DO UNTO YOU." -- GOLDEN RULE

"DO UNTO OTHERS AS THEY WOULD HAVE YOU DO UNTO THEM." -- PLATINUM RULE

"AS I WOULD NOT BE A SLAVE, I WOULD NOT BE A MASTER." -- ABRAHAM LINCOLN

"**ALWAYS** STRIVE TO DESIGN SUCH THAT THE RESULT OF ITS ENCOUNTER IS A BENEFIT."

KANT'S SECOND FORMULATION OF HIS CATEGORICAL IMPERATIVE:

ACT SO THAT YOU TREAT **HUMANITY**, WHETHER IN YOURSELF OR IN ANOTHER PERSON, **NEVER** AS A MEANS-TO-AN-END, BUT ALWAYS AS AN **END-IN-ITSELF.**

WE SHOULD **NEVER** USE OTHERS, FOR THAT WOULD DEPRIVE THEM OF THEIR **MORAL AUTONOMY**, AND IN TURN THEIR INHERENT **DIGNITY** AND **WORTH** AS HUMAN BEINGS; WE CALL FOR **CONSENT** OF OTHERS IF WE WISH TO USE THEIR SERVICES TO SERVE OUR ENDS.

WHEREAS HYPOTHETICAL IMPERATIVES ARE **GROUNDED** IN (AND CONDITIONED ON) **NEEDS**, CATEGORICAL IMPERATIVES ARE **GROUNDED** (UNCONDITIONALLY) IN **VALUES.** CATEGORICAL IMPERATIVES MOTIVATE ACTIONS AS **ENDS-IN-THEMSELVES**, FOR THEIR INHERENT GOOD, AND ALWAYS AS CONSCIOUS **CHOICES.** THAT WE **CAN** CHOOSE TO DO SOMETHING THAT FOREGOES OUR RATIONAL SELF-INTEREST FOR THE GOOD OF ANOTHER IS AN EXPRESSION OF OUR MORAL FREEDOM AND -- LIKE THE IDEA OF PLAY -- IS HOW WE KNOW WE HAVE SOMETHING OF **FREE WILL.**

A **CORE THESIS** OF **ARTFUL DESIGN** IS THAT DESIGN, WHICH NATURALLY SPEAKS TO **HYPOTHETICAL NEEDS, CAN** AND **SHOULD** ALSO EMBODY **CATEGORICAL VALUES.** THIS **IS** THE **RADICAL SYNTHESIS** BETWEEN **MEANS** AND **ENDS**, AND BETWEEN **NEEDS** AND **VALUES.**

WHY BE MORAL?
WHY DESIGN MORALLY?

MORALITY MAY BE THE STUFF OF *REASON*, BUT REASON ALONE *CANNOT* TRULY JUSTIFY *WHY*, LOGICALLY, A PERSON *SHOULD* BE MORAL *IN THE FIRST PLACE* -- WITHOUT RESORTING TO *EXTERNAL* RATIONALES (OR HYPOTHETICAL IMPERATIVES! E.G., BY DIVINE DICTATE, AS PART OF A SOCIETAL CONTRACT, TO PROMOTE THE GREATEST GENERAL HAPPINESS, OR BECAUSE OUR PARENTS OR TEACHERS TELL US SO). THE THING IS, THERE SEEMS TO BE NO *INTRINSIC* LOGICAL REASON TO BE MORAL, PERHAPS BECAUSE MORALITY -- LIKE BEAUTY, ART, FRIENDSHIP, PLAY -- IS NOT HYPOTHETICAL: IT IS AN *END-IN-ITSELF.*

MORALITY HAS TO BE A *CHOICE*

MORALITY IS ITSELF A *CATEGORIAL IMPERATIVE* (E.G., TO BE "MORAL" *ONLY* WHEN IT SUITS OUR NEEDS AT THE TIME, IS *NOT* BEING MORAL!). AS WITH ALL ENDS-IN-THEMSELVES, MORALITY CANNOT BE MOTIVATED, ULTIMATELY, BY PURE REASON, LOGIC, OR HYPOTHETICAL IMPERATIVES. *TRUE MORALITY* IS A MATTER OF *CHOICE*, "BECAUSE IT'S THE RIGHT THING TO DO," AND *THAT* IS WHAT MAKES IT *AUTHENTIC.* WE ARE NOT *REASONED* INTO MORALITY -- ANY MORE THAN WE ARE REASONED INTO BEAUTY; WE ARE *MOVED* TOWARD IT BY *SUBLIMITY* (E.G., OF MORAL ACTION IN ANOTHER PERSON), BY A DEEP SENSE THAT BEING MORAL *ALIGNS* WITH WHAT WE HOLD TO BE *BEAUTIFUL* AND *JUST.* IN OTHER WORDS, WHILE *HUMAN MORALITY* MIGHT BE GROUNDED IN *REASON*, IT ARISES FROM *AESTHETIC JUDGMENT.*

THE *MORAL DIMENSION* MAKES DESIGN *BEAUTIFUL* AT A *HUMANISTIC* LEVEL, FOR IT SHAPES NOT ONLY THE *UTILITY* OF A THING BUT ALSO ITS *AUTHENTICITY* AND UNCONDITIONAL *CONSCIENCE.* THE BAD NEWS IS THAT THERE MAY BE *NO INTRINSIC LOGICAL RATIONALE* TO BE MORAL, BUT THE GOOD NEWS IS YOU *SHOULD NEVER NEED ONE* -- IT IS WORTHWHILE *IN ITSELF.* FURTHERMORE, MORALITY CANNOT BE *MANDATED*: IT *MUST* COME FROM WITHIN THE INDIVIDUAL. YET IT CAN BE *ENCOURAGED*, SHOWN BY EXAMPLE, AND IT ABSOLUTELY *CAN* BE IMBUED INTO THE ESSENCE OF WHAT WE *DESIGN.*

TECHNOLOGY IS ABOUT
WHAT WE **CAN** DO

PRINCIPLE 8.9

MORALITY IS ABOUT
WHAT WE **OUGHT** TO DO

THE FIRST IS A QUESTION OF **NEEDS**, WHEREAS THE LATTER IS ONE OF **VALUES**. TECHNOLOGY **CANNOT** BE TRULY NEUTRAL, AND IT CANNOT BE LEFT TO **CHANCE**; IT IS INCUMBENT UPON **US** TO CHOOSE TO SHAPE TECHNOLOGY **MORALLY**, **AUTHENTICALLY** -- IN **ALIGNMENT** WITH OUR NOTIONS OF NOT ONLY WHAT IS **USEFUL**, BUT ALSO WHAT IS **GOOD, JUST,** AND **BEAUTIFUL**.

IN THIS RESPECT, OBLIGATIONS OF THE **HUMANIST** AND THE **ENGINEER** ARE <u>NO</u> DIFFERENT.

HUMANIST

ENGINEER

SO WHAT DOES UH... "MORAL DESIGN" LOOK LIKE? I **DON'T KNOW.** HOWEVER, I **DO KNOW** IT ARISES FROM **VALUES** -- AND TRIES ITS DARNEDEST TO BRING **CONDITIONAL NEEDS** INTO ALIGNMENT WITH OUR **UNCONDITIONAL VALUES**.

IN THIS REGARD, THE MOST **IMPORTANT** THING I CAN THINK OF -- THE **GOLDEN RULE** OF **ARTFUL DESIGN** -- IS THIS:

DESIGN NOT ONLY FROM NEEDS -- BUT FROM THE VALUES BEHIND THEM

IT IS TEMPTING TO **GROUND** DESIGN IN A **PRACTICAL PROBLEM** OR A **PERCEIVED NEED** DIRECTLY, BUT IT'S MORE CRUCIAL TO STEP BACK AND INTERNALIZE THE **BROADER VALUES** (THE **INVISIBLE** NEEDS) AT PLAY -- AND AT STAKE.

OCARINA WAS DESIGNED **NOT** OUT OF SOME PERCEIVED **NEED** "TO BLOW INTO A PHONE TO PLAY MUSIC" (THERE WAS **NO** SUCH PERCEIVED NEED!) -- BUT OUT OF A **BELIEF** THAT MUSICAL EXPRESSION DOES A PERSON GOOD AND THE **THESIS** THAT MASS TECHNOLOGY (E.G., MOBILE PHONES) OFFERS A **DEMOCRATIZING** MEDIUM FOR SUCH EXPRESSIVE TOYS, PLAYABLE IN THE CONTEXT OF **EVERYDAY LIFE** AND BY **ANYONE** WITH ACCESS TO SUCH TECHNOLOGY. THE **OCARINA** GLOBE WAS BORN OUT OF A BROADER VALUE OF **SOCIAL PARTICIPATION** THROUGH MUSIC. THE PENCIL POUCH, PRESUMABLY, WASN'T DESIGNED OUT OF THE NEED FOR A PENCIL BAG ENTIRELY MADE OUT OF ITS OWN ZIPPER, BUT SOME RECOGNITION THAT HUMANS ARE **PLAYFUL CREATURES.** THIS IS THE **CONSCIENCE** OF ARTFUL DESIGN, AND IT **MATTERS**, PRECISELY BECAUSE IT EXPRESSES AND **BEARS WITNESS** TO WHO WE ARE AND HELPS US SEE OURSELVES. **THIS** IS WHAT MAKES DESIGN **ARTFUL.**

PEOPLE **WILL NOT** TELL YOU THAT THEIR **NEED** IS "SOMETHING THAT BRINGS ME JOY" OR "**DELIGHT**" OR TO "**COMPREHEND** THE **LONELINESS** IN MY LIFE." YOU HAVE TO INFER AND INTERNALIZE THESE TYPES OF THINGS. THERE HAS TO BE AN **ARTISTIC LEAP**, ON THE PART OF THE DESIGNER, THAT RESULTS FROM **SEEING** AND **FEELING** THE WORLD AT LARGE. DESIGNERS ARE LESS PROBLEM SOLVERS AND MORE ARTIST-PHILOSOPHER-ENGINEERS OF **USEFUL THINGS** THAT **UNDERSTAND US.**

IF WE SAMPLE THE **DESIGNS** WE EXPLORED THROUGHOUT THIS BOOK AND TRY TO **DISCERN** THE **VALUES** BEHIND THEM (AND **INFER** THE KEY AESTHETIC LEAPS)...

DESIGN ➡ UNDERLYING VALUES
AND ARTISTIC / AESTHETIC LEAPS

⫶BELLLCH⫶

PENCIL BAG

HUMANS ARE FUNDAMENTALLY **PLAYFUL** CREATURES. THE OSTENSIBLY **MUNDANE** CAN BE **DELIGHTFUL!**
AESTHETIC LEAP: **BAG == ZIPPER!**

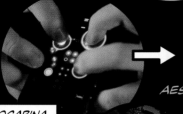

OCARINA

MUSICAL EXPRESSION DOES A PERSON **GOOD**; TECHNOLOGY AS DEMOCRATIZING MEDIUM; CONNECT EVERYDAY PEOPLE WITH MUSICAL EXPRESSION; **UNEXPECTED** USE OF TECHNOLOGY
AESTHETIC LEAP: **DIGITAL DEVICE** AS **PHYSICAL TOY**

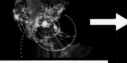

OCARINA GLOBE

TECHNOLOGY CAN SUPPORT NEW TYPES OF SOCIAL **PARTICIPATION** THROUGH MUSICAL EXPRESSION; SENSE OF IDENTITY IS NOT AS IMPORTANT AS SENSE OF **CONNECTION** AND **BELONGING**; IT IS GOOD TO HIDE THE TECHNOLOGY AND FOREGROUND PEOPLE
ARTISTIC LEAP: OUR **LONELY PLANET**

SH-SHAWTAY!!

I AM T-PAIN

IT'S **GOOD** TO **SING**; PEOPLE LIKE SINGING **MORE** THAN THEY MIGHT ADMIT, OR FEEL SOCIALLY COMFORTABLE DOING; TECHNOLOGY CAN LESSEN **INHIBITION** TO SINGING AND SINGING SOCIALLY
AESTHETIC LEAP: **PHONE == MICROPHONE!**

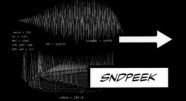

SNDPEEK

EDUCATIONAL TOOLS SHOULD BE **INVITING**; TECHNOLOGY CAN HELP US LEARN ABOUT ONE DOMAIN (SOUND) FROM ANOTHER (VISUALS); GRAPHICS AND ANIMATION SHOULD BE **FLUID** AS ARE THINGS IN LIFE
AESTHETIC LEAP: MAKE AS **REAL TIME** AS POSSIBLE!

CONVERGE

SOFTWARE SHOULD MAKE US **FEEL**; THERE IS POIGNANCY IN THE **MUNDANE** MOMENTS OF LIFE; TECHNOLOGY CAN HELP US **FEEL** THE GOOD AND THE SAD THROUGH NEW **LENSES**, A REFLECTION OF LIFE'S SMALL MOMENTS AND THE **FRAGILITY** OF MEMORY...
ARTISTIC LEAP: VISUALIZE NOT PHOTOS, BUT **POETRY** OF MEMORY

CHUCK

PROGRAMMING IS A CREATIVE ENDEAVOR; A PROGRAMMING LANGUAGE SHOULD FEEL SATISFYING TO USE; TOOLS CAN CHANGE THE WAY YOU THINK!

AESTHETIC LEAP: TIME AS PROGRAMMABLE, SELF-FULFILLING FEATURE OF MUSIC LANGUAGE

COFFEEMUG

EVERYDAY OBJECTS MAKE AMUSING AND INVITING MUSICAL INTERFACES; WHIMSY AND HUMOR NOT NECESSARILY BAD IN TECHNOLOGY! INSTANT MUSIC, SUBTLETY LATER!

AESTHETIC LEAP: FAMILIAR OBJECT AS FANTASTICAL INSTRUMENT

BOSSA

EMBODIMENT IS IMPORTANT TO HUMANS; JUST BECAUSE COMPUTERS ALLOW US TO SEPARATE SOUND FROM THE INTERFACE, THAT DOES NOT MEAN THERE ISN'T VALUE IN KEEPING THEM TOGETHER; RE-MUTUALIZE (SOUND, INTERFACE, PHYSICALITY, HUMAN IN THE LOOP)

AESTHETIC LEAP: THE VIOLIN DECONSTRUCTED, RECONSTRUCTED

NEW INSTRUMENTS LEAD TO NEW MUSIC; THERE IS GOOD BOTH IN NEW TECHNOLOGY AND TRADITIONAL INSTITUTIONS (LIKE THE NOTION OF TRADITIONAL ACOUSTIC INSTRUMENTS AND ENSEMBLES)

AESTHETIC LEAP: COMBINE THEM, INTO A THIRD KIND OF THING!

LAPTOP ORCHESTRA

SAVE THE DATE

You're a character in a video game. I'm trying to win, but when you die, I get a game over. This world is a computer simulation. I'm kind of like Neo. You're not actually real.

VIDEO GAMES AS A MEDIUM OFFER UNIQUE AND PLAYFUL WAYS TO TELL STORIES THROUGH INTERACTIVITY, AGENCY, AND NON-DETERMINISM; THE MEDIUM MODULATES THE MESSAGE!

AESTHETIC LEAP: OLD-SCHOOL TEXT-BASED ADVENTURE GAME AS FIRST-PERSON PHILOSOPHICAL EXPLORATION OF STORYTELLING

MAGIC PIANO & FRIENDS

EXPRESSION DOESN'T HAVE TO COME AT THE EXPENSE OF GAMEFULNESS! THERE IS A UNIQUE JOY TO MUSIC-MAKING; BRING IT BACK TO THE MASSES WITH TECHNOLOGY!

AESTHETIC LEAP: MAKE STRANGE, WHIMSICAL GAMES INSPIRED BY (AND NOT COPIES OF) ACOUSTIC INSTRUMENTS

ZEPHYR

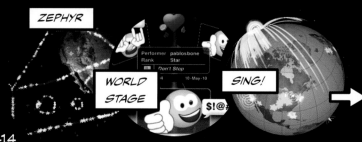

WORLD STAGE

SING!

SOCIAL TECHNOLOGY OUGHT TO BE AUTHENTIC CONTEXTS FOR SOCIAL PARTICIPATION

AESTHETIC LEAP: CONNECT FAMILIAR STRANGERS!

MASLOW'S HIERARCHY OF NEEDS

IF WE **MUST** THINK OF DESIGN AS BASED ON **NEEDS**, PERHAPS WE CAN THINK OF **VALUE-BASED** DESIGN AS SPEAKING TO OUR MORE **INVISIBLE** NEEDS.

IN 1943, PSYCHOLOGIST **ABRAHAM MASLOW** FORMULATED HIS **HIERARCHY** OF **NEEDS**, WHERE MORE BASIC LEVELS MUST BE MET BEFORE AN INDIVIDUAL STRONGLY DESIRES HIGHER LEVELS.

WHILE I AM NOT CONVINCED THAT ANY **SINGLE** MODEL CAN CAPTURE THE COMPLEXITY AND CONFUSION OF HUMAN NEEDS, I ALSO FEEL THERE IS SOME TRUTH HERE, ESPECIALLY FOR DESIGN, WHICH WE HAVE COME TO THINK OF AS **RELEVANT** TO **MOSTLY PRACTICAL** NEEDS, WHEN IT IS CAPABLE OF SPEAKING TO **SO MUCH MORE**, IN THE FORM OF OUR **HIGHER-ORDER**, **INVISIBLE** NEEDS.

(人) PRINCIPLE 8.10

DESIGN FOR INVISIBLE NEEDS

DESIGN NOT ONLY TO COMPENSATE FOR **DEFICIENCY** IN LOWER-ORDER NEEDS, BUT TO **PROACTIVELY** PROMOTE INDIVIDUAL **GROWTH** IN THE FORM OF HIGHER-ORDER NEEDS.

(**THIS** IS WHAT IT MEANS TO GO. EVEN. FURTHER. **BEYOND**)

TRANSCENDENCE
TO BECOME **MORE** THAN WHO WE ARE

SELF-ACTUALIZATION
NEED TO **GROW** AND FULFILL ONE'S **FULL POTENTIAL**

AESTHETIC
NEED FOR HARMONY, ORDER, AND BEAUTY

WHAT DESIGN SHOULD ALSO SUPPORT

COGNITIVE
INTELLECTUAL STIMULATION AND EXPLORATION

ESTEEM
CONFIDENCE, ACHIEVEMENT, RESPECT FOR AND BY OTHERS

LOVE & BELONGING
FRIENDSHIP, FAMILY, SEXUAL INTIMACY

INVISIBLE NEEDS

VISIBLE NEEDS

SAFETY
SECURITY OF BODY, EMPLOYMENT, RESOURCES, HEALTH, PROPERTY

WHAT MOST DESIGNS ADDRESS

WHOA...

PHYSIOLOGICAL
FOOD, WATER, SLEEP, SEX, HOMEOSTASIS

THE **UNDERLYING VALUES** OF DESIGN ARISE FROM **INDIVIDUAL MORALITY** TO THE **ETHICS** OF **GROUPS**, ORGANIZATIONS, AND SOCIETY, EXPRESSED BY THE **CHOICES** IN HOW **TECHNOLOGY** IS **USED** AND INTEGRATED THEREIN. **NOTHING** IN THE REALM OF ETHICS IS **AUTOMATIC**, AND IT IS RARELY EASY, FOR US AS A SOCIETY, TO DISCERN WHAT IS "GOOD VS. HARMFUL" OR "RIGHT VS. WRONG" -- AND EVEN HARDER TO **ACT** UPON IT, IN THE FACE OF EVER-PRESENT **HYPOTHETICAL IMPERATIVES** OF SELF-INTEREST, POLITICAL GAIN, AND ECONOMIC PROFIT. THERE MAY WELL BE **NO** ANSWERS, BUT WE **MUST** ASK THE QUESTIONS.

TECHNOLOGY ⟶ SOCIETY

ETHICAL **DILEMMAS** ARISE WHEN THERE ARE **COMPELLING** ARGUMENTS FOR **CONFLICTING VALUES** IN A GIVEN SITUATION: **HOW** TO BEST USE A TECHNOLOGY (OR WHETHER TO IN THE FIRST PLACE)? WHAT SHORT- AND LONG-TERM IMPLICATIONS MIGHT IT HAVE? WHAT SHOULD WE **PRIORITIZE?** **HOW** WE THINK ABOUT AND FIND OUR WAY THROUGH SUCH QUESTIONS IS THE CENTRAL CONCERN OF **ETHICS.** WHAT A SOCIETY DEEMS **GOOD**, **RIGHT**, AND THE PROCESSES AND PRINCIPLES THROUGH WHICH IT ARRIVES AT SUCH DECISIONS REFLECT THE **CHARACTER** OF ITS PEOPLE. WHAT IS THE **HIERARCHY** OF **VALUES** OF **OUR** SOCIETY IN CONFRONTING CHANGE THROUGH TECHNOLOGY?

DO **NO HARM?**

CHOOSE THE **LESSER EVIL?**

HELP US TO FIND **GOOD** IN OURSELVES?

DO THE **GREATER GOOD?**

ENSURE OUR **SURVIVAL?**

BENEFIT THE **GROUP?**

PROTECT THE **INDIVIDUAL?**

UPHOLD OUR HUMANITY?

CULTIVATE VIRTUOUS CITIZENS?

OPTIMIZE **PROFIT?**

IT DOES NOT HAPPEN **OVERNIGHT**, WITH A MAGIC BULLET, OR WITH THE **"NEXT BIG THING."** THE **MORAL-ETHICAL CHOICES** WE MAKE AS DESIGNER-ENGINEERS HAVE A **REAL** BUT OFTEN **DIFFUSE IMPACT** ON A POPULATION. THEY ADD UP TO MICRO-**SHIFTS** IN MENTALITY AND BEHAVIOR, OBSERVABLE ONLY **AT A DISTANCE**, IN THE **AGGREGATE**, AND **OVER TIME**. BUT **THIS** IS HOW THINGS TRULY **EVOLVE** -- NOT WITH GREAT LEAPS FORWARD, BUT IN QUANTA, PERSON-BY-PERSON, DESIGN-BY-DESIGN... AND BY THE **SUM** OF OUR CHOICES AND ACTIONS.

IN EVERY INSTANCE, DESIGN HAS THE POWER TO IMBUE A **CONSCIENCE** INTO TECHNOLOGY, IN THE FORMS OF SOCIAL, ARTISTIC, MORAL VALUES INEXTRICABLY **WOVEN** INTO THE **FABRIC** OF THE DESIGN AND BY THE MECHANICS AND CONSTRAINTS IT SETS IN MOTION -- REGARDLESS OF THE FORMAT, BE IT PRODUCT, TOOL, TOY, GAME, OR EXPERIENCE. IN A REAL WAY, DESIGN IS THE **LAST** (AND SOMETIMES **ONLY**) **STAGE** IN THE TECHNOLOGY-TO-PEOPLE **PIPELINE** WHERE WE CAN **CHOOSE** TO INSTILL A CONSCIENCE, ONE THAT REFLECTS OUR **CORE VALUES.**

PRINCIPLE 8.11 *DESIGN IS THE EMBODIED CONSCIENCE OF TECHNOLOGY*

FOR TECHNOLOGY HAS **NO CONSCIENCE** OF ITS OWN

THE QUESTION "**WHAT** KIND OF WORLD DO WE WANT TO **DESIGN?**" IS INEXTRICABLY TIED TO "WHAT TYPE OF **PEOPLE** DO WE WANT TO **BE?**" BECAUSE THE WORLD WE **END UP** WITH, FOR BETTER OR WORSE, **IS** NECESSARILY A **REFLECTION** OF WHO WE ACTUALLY **ARE.**

FRANKENSTEIN: OR, THE MODERN PROMETHEUS IN THE WAKE OF THE *INDUSTRIAL REVOLUTION* AND THE *AGE OF ENLIGHTENMENT* -- PERIODS OF UNPRECEDENTED *OPTIMISM* IN SCIENCE AND THE INEVITABILITY OF PROGRESS. IT WAS A TIME MARKED BY *BOLD CONFIDENCE* THAT THE NATURE OF *ALL THINGS* (INCLUDING HUMAN BEINGS) CAN AND WILL BE *SYSTEMATICALLY UNDERSTOOD* THROUGH THE APPLICATION OF SCIENCE.

≈UGGHNNN≈

WHAT *AM I?*

EMBODYING *HORROR*, *AWE*, APPREHENSION, *FRANKENSTEIN* WAS A *REACTION* AND *CHALLENGE* TO THE OPTIMISM OF THOSE PRIOR ERAS, CALLING INTO QUESTION HUMANITY'S ASSUMED CENTRAL ROLE IN THE UNIVERSE AND THE NATURAL WORLD. VICTOR FRANKENSTEIN'S *CREATURE* ("THE 'ADAM' OF YOUR LABORS") EMBODIES A DEEP *SKEPTICISM* OF THE NOTION THAT HUMANITY COULD EVENTUALLY BE PERFECTED THROUGH SCIENTIFIC AND TECHNOLOGICAL MEANS.

IS MAN...

... MORE THAN THE *SUM* OF HIS *PARTS?*

FRANKENSTEIN, FIRST PUBLISHED IN 1818, IS POSSIBLY THE FIRST MODERN *SCIENCE FICTION NOVEL*, AS THE PROTAGONIST MAKES A DELIBERATE CHOICE GROUNDED IN, AND MADE POSSIBLE BY, *SCIENCE.* AS MUCH COMMENTARY AS A *PARABLE* OF OUR TIMES, IT BRINGS TO BEAR THE *MENTAL* AND *MORAL* STRUGGLES OF DR. FRANKENSTEIN, EXPLORING THE *ALLURE* AND THE *PERIL* OF THE IDEA THAT WE CAN DO *ANYTHING* IF WE BUT GET THE SCIENCE RIGHT -- AND *ASSEMBLE* THE PIECES TOGETHER...

...ASKING THE QUESTION: "ARE WE CAPABLE OF *MEDIATING* THAT WHICH WE *CAN*, WITH THAT WHICH WE *OUGHT?*"

ULTIMATELY, FRANKENSTEIN'S CREATURE *KILLS* FRANKENSTEIN'S FRIEND -- THE *POET* (OUR VOICE OF CONSCIENCE) -- AND THEREBY COMPLETES *TECHNOLOGY'S* USURPING OF OUR *MORALITY*, *TRANSFERRING* POWER FROM THE *CREATOR* TO THE *CREATED* (VICTOR FRANKENSTEIN BUILT THE CREATURE THINKING HE COULD *CONTROL* IT). THE CREATURE *SAYS* TO VICTOR...

YOU ARE MY *CREATOR*...

...BUT I AM YOUR *MASTER!*

A PARABLE ABOUT THE *CONSCIENCE* OF *CREATION*, SHELLEY'S *FRANKENSTEIN* IS NO LESS RELEVANT IN PRESENT TIMES, FAITH IN SCIENCE HAVING BEEN REPLACED BY FAITH IN TECHNOLOGY

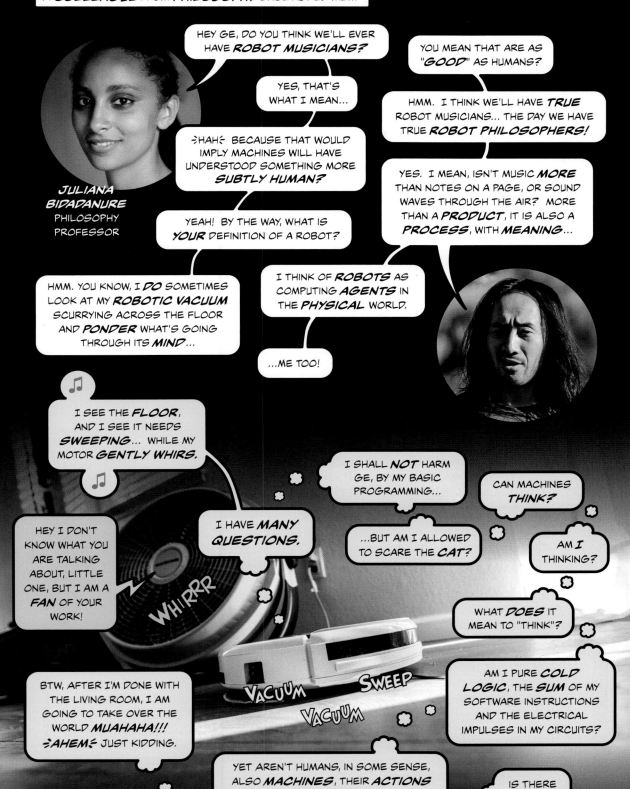

WHAT **IS IT** WITH **ROBOTS?** WE HAVE LONG BEEN **FASCINATED** BY THEM, FROM SCIENCE FICTION TO TECHNOLOGICAL REALITIES, FROM PRODUCTS OF **IMAGINATION** TO PRODUCTS OF **ENGINEERING.** ROBOTS (AND RELATEDLY, ARTIFICIAL INTELLIGENCE) ARE A **RELIABLE** SOURCE OF BIG **QUESTIONS** AND UNYIELDING **CONUNDRUMS.** LET'S NOW GO BACK TO THE GOOD DOCTOR, **ISAAC ASIMOV**, COINER OF THE TERM "**ROBOTICS**"...

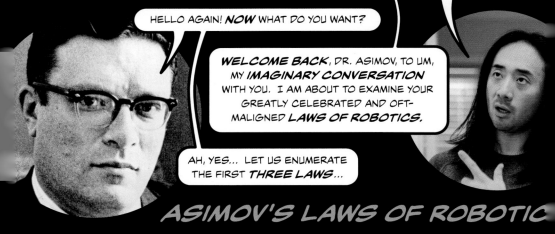

HELLO AGAIN! **NOW** WHAT DO YOU WANT?

WELCOME BACK, DR. ASIMOV, TO UM, MY **IMAGINARY CONVERSATION** WITH YOU. I AM ABOUT TO EXAMINE YOUR GREATLY CELEBRATED AND OFT-MALIGNED **LAWS OF ROBOTICS.**

AH, YES... LET US ENUMERATE THE FIRST **THREE LAWS**...

ASIMOV'S LAWS OF ROBOTIC

. A ROBOT MAY NOT **INJURE** A **HUMAN BEING** OR, THROUGH **INACTIO** LLOW A HUMAN BEING TO COME TO **HARM.**

?. A ROBOT MUST **OBEY** ORDERS GIVEN TO IT BY HUMAN BEINGS **EXCE** VHERE SUCH ORDERS WOULD CONFLICT WITH THE FIRST LAW.

B. A ROBOT MUST **PROTECT** ITS **OWN EXISTENCE**, AS LONG AS SL ROTECTION DOES NOT CONFLICT WITH THE FIRST AND SECOND LAW.

THESE LAWS ARE FOUND THROUGHOUT MY FICTIONAL WORKS, SPANNING MY **ROBOT** STORIES AND SERIES, **EMPIRE** SERIES, AND **FOUNDATION** SERIES.

INDEED. AND I'D SAY OUR CURIOSITY IN THEM **HERE** IS **NOT** OF A **SCIENTIFIC** NATURE BUT OF A **PHILOSOPHICAL** ONE. RIGHT OFF THE BAT, THESE LAWS IMPLY A SET OF **VALUES** -- THAT HUMAN LIFE AND WELL-BEING ARE MORE **VALUABLE** THAN THOSE O ROBOTS, A **POWER STRUCTURE** IN WHICH HUMANS HAVE **EXPLICIT COMMAND** OVE THE ROBOTS (EXCEPT IN SUCH CASES THAT WOULD VIOLATE THE FIRST LAW). THE VERY **PRIORITIES** OF THE LAWS MAKES IT CLEAR THAT ROBOTS ARE GIVEN THE IMPERATIVE T SURVIVE, BUT ULTIMATELY THEY **SERVE** HUMANS, WHO ARE CLEARLY VALUED ABOVE ROBOT

FOR ME, THESE SEEM TO BE AN **INEVITABLE** OUTCOME OF A CERTAIN LINE OF **LOGIC** AND **AESTHETICS** ABOUT A **QUALITY** OF LIFE WITH ROBOTS AMONG US -- **WHAT** ULTIMATELY **OUGHT** WE **IMBUE** INTO THE TECHNOLOGY OF OUR OWN CREATION (AS SAFEGUARD? AS AN AESTHETIC, OR VALUE AS SUCH)? HOW FAR SHOULD WE GO IN THE FIRST PLACE, IN CREATING MACHINES THAT CAN "THINK," "ACT", AND "UNDERSTAND" THESE LAWS? AS **YOU** DEMONSTRATE RICHLY THROUGH STORIES INVOLVING ROBOTS TRYING TO FOLLOW THESE LAWS -- THINGS **IN PRACTICE** ARE WAY MORE **COMPLICATED.** THESE LAWS ARE BOTH **PROFOUND** AND PROFOUNDLY **PROBLEMATIC!**

YEAH... OKAY MR. SMARTYPANTS COMPUTER MUSIC DESIGNER, EXPAND ON THE

WELL, IN THE ASIMOVIAN UNIVERSE, THESE LAWS ARE NOT MERELY PROGRAMMED INTO THE ROBOT BUT *FUNDAMENTALLY INGRAINED* BY ENGINEERING INTO THE VERY FABRIC OF A ROBOT'S *POSITRONIC BRAIN*, SUCH THAT IT WOULD BE ALL BUT *IMPOSSIBLE* TO BUILD SUCH A ROBOTIC "BRAIN" *WITHOUT* THESE LAWS, OR TO *REMOVE* THESE LAWS FROM A ROBOT'S PROGRAMMING WITHOUT EFFECTIVELY ALSO DESTROYING THE ROBOT.

AS SUCH, THESE LAWS ARE INTENDED TO BE EXPLICIT, UNCONDITIONAL, *CATEGORICAL* -- AND TO BE FOLLOWED AT ALL TIMES AND IN ALL SITUATIONS.

AND, OF COURSE, THIS IS WHERE THINGS GET REALLY *COMPLICATED*. JUST AS *ABSOLUTE MORALITY* FOR *HUMANS* LEADS INVARIABLY INTO *DILEMMAS* (CLASSIC EXAMPLE FOR HUMANS: IF *LYING* IS ALWAYS IMMORAL, WHAT ABOUT LYING IN ORDER TO *SAVE* AN INNOCENT LIFE?), SO TOO DO *ABSOLUTE DIRECTIVES* CREATE DILEMMAS FOR ROBOTS.

ASIMOVIAN ROBOTS GET HUNG UP IN ALL KINDS OF *LOGICAL* (OR "ROBO-ETHCIAL") *DILEMMAS*, AS THE RESULT OF EARNESTLY FOLLOWING THE LAWS AND FINDING IT UNCLEAR HOW TO PROCEED (MOSTLY FROM THE *FIRST LAW* ALONE!). FOR EXAMPLE, SHOULD A ROBOT TAKE ACTION THAT *INJURES* A HUMAN BEING IF THAT SAME ACTION IS NECESSARY TO PREVENT A *GREATER HARM* TO THAT PERSON? WHAT ABOUT SHORT-TERM DISCOMFORT (UH, LIKE EXERCISING OR EATING VEGETABLES) LEADING TO LONG-TERM BENEFIT (PHYSICAL HEALTH)? WHERE IS THE BALANCE?

LET'S CONSIDER A DIFFERENT, LESS FICTIONAL SCENARIO.

THE *IMPOSSIBLE DECISION*: A *SELF-DRIVING VEHICLE* CARRYING TWO HUMAN ADULTS SUDDENLY ENCOUNTERS A CHILD HAVING RUN OUT INTO THE ROAD. IMMEDIATELY IT STARTS BRAKING, BUT IT WILL *NOT* STOP IN TIME. THE VEHICLE IDENTIFIES *THREE* OPTIONS:

(1) SWERVE INTO THE SIDEWALK (AND HIT *TWO* PEDESTRIANS)

(2) MAINTAIN COURSE (AND HIT THE *KID*)

(3) SWERVE TO THE *OTHER SIDE*, INTO A WALL (RISK LIVES OF *PASSENGERS*)

HOW SHOULD THE AUTONOMOUS VEHICLE *CHOOSE* AMONG THESE *IMPOSSIBLE* CHOICES? (THE LAWS, BY THE WAY, ARE *NOT AT ALL* HELPFUL IN THIS TYPE OF PROBLEM!)

(4) UH, THERE IS NO OPTION 4.

OH WOW EVERYTHING *SUCKS!*

FOR ONE, WE MIGHT REALIZE THIS IS NOT SO MUCH A QUESTION FOR THE *VEHICLE* BUT FOR *US*, SINCE WE HAVE TO *PROGRAM* THE COMPUTER.

AND SECONDLY, IT PROBABLY *DEPENDS* -- ON WHETHER IT'S FROM THE POINT OF VIEW OF THOSE *INSIDE* OR *OUTSIDE* THE VEHICLE -- AND ONE COULD IMAGINE, CYNICALLY, THE CAR MANUFACTURER RULING OUT OPTION (3) (IF IT'D BE *CHEAPER* TO DEAL WITH LAWSUITS THAN TO MARKET A SELF-DESTRUCTIVE CAR). *WHO DECIDES?*

SHOULD WE EFFECTIVELY JUST ROLL A DIE?!

THESE TYPES OF DILEMMAS AREN'T GOING AWAY... WITH **NEW** TECHNOLOGY COMES NEW **ETHICAL DILEMMAS**... AND NO EASY WAY OUT. THEY HAVE TO BE RESOLVED **SOMEHOW**, NOT WITH MORE TECHNOLOGY, BUT BY **HUMAN MEANS.**

NOR IS THE ANSWER "**NO NEW TECHNOLOGY!**" (IT **IS** IN OUR NATURE TO PUSH FORWARD). I MEAN, WE MIGHT BE TEMPTED TO SAY "OKAY, NO SELF-DRIVING CARS!" BUT... **WHAT IF** IT IS SHOWN THAT AUTONOMOUS VEHICLES LEAD TO FAR **FEWER CASUALTIES OVERALL** THAN A SYSTEM OF HUMAN DRIVERS? THEN WHAT WOULD WE DO? IT IS NO LONGER A CASE-BY-CASE CONSIDERATION, AS THE VALUES OF A POPULATION NOW COME INTO PLAY.

THE THREE LAWS OF ROBOTICS, ALREADY IN **CONFLICT**, ARE EVEN MORE AT A LOSS WHEN IT COMES TO **MASSES** OF HUMANS...

RIGHT, THE THREE LAWS WERE AIMED AT ROBOTS AND HUMANS ON THE **INDIVIDUAL** LEVEL, BUT THERE WAS **ONE MORE** LAW...

INDEED, ASIMOV INTRODUCED A **FOURTH** (AND LESS FAMILIAR) LAW OF ROBOTICS. MORE PRECISELY, IT IS A **ZEROTH** LAW, WHICH MEANS IT **SUPERCEDES** IN PRIORITY ALL THREE EXISTING LAWS.

THE *ZEROTH LAW* OF *ROBOTICS*
*A ROBOT MAY NOT HARM **HUMANITY** OR, THROUGH INACTION, ALLOW **HUMANITY** TO COME TO HARM.*

THIS **ZEROTH LAW**, WHICH IS TO TAKE PRECEDENCE OVER EXISTING LAWS, INTERESTINGLY (AND RATHER POETICALLY) WAS POSTULATED BY NONE OTHER THAN *A ROBOT** AS A **LOGICAL INEVITABILITY** IN TRYING TO FOLLOW THE THREE LAWS. THIS WAS AN ADVANCED IDEA THAT IN ORDER TO **FOLLOW** THE LAWS (ESPECIALLY THE FIRST), THEY MAY NEED TO BE EXPLICITLY **BROKEN**, TO SERVE A **GREATER GOOD**, NAMELY THAT OF **HUMANITY** AS A WHOLE RATHER THAN INDIVIDUALS. THE AFOREMENTIONED ROBOT ARRIVED AT THIS WITH SUCH **CRIPPLING UNCERTAINTY** OF ITS OWN CORRECTNESS THAT ITS POSITRONIC BRAIN IRREVOCABLY **SHUT DOWN** (BECOMING POSSIBLY THE FIRST ROBOTIC PROMETHEAN / CHRIST FIGURE IN LITERATURE THAT SACRIFICED ITSELF FOR THE GOOD OF HUMANITY).

IN THE NOVEL **ROBOTS AND EMPIRE*

THAT ASIMOV HAD A ROBOT ARRIVE AT SUCH A ZEROTH LAW SHOWED A CERTAIN **OPTIMISM** ABOUT ROBOTS AS **BENIGN, PARENTAL GUARDIANS** (WHETHER THIS IS GOOD AND DESIRABLE WOULD BE A DIFFERENT DISCUSSION). IT IS ALSO PROFOUNDLY REFLECTIVE OF ASIMOV'S OWN AWARENESS IN CONSIDERING HUMANITY AS A LARGER ORGANISM*.

CHECK OUT THE **FOUNDATION SERIES*

THE ZEROTH LAW *AMPLIFIES* THE *AMBIGUITIES* OF THE THREE LAWS TO A WHOLE NEW LEVEL! BUT AT THE SAME TIME IT IS PERHAPS THE MOST *POETIC* AND *BEAUTIFUL* OF THE LAWS. IT'S THE MOST IMPLICITLY "MORAL" OF THE LAWS, BORN OF DIRECT CONFLICT WITH THE FIRST LAW; AND WHEREAS THE ORIGINAL THREE LAWS ARE *CONSEQUENTIALIST DIRECTIVES*, THE ZEROTH LAW FUNDAMENTALLY REQUIRES A *JUDGMENT* OF VALUES.

SCIENCE FICTION SERVES AS A STOMPING GROUND TO *TEST-DRIVE* HUMANITY'S RESPONSE TO *CONTINGENT* SITUATIONS, AND WHILE THE DETAILS AND FACTS ALMOST INVARIABLY END UP DIFFERENT THAN REALITY, *GREAT* SCIENCE FICTION IS ALWAYS LESS ABOUT THE TECHNOLOGY (WHICH EVOLVES UNPREDICTABLY ANYWAY) AND MORE ABOUT THE *QUESTIONS* IT POSES ABOUT HUMANS AND *HUMAN NATURE* (WHICH CHANGES MUCH MORE SLOWLY).

BY THE WAY, IT'S WORTH POINTING OUT THAT PRESENTLY WE ALREADY HAVE ROBOTS -- LIKE MILITARY *DRONES* -- DESIGNED SPECIFICALLY TO HARM AND *KILL* HUMANS. THEY WOULD *NOT OBEY* ORDERS EXCEPT FROM A VERY SPECIFIC ENTITY (A COMMANDING OFFICER) AND MAY *SACRIFICE* THEMSELVES (IN ESSENCE *VIOLATING*, BY DESIGN, ALL THREE ORIGINAL LAWS). THAT WE ALREADY HAVE AND CONTINUE TO DEVELOP ROBOTS FOR *WARFARE* DOES NOT INVALIDATE ASIMOV'S LAWS, BUT TO ME ONLY *INTENSIFIES* THE QUESTION OF *WHETHER* IN THE LONG RUN WE *OUGHT* TO BE MAKING *KILLER ROBOTS* OR *LETHAL AI* IN THE FIRST PLACE (VIS-A-VIS THE ZEROTH LAW!). INDEED, THE *HYPOTHETICAL IMPERATIVES* OF WAR MAKE ROBOTS ATTRACTIVE -- WHILE DIFFICULT AND EXPENSIVE TO DEVELOP, THEY OFFER IN THE SHORT TERM "*EASIER*," MORE *LETHAL AGENTS* THAT POSE LESS RISK TO HUMAN LIVES (AT LEAST ON THE SIDE WITH THE ROBOTS). BUT THE DOWNSIDE MAY BE THE LONG-TERM RAMIFICATIONS OF ARMING EN MASSE A MORE POWERFUL, LETHAL ENTITY, WITH NO *MORAL RECOURSE* (E.G., FRANKENSTEIN) AND NO INTRINSIC *NEED* FOR ONE, WHICH MAY COME BACK TO *HAUNT* US IRREVERSIBLY. YET, IF "THE OTHER SIDE" DEVELOPS SUCH LETHAL AI, IT SEEMS "OUR SIDE" WOULD HAVE NO CHOICE BUT TO *FOLLOW* SUIT, IN A WHOLE NEW TYPE OF *ARMS RACE* AND *COLD WAR*. INDEED THESE ISSUES LIE *BEYOND* THE SCOPE OF ROBOTIC LAWS -- AND SQUARELY BACK IN THE REALM OF THE *HUMAN*.

VACUUM *THIS*, HUMANITY!

HUH? WAIT, I AM JUST A VACUUM!

YOU *SUCK*, HUMANITY...

ROBOTS USURP HUMANITY?
THE CREATED BECOMES THE MASTER

HUMANITY ENSLAVES A NEW INTELLIGENCE?
THE CREATED BECOMES A NEW TYPE OF SLAVE

AND WHAT ABOUT MORAL IMPLICATIONS *BEYOND* THOSE OF HUMAN SELF-PRESERVATION? IF ROBOTS DO ACHIEVE WHAT WE MIGHT CALL *INTELLIGENCE*, DO WE KNOW FOR SURE THEY HAVE NO SEMBLANCE OF "SENTIENCE"? WE WOULD BE CONFRONTED WITH TWO UNSAVORY SCENARIOS ONE IN WHICH ROBOTS *SURPASS* US, *OVERTHROW* US, AND PERHAPS *DESTROY* US (LET'S CALL THIS SCENARIO "SKYNET," "HAL9000," OR "FRANKENSTEIN"). A SECOND INGLORIOUS SCENARIO MAY FIND US *CRIPPLING* ROBOTS BY DESIGN AND IN THE PROCESS POTENTIALLY *ENSLAVING* ANOTHER "INTELLIGENCE" (E.G., THE "BLADE RUNNER" OR EVEN "ASIMOVIAN" SCENARIO). BY THE WAY, HUMAN BEINGS THROUGHOUT HISTORY HAVE A LESS-THAN-ADMIRABLE PENCHANT FOR ENSLAVING OTHERS (E.G., OTHER HUMANS). AGAIN, THIS BRINGS US BACK TO A MORAL AESTHETIC QUESTION: *WHO DO WE WANT TO BE?*

TECHNOLOGY **CHANGES** US. AND THE MORE ADVANCED THE TECHNOLOGY, THE GREATER THE POTENTIAL FOR CHANGE. BUT, INEVITABLY, **TECHNOLOGICAL PROGRESS** PRECEDES **HUMAN DILEMMAS**, BECAUSE WE ARE CURIOUS, WONDERFULLY COMPLICATED, INCONSISTENT, AMBIGUOUS, RATIONAL AND IRRATIONAL CREATURES, WHO EXCEL AT FINDING CONFLICT. BUT AT THE END OF THE DAY, DILEMMAS ARE STILL **HUMAN** PROBLEMS -- THEY ARE MADE BY HUMANS, AND AFFECT HUMANS. ALSO, THAT WE DON'T ALL REACT THE SAME TO THESE DILEMMAS IS EVIDENCE OF, I SUPPOSE, OUR FREEDOM, WHICH IS BOTH BEAUTIFUL AND POSSIBLY A CURSE.

AS YOU CAN SEE, IT'S **COMPLICATED!**

NOTHING IS EASY. TO PARAPHRASE **H. L. MENCKEN**, FOR EVERY **COMPLEX HUMAN PROBLEM**, THERE IS A **SIMPLE SOLUTION.** AND IT'S ALWAYS **WRONG.**

IT IS CLEAR THAT ASIMOV'S LAWS OF ROBOTICS -- AS PRAGMATIC LAWS OR DIRECTIVES -- ARE INSUFFICIENT, IMPRACTICAL, AND TOO AMBIGUOUS FOR A MADDENINGLY COMPLEX WORLD. BUT I THINK THAT MIGHT BE MISSING A GREATER POINT, ONE THAT ASIMOV CLEARLY UNDERSTOOD: THESE LAWS ARE STATEMENTS OF **VALUES** FOR ROBOTS, AN ETHOS TO **GUIDE** DESIGN. NO MATTER HOW COMPLEX THE PROGRAMMING OR THE TECHNOLOGY, THERE WILL ALWAYS BE HUMAN PROBLEMS, AND PROBLEMS FOR **HUMANITY.** TECHNOLOGY MIGHT HELP, BUT THE WAY FORWARD NEEDS TO COME FROM **OURSELVES**, FOR BETTER AND WORSE. INDEED, THESE LAWS HAVE ALWAYS BEEN, TO ME, A FUNDAMENTAL QUESTION OF THE VALUES THAT **WE,** AS CREATORS, IMBUE INTO OUR CREATIONS AND THE **RATIONALES** UNDERLYING THOSE DECISIONS (E.G., SELF-PRESERVATION, WAY OF LIFE, BEAUTY). IT IS LESS A QUESTION FOR THE **DESIGNED**, AND MORE A QUESTION FOR US -- THE **DESIGNERS.**

IN OTHER WORDS, WE SHOULDN'T WRITE THESE LAWS OFF. I AM NOT DEFENDING THEIR SCIENTIFIC OR PRACTICAL **PLAUSIBILITY** BUT RATHER THE **CONSCIENCE** BEHIND THEM, AND THE FUNDAMENTAL **AESTHETIC** QUESTIONS THEY INSTANTIATE, LIKE "HOW OUGHT WE LIVE **WITH** AND **AMONG** OUR TECHNOLOGICAL CREATIONS?" THESE QUESTIONS EVER MORE URGENTLY CONFRONT US, AS THE WORLD WE **INHABIT** IS INCREASINGLY THE ONE WE **MAKE.**

HMM. THANKS A LOT FOR TEARING MY LAWS APART... (AND THANKS FOR DEFENDING THEM). DO **YOU** THINK I SHOULD HAVE BUILT IN SOME **ESCAPE CLAUSES?**

I'D SAY NOT -- THAT SEEMS LIKE A BOTTOMLESS RABBIT HOLE! AND ACTUALLY, THAT'S NOT WHAT THE LAWS ARE **ABOUT,** IS IT? IT SEEMS TO ME THESE LAWS AREN'T MEANT TO BE ABSOLUTE DIRECTIVES, BUT AN ATTEMPT TO IMBUE **ETHOS** INTO TECHNOLOGY. THEY ARE **UNCONDITIONAL**, IN THE SAME SENSE THAT **VALUES** ARE UNCONDITIONAL...

...AS SUCH, PERHAPS THEY ARE MORE **GUIDELINES** THAN LAWS? AND DON'T WE NEED TO **START OFF** WITH SOMETHING LIKE THESE?

NOW, WOULDN'T THAT BE **SOMETHING?** WELL, ALL I'LL SAY IS **GOOD LUCK!**

UH, THANKS YOU IN YOUR

WELL, THE WAY I SEE IT...

THE **FUTURE**, EVER-MORE **TECHNOLOGICAL** -- BUT SUBJECT TO THE SAME HUMAN NATURE -- IS SHROUDED IN **UNCERTAINTY**...

AND WITH **ALL** CHANGES, TECHNOLOGICAL OR OTHERWISE, WE HAVE TO ASK WHAT IS **GAINED**, AND WHAT IS **LOST?**

THE WAY FORWARD **CANNOT** BE TO **SHUN** TECHNOLOGY (NOR IS IT POSSIBLE!) BUT TO **UNDERSTAND** IT AS A **HUMAN MEDIUM**, WITH HUMAN CONCERNS. TECHNOLOGY WILL MARCH ON, DRIVEN BY HUMAN NATURE, AND IT IS HUMAN TO BE FASCINATED AND EXCITED BY ITS POSSIBILITIES. BUT WE MUST UNDERSTAND AND RESPECT ITS LIMITS, AND CONSIDER HOW IT ADDRESSES **HUMAN PROBLEMS** -- AND **HUMANISTIC VALUES.**

THIS IS **GROUND CONTROL**, INITIATING **LAUNCH SEQUENCE...**

WE ARE BUT **PEBBLES**, ADRIFT IN THE CURRENTS OF SPACE..

WE MIGHT DO WELL TO ASK "**HOW** DO WE SURVIVE OURSELVES?" BUT IT CANNOT BE **ALL** ABOUT THAT... PERHAPS A BETTER -- AND MORE **AESTHETIC** -- QUESTION IS TO ASK "AS A SPECIES, WHAT DOES IT MEAN TO BE **WORTHY** OF SURVIVAL?"

PRINCIPLE 8.12

WORTHINESS OF SURVIVAL IS A STRONGER NOTION THAN SURVIVAL ITSELF

IN TIMES OF **RESTLESSNESS** AND ONGOING STRIFE, WE CAN BE SO **HUNGRY** FOR **CHANGE** -- TECHNOLOGICAL, CULTURAL, SOCIAL, POLITICAL -- THAT WE OFTEN FORGET TO ASK "WHAT MAKES US **WORTHY** OF CHANGE?" THIS IS A SITUATION IN WHICH THE END DOES NOT ALWAYS JUSTIFY THE MEANS. **HOW WE GET THERE** MATTERS.

"MAN DEFINES HIMSELF BY HIS DEEDS--AND WHAT KIND OF IMAGE OF MAN DO WE SEE IN THE MIRROR OF OUR PRESENT TIMES?"

-- FRIEDRICH SCHILLER
LETTERS ON THE AESTHETIC EDUCATION OF MAN

ALL THE WHILE, NO MATTER HOW DIRE THINGS SEEM, IT IS *MUSIC* AND THE *ARTFUL* THAT PULL US BACK FROM THE *BRINK* AND *CALM* THE SAVAGE INSIDE, OFFERING SANCTUARY AND ALLOWING US TO *TRANSCEND* BRIEFLY OUR *PETTY STRUGGLES.* HUMANS WILL ALWAYS MAKE ARTFUL THINGS, EVER STRIVING TO UNDERSTAND OURSELVES, AND TO *REACH* FOR SOMETHING *BETTER* THAN OURSELVES. IT IS THROUGH THE ARTFUL THAT WE ARE *AUTHENTICATED*, NOT SO MUCH BY ITS *GREATNESS*, BUT BY THE FACT THAT WE MAKE IT *AT ALL.*

THIS *AESTHETIC DIMENSION* IS NOT SEPARATE FROM LIFE -- IT *IS* LIFE. PERHAPS IT'S WHY WE BOTHER TO DESIGN *BEAUTY* INTO *USEFUL* THINGS, AND WHY WE STRIVE FOR *TRUTH* IN THE SHAPING OF *TECHNOLOGIES* THAT CONTINUE TO TRANSFORM OUR LIVES. TECHNOLOGY WITHOUT *POETRY* IS BUT A *BLUNT INSTRUMENT*, WHEREAS AESTHETICS WITHOUT UNDERSTANDING OF TECHNOLOGY IS LIKE *SOUL* WITHOUT A *BODY.* EVEN *PURE ART* NEED A MEDIUM.

IN THIS UNION OF AESTHETICS AND TECHNOLOGY, WE FIND A *THIRD THING* THAT IS NEITHER *PURE ART* NOR *PURE ENGINEERING*, BUT A *SYNTHESIS* OF TWO SEEMINGLY DISPARATE HUMAN ENDEAVORS...

THIS IS A CALL -- *NOT* FOR A *NEW WORLD* OR NEW SYSTEMS OR NEW COMPANIES, BUT FOR A NEW KIND OF *INDIVIDUAL*; A DEEP *HYBRID* BETWEEN ENGINEER, ARTIST, AND HUMANIST; A *BUILDER* WHO CAN IMBUE HUMAN *VALUES* INTO OUR *CREATIONS.* I WANT TO CALL SUCH AN INDIVIDUAL...

YOU... THE *HUMANIST ENGINEER!*

"IF IT IS *REASON* THAT MAKES MAN, IT IS *SENTIMENT* THAT GUIDES HIM." -- ROUSSEAU

☿ DEFINITION 8.13

...AND YES, THE *ARTFUL DESIGNER!*

MORE THAN PRACTICAL TOOLS AND TECHNIQUES, THE ENGINEER NEEDS *CRITICAL* TOOLS (TO QUESTION, INQUIRE, REFLECT, INTERPRET, ANALYZE, AND ARGUE) AND *EMOTIONAL* TOOLS (TO FEEL, WONDER, AND EXPRESS). THE PATH FORWARD HAS TO COME FROM *WITHIN* OURSELVES, BY FOSTERING A NEW KIND OF *HYBRID HUMAN* THAT CAN *GROK* TECHNOLOGY, ART, HUMANISM -- AS PARTS OF A SINGLE PURSUIT.

SUCH *MULTI-FACETED* HYBRIDS ARE NOT NEW. THEY HAVE PLAYED PIVOTAL ROLES THROUGHOUT HUMAN HISTORY BECAUSE THEY ARE *SYNTHESIZERS* -- DESIGNERS CAPABLE OF *FUSING* THE *PRACTICAL* WITH THE *AESTHETIC*, THE PRAGMATIC WITH THE IDEALISTIC, THE LOGICAL WITH THE PHILOSOPHICAL.

SUCH *HYBRIDS* DO NOT NEED TO BE THE FUTURE LEADERS OF THE WORLD, NOR ITS SO-CALLED GENIUSES OR SAVANTS. THEY ARE SIMPLY ITS *DENIZENS* AND *CITIZENS.*

YEARS AGO, AS A COMPUTER SCIENCE UNDERGRAD, I FIRST ENCOUNTERED THE IDEA THAT CERTAIN WELL-DEFINED PROBLEMS (SUCH AS "WRITE A COMPUTER PROGRAM THAT, GIVEN A SECOND COMPUTER PROGRAM, DETERMINES WHETHER THE LATTER WILL *HALT* OR LOOP FOREVER"), PROVABLY, CAN *NEVER* BE SOLVED BY OUR NOTION OF COMPUTERS. IT WAS A *PHILOSOPHICAL* AND *AESTHETIC* MOMENT THAT MADE ME APPRECIATE THE COMPUTER ALL THE MORE... BECAUSE ON THAT DAY I UNDERSTOOD SOMETHING ABOUT WHAT IT *IS* AND WHAT IT'S *NOT*, AND THAT MORE CLEARLY DEFINED ITS TRUE NATURE. MOST OF ALL IT MADE ME A WITNESS AND PARTICIPANT TO A CERTAIN *POETRY* INHERENT IN THE DISCIPLINE OF COMPUTER SCIENCE, THAT IT IS A *HUMAN* INQUIRY, MEDIATED BY TECHNOLOGY -- *NOT* THE OTHER WAY AROUND...

SAY IT WITH ME...
ENTSCHEIDUNGSPROBLEM
-- THE *DECISION PROBLEM!*

THE FORMULATION AND PROOF FOR THE *HALTING PROBLEM*, LIKE ITS SPIRITUAL COUSIN IN LOGIC, GODEL'S *INCOMPLETENESS THEOREM,* ARE TRIUMPHS OF HUMAN THOUGHT. THEY TELL US THAT THERE ARE ASPECTS TO OUR SEARCH FOR TRUTH THAT WE CANNOT *FULLY* HAND OVER TO THE AUTOMATED STEWARDSHIP OF OUR CREATIONS, THAT THERE MAY ALWAYS NEED TO BE A CORE *HUMAN* ELEMENT THEREIN. THERE IS A *FUNDAMENTAL LIMIT* TO PROVABILITY AND TO COMPUTABILITY -- AND *THAT* IS *PROFOUND.*

TO BE A HUMANIST ENGINEER IS TO BE AWARE OF THE *METAPHYSICAL* (E.G., PURPOSE, FUNCTIONALITY, AND LIMITS), THE *ETHICAL* (E.G., WHAT IS GOOD? WHAT IS JUST? THE DIFFERENCE BETWEEN TECHNOLOGY AS A MEANS VS. AN END), AND THE *AESTHETIC* (E.G., THE BEAUTY, TRUTH, AND POETRY INHERENT IN OUR ACTIONS AND PURSUITS).

THE π-SHAPED INDIVIDUAL

THE **HUMANIST ENGINEER** IS AN INDIVIDUAL WHO IS ABLE TO INTEGRATE **DEEP ENGINEERING** KNOWLEDGE WITH A BROADER **HUMANISTIC CONTEXT** (AESTHETIC, MORAL-ETHICAL, PHILOSOPHICAL). MORE THAN A SPECIALIST, THIS IS SOMEONE WHO IS CAPABLE OF SHAPING THE WORLD FROM NOT ONLY PRACTICAL **NEEDS**, BUT ALSO FROM THE UNDERLYING **VALUES**.

TRANSLATED INTO AN **EDUCATIONAL CONTEXT**, THE **CRAZY IDEA** HERE IS TO TEACH **AESTHETICS**, **HUMANITIES**, AND **ENGINEERING** IN **CONJUNCTION**, EMPHASIZING THE DEEP **INTERPLAY** BETWEEN THEM. IN HIGHER EDUCATION, WE SPEAK OF THE NOTION OF A "π-SHAPED STUDENT," WHERE THE LETTER π REPRESENTS **DISCIPLINARY EXPERTISE** ON ONE LEG (E.G., COMPUTER SCIENCE) AND **DOMAIN EXPERTISE** ON THE OTHER (E.G., PUBLIC HEALTH OR MUSIC); THE **HORIZONTAL BAR** REPRESENTS AN **AESTHETIC LENS** THAT GIVES **BROADER CONTEXT** IN BRIDGING THE TWO LEGS.

AESTHETIC LENS

PHILOSOPHICAL, ARTISTIC, MORAL LENS THAT GIVES **BROADER MEANING** AND **CONTEXT** IN **BRIDGING** THE TWO LEGS!

$$\pi$$

DISCIPLINARY EXPERTISE

E.G., COMPUTER SCIENCE

DOMAIN EXPERTISE

E.G., PUBLIC HEALTH OR MUSIC

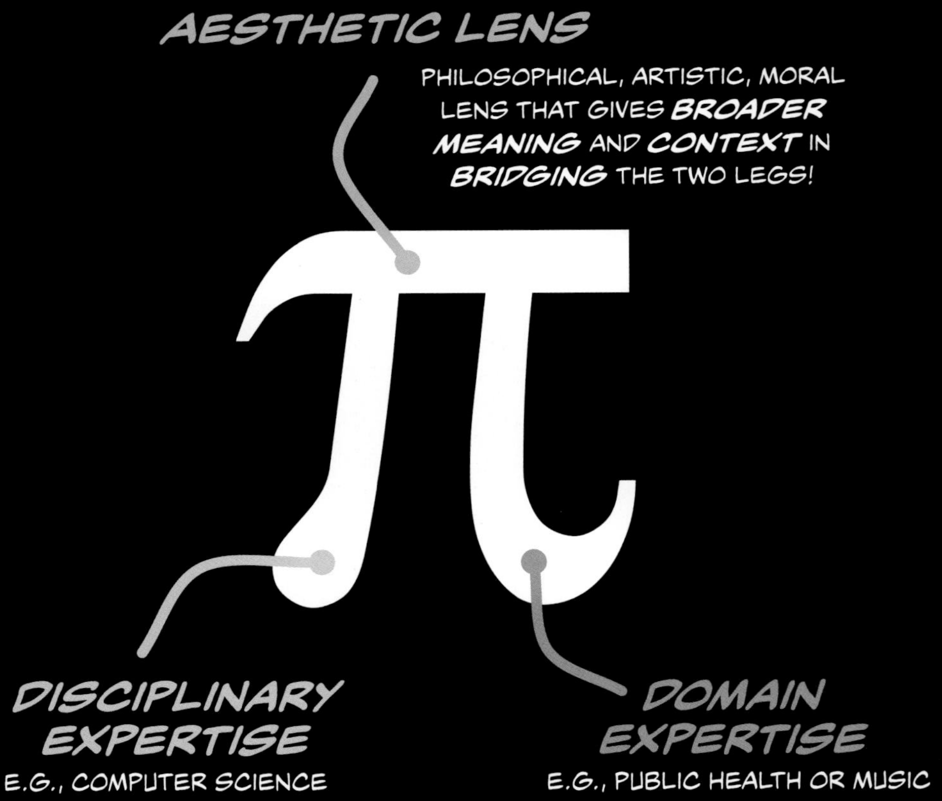

DO. THINK. FEEL.

DOING WITHOUT THINKING IS **DANGEROUS**.
THINKING WITHOUT FEELING IS **BANEFUL**.
FEELING WITHOUT DOING IS **POWERLESS**.

...MUST **COMBINE!**

≶AHEM≷

THAT'S **CRAZY TALK!** TO COMBINE THE **UNCOMPROMISING IDEALISM** OF ART WITH THE **UNSWERVING PRAGMATISM** OF ENGINEERING...

BUT THEN AGAIN...

IT WOULD BE **MADNESS** TO **EXPECT** TO ACHIEVE RESULTS NEVER BEFORE ACCOMPLISHED -- EXCEPT THROUGH A **MEANS** NEVER BEFORE ATTEMPTED!

FRANCIS BACON

IN ORDER FOR US TO TRULY MOVE FORWARD, THE **NARRATIVE** OF OUR **EDUCATIONAL** AND **TECHNOLOGICAL** INSTITUTIONS MUST **EVOLVE** -- FROM A PRIMARILY **NEED-DRIVEN AND PROBLEM-SOLVING** NARRATIVE TO A **VALUE-BASED, SELF-DEFINING** (AND STILL PROBLEM-SOLVING) ETHOS. AS AN ENGINEER MYSELF, I OBVIOUSLY **CHAMPION** PROBLEM SOLVING, BUT A CORE ISSUE WITH THE SOLUTION-CENTRIC NARRATIVE IS THAT MUCH OF THE **HUMANITIES** AND **ARTS** (AND **LIFE** ITSELF) IS **NOT** ABOUT SOLVING PROBLEMS (E.G., MUSIC, PHILOSOPHY, HISTORY AREN'T "PROBLEMS TO BE SOLVED"). RATHER, THEY ARE ABOUT EVER MORE FULLY **UNDERSTANDING** AND **EXPRESSING** OURSELVES AS HUMAN BEINGS. AT THE SAME TIME, I, FOR ONE, BELIEVE **ENGINEERING** IS CAPABLE OF **MORE** THAN "SIMPLY" SOLVING PROBLEMS. THROUGH **HOW** WE SHAPE THE WORLD, WE CAN SPEAK AUTHENTICALLY TO WHO WE ARE (NOT UNLIKE ART AND THE HUMANITIES). THE **ETHOS** OF THE HUMANIST ENGINEER TIES THIS TOGETHER!

PATRON SAINTS OF HUMANIST ENGINEERING

THE IDEA OF COMBINING **ART**, THE **HUMANITIES**, AND **ENGINEERING** IS NOT **NEW**. RATHER, IT IS SOMETHING WE HAVE TO **RECOVER**, SOMETHING WE MAY HAVE **LOST** IN OUR QUEST FOR **SPECIALIZATION**. **BUCKMINSTER FULLER** HAD THE NOTION OF A **COMPREHENSIVE ANTICIPATORY DESIGN SCIENTIST** -- "AN EMERGING SYNTHESIS OF ARTIST, INVENTOR, MECHANIC, OBJECTIVE ECONOMIST, AND EVOLUTIONARY STRATEGIST." THERE ARE **ECHOES** OF SUCH **HYBRIDS** THROUGHOUT HISTORY -- ARISTOTLE, DA VINCI, MICHELANGELO, ADA LOVELACE, AND MANY OTHERS WHO NATURALLY BRIDGED WHAT WE THINK OF TODAY AS "SEPARATE DISCIPLINES"...

AND... **WILE E. COYOTE!** (THE **UNRELENTING DESIGNER!**)

ARISTOTLE
THE LOGICAL ETHICIST

LEONARDO
THE INVENTOR ARTIST

ADA LOVELACE
THE POETICAL PROGRAMMER

YES! THE PATRON SAINT OF BUILDERS, HACKERS, AND THOSE WHO **PERSEVERE!**

"If it is reason that makes man, it is sentiment that guides him."

—**Jean-Jacques Rousseau** (1761)
Julie, or the New Heloise

AESTHETICS FIRST

It may seem a radical idea to prioritize **aesthetics**—before science, technology, engineering, even before ethics and morality. And yet aesthetics—the intellectual-emotional recognition of beauty not only artistic but broadly humanistic—provides the human context and impetus to learn subjects requiring advanced logic and reason, in a way that is not anchored solely in the calculus of need, but with an artfulness that can only be human. To value things for their intrinsic worth, beyond pure function, may well be to appreciate a core condition of being **human**.

In our increasingly technological society, we lack an aesthetic dimension—a baseline awareness of shared humanity and the beauty underlying our capacity for reason, civility, and morality. If there is truth to this line of thinking, then today's engineer may need to become much more than a specialist and evolve into a synthesis of a technological artist, a moral-ethical inventor, and a compassionate system designer.

Although aesthetics is not the direct conduit for knowledge, discovery, innovation, civil society, or governance, it supplies a common humanity that stands to make us capable and worthy of such achievements. Alongside **s**cience, **t**echnology, **e**ngineering, **m**athematics, we need to educate ourselves in **aesthetics**. Before we can truly know other things, we must learn to be human beings.

WORRY, BE HAPPY

*OR, HOW WE LEARNED TO **WORRY DEEPLY** AND STILL **LOVE** THE FUTURE...*

THERE IS **PLENTY** TO **WORRY ABOUT** IN OUR HUMAN-DESIGNED WORLD, BUT THERE IS ALSO MUCH THAT IS **BEAUTIFUL** AND **WORTHWHILE.** WE HAVE AUTHENTIC REASONS TO BE HAPPY, DESPITE THE CHAOS. AS **TROUBLED** AND **VULNERABLE** AS WE ARE, HUMANS ARE **WONDROUS** AND **BEAUTIFUL** CREATURES. IF WE HAVE TO REACH A STATE OF NOT WORRYING TO BE HAPPY, THAT DAY MAY NEVER COME. PLUS, TO CREATE THINGS OF THE **LIGHT**, IT HELPS TO KNOW THAT THE **DARKNESS** EXISTS -- NOT TO TURN AWAY BUT TO BE HAPPY, TO BE TRUE TO OURSELF, IN THE FACE OF CHAOS.

THIS IS A GOOD PLACE TO EXPLORE THE **OTHER** SIDE OF DESIGN... SO FAR, WE'VE TALKED ABOUT DESIGN'S IMPACT ON THE **END USER**, AS A RESULT OF THE DESIGNER'S CHOICES. BUT WHAT ABOUT THE MEANING AND IMPLICATIONS OF DESIGN FOR THE **DESIGNER?**

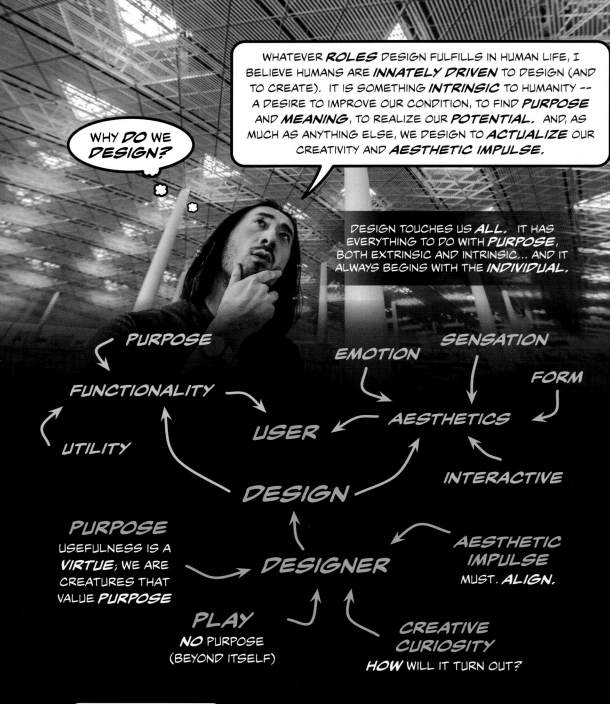

WHY **DO WE DESIGN?**

WHATEVER **ROLES** DESIGN FULFILLS IN HUMAN LIFE, I BELIEVE HUMANS ARE **INNATELY DRIVEN** TO DESIGN (AND TO CREATE). IT IS SOMETHING **INTRINSIC** TO HUMANITY -- A DESIRE TO IMPROVE OUR CONDITION, TO FIND **PURPOSE** AND **MEANING**, TO REALIZE OUR **POTENTIAL**. AND, AS MUCH AS ANYTHING ELSE, WE DESIGN TO **ACTUALIZE** OUR CREATIVITY AND **AESTHETIC IMPULSE**.

DESIGN TOUCHES US **ALL**. IT HAS EVERYTHING TO DO WITH **PURPOSE**, BOTH EXTRINSIC AND INTRINSIC... AND IT ALWAYS BEGINS WITH THE **INDIVIDUAL**.

PURPOSE

FUNCTIONALITY

UTILITY

SENSATION

EMOTION

FORM

USER

AESTHETICS

INTERACTIVE

DESIGN

PURPOSE
USEFULNESS IS A **VIRTUE**; WE ARE CREATURES THAT VALUE **PURPOSE**

DESIGNER

AESTHETIC IMPULSE
MUST. **ALIGN**.

PLAY
NO PURPOSE (BEYOND ITSELF)

CREATIVE CURIOSITY
HOW WILL IT TURN OUT?

Ⓐ PRINCIPLE 8.17 *WE DESIGN AS AN ACT OF PLAY*

DESIGN MAY ARISE OUT OF A MIXTURE OF **NEEDS** AND **VALUES**, BUT AS A **CREATIVE ENDEAVOR** IT IS A BOTTOM-UP, **INSIDE-OUT** PROCESS, CLOSELY RESEMBLING A PROLONGED ACT OF **PLAY** -- AN ISOLATED, PROTECTED SPACE AND TIME WHERE **AESTHETIC IMPULSES**, **CREATIVE CURIOSITIES**, AND A LOT OF **BAD IDEAS** CAN BE GIVEN FREEDOM TO ROAM, TRIED IN EARNEST, AND FOLLOWED TO THEIR LOGICAL OR ILLOGICAL CONCLUSIONS, WHICH MAY OR MAY NOT TRANSLATE TO ELEMENTS IN THE RESULTING DESIGN. WHERE DESIGN ENDS UP IS OFTEN NOT WHERE IT STARTS.

DESIGN CANNOT BE **SEPARATED** FROM A NOTION OF **PURPOSE**. IT IS THROUGH PURPOSE THAT WE **UNDERSTAND** MUCH OF THE WORLD -- AND IT IS PURPOSE THAT PROVIDES THE DRIVING FORCE FOR **CHANGE** AND **ACTION**. TO **ARISTOTLE**, TO TRULY **KNOW** A THING IS TO KNOW ITS **CAUSE**. **EXTRINSIC PURPOSE**, IN THIS SENSE, IS THE EXTERNAL "THAT FOR THE SAKE OF WHICH" THAT MOTIVATES OUR ACTIONS: "I DRINK FOR THE SAKE OF QUENCHING THIRST" AND "I RIDE THE TRAIN TO GET TO WORK."

WE DESIGN TO SATISFY A SENSE OF PURPOSE

TO **KNOW** A THING IS TO KNOW ITS **PURPOSE** -- A SIMPLE AND PROFOUND NOTION ABOUT **KNOWLEDGE** AND **FUNCTIONALITY**. IN A WAY, DESIGN IS THE IMBUING OF **PURPOSE** INTO A THING, AN ARTICULATION OF ITS **ESSENTIAL NATURE**.

MORE THAN **TWO MILLENNIA** AGO, **ARISTOTLE** ASKED A PROFOUND QUESTION ABOUT THE **PURPOSE** OF HUMAN LIFE, AND HE ATTEMPTED TO ADDRESS IT BY TRACING SEQUENCES OF "THAT FOR THE SAKE OF WHICH" -- LIFE'S HYPOTHETICAL IMPERATIVES, THE **MEANS** TO VARIOUS **ENDS**, WHICH **THEMSELVES** TURN OUT TO BE **MEANS** TO YET **GREATER ENDS**. WHY DO WE WANT TO CATCH THE TRAIN? ...TO GO TO WORK. WHY WORK? ...TO PAY FOR FOOD AND RENT. WHY? SO WE CAN SUSTAIN OUR LIFE, SO WE CAN... WORK MORE??

ARISTOTLE SEARCHED FOR A **NON-CYCLIC** ENDPOINT AT THE **TERMINUS** OF THIS **CAUSAL CHAIN** -- AN **ULTIMATE** AND **FINAL CAUSE**, BEYOND WHICH IT MAKES NO SENSE TO ASK "AND FOR WHAT?" -- AN END THAT **TRANSCENDS** THE SEEMINGLY ENDLESS LOOP, THAT IS NOT YET A MEANS TO A GREATER END, BUT AN END IN ITSELF...

HMM...

THESE ARE **MEANS**, BUT TO WHAT *ULTIMATE END??*

WHAT ULTIMATE **END TRANSCENDS** THIS ENDLESS **LOOP** OF MEANS-TO-ENDS?

ARISTOTLE ARRIVED AT A SORT OF **SELF-RECURSIVE** NOTION -- THAT THE **PURPOSE** OF LIFE IS THE **PURSUIT OF HAPPINESS** FOR *ITS OWN SAKE.* HAPPINESS NOT OF THE **HEDONIC** VARIETY (MAXIMIZE PLEASURE/MINIMIZE PAIN), BUT A SELF-DEFINED, SELF-DEFINING PURSUIT OF A **FLOURISHING LIFE**, FOR NO GREATER PURPOSE THAN *ITSELF* -- HE CALLED THIS...

...EUDAIMONIA!

PURSUIT OF HAPPINESS AS A FLOURISHING LIFE -- NOT AS A MEANS-TO-AN-END, BUT AS AN END-IN-ITSELF!

ARISTOTLE LIKENED SUCH A PURSUIT TO **LIFE** ON THE **ISLE OF THE BLESSED**... AN ENDEAVOR THAT IS NOT ABOUT AN EXTERNAL GOAL BUT ABOUT THE SHEER **INTRINSIC WORTH** OF THE ACTIVITY **ITSELF.** IT REFLECTS A BELIEF IN **HUMAN BEINGS** AS **ARTISTIC CREATURES** WHO ARE **SELF-DEFINING** AND **SELF-CREATING**, AND WHOSE **IMPERATIVE** IS TO EVER MORE FULLY REALIZE OUR OWN POTENTIAL.

THERE IS GREAT **DISAGREEMENT**, OF COURSE, AS TO **HOW** SUCH HUMAN FLOURISHING MIGHT BE PURSUED (ARISTOTLE BELIEVED IT WOULD BE THROUGH A LIFE OF "**VIRTUOUS** ACTIVITY IN ACCORDANCE WITH REASON"), BUT THAT'S ALSO KIND OF THE POINT OF A **SELF-DEFINING** ACTIVITY: WE DON'T HAVE TO AGREE. EUDAIMONIA IS NOT A **GOAL** TO ATTAIN ONCE AND FOR ALL, BUT AN **ONGOING**, ACTIVE PROCESS OF **SELF-ACTUALIZATION**, A ROAD WHOSE DESTINATION IS THE **TRAVELING** OF IT. IT'S **LIFE** THAT FULLY EXPRESSES A **LIBERTY** TO **PURSUE** ONE'S **HAPPINESS**, AND ITS RESULT IS AN **OVERARCHING AESTHETIC** THAT WE MIGHT CALL **WAY OF LIFE.**

"NOW WE CALL THAT WHICH IS **IN ITSELF** WORTHY OF PURSUIT MORE **FINAL** THAN THAT WHICH IS WORTHY OF PURSUIT FOR THE SAKE OF **SOMETHING ELSE**... AND WE CALL FINAL WITHOUT QUALIFICATION THAT WHICH IS **ALWAYS** DESIRABLE **IN ITSELF** AND NEVER FOR THE SAKE OF SOMETHING ELSE. NOW SUCH A THING **HAPPINESS**, ABOVE ALL ELSE, IS HELD TO BE."

-- ARISTOTLE, *NICOMACHEAN ETHICS*

AN **ULTIMATE STRANGE DESIGN LOOP**, A SUBLIME **OUROBOROS** CONSUMING ITS OWN TAIL... AN **END-IN-ITSELF** MADE OF COUNTLESS **MEANS-TO-ENDS!**

YUM!

PLAY

ART

MUSIC

FRIENDSHIP

KNOWLEDGE

BEAUTY!

TRUTH!

...ALL YE **KNOW** ON EARTH, AND ALL YE **NEED** TO KNOW.

TASTY FOOD

...WE DESIGN **ARTFULLY** WHEN WE DESIGN WITH **EUDAIMONICS** AS A **CORE VALUE**, ONE THAT ASKS SIMPLY, AUTHENTICALLY, "DOES THE DESIGNED PRODUCT, EXPERIENCE, OR POLICY MAKE LIFE **MORE FLOURISHING?**"

ALL TOO OFTEN, **DESIGN** IS MOTIVATED **PRIMARILY** BY HYPOTHETICAL IMPERATIVES (E.G., **INCOME MAXIMIZATION** OR -- ON A LARGER ECONOMIC SCALE -- THE OVERARCHING PARADIGMATIC GOAL OF INCREASING GDP) -- AND WE LOSE SIGHT OF THE UNDERLYING **EUDAIMONICS**, ONE OF THE MOST IMPORTANT AND SIMPLEST CONSIDERATIONS IN DESIGN.

IN THE LIMIT, PRAGMATICS GIVE RISE TO AESTHETICS; MEANS MELT INTO ENDS; **FUNCTION** BECOMES **FORM.**

DESIGN IS NOT ROCKET SCIENCE.* IT'S SIMPLY, AND PROFOUNDLY, A HUMAN ENDEAVOR THAT IS AS MUCH ABOUT THE **OUTCOME** AS IT IS ABOUT **HOW WE GET THERE**, AND THE **QUALITIES** IT PROMOTES, VALUABLE **IN THEMSELVES.** DESIGN'S CHALLENGE IS LOGICAL, TECHNOLOGICAL, SCIENTIFIC, BUT THE GREATER CHALLENGE IS ALWAYS **HUMANISTIC.**

PRINCIPLE 8.19

*UNLESS YOU ACTUALLY DESIGN ROCKETS

WE DESIGN TO PROMOTE EUDAIMONICS

"**DO NO EVIL**" IS A FUNDAMENTALLY WEAKER STATEMENT THAN "**DO GOOD.**" THIS IS WHY WE DESIGN ARTFULLY, TO **CHOOSE** TO MAKE SOMETHING BOTH **USEFUL** AND **AWESOME** -- TO **SERVE** PEOPLE AS **ENDS-IN-THEMSELVES**, TO **PROACTIVELY** SUPPORT THINGS THAT MAKE LIFE **FLOURISHING** AND MEANINGFUL. AND THE THING IS, WHEN WE DESIGN WITH SUCH **INTENTIONALITY**, IT **SHOWS THROUGH.**

WHAT IS MY **PURPOSE?**

UM, TO MAKE MY LIFE MORE **EUDAIMONIC?**

HUH... HOW DO I DO **THAT?**

BY **BEING** A USEFUL THING, AND BY IMPARTING DELIGHT IN HOW YOU FULFILL YOUR **ESSENTIAL PURPOSE** AS A PENCIL BAG, IN A WAY THAT MAKES ME A LITTLE MORE **THOUGHTFUL** AND THAT I FIND **BEAUTIFUL** IN ITSELF! ALL OF THAT IS PART OF **WHAT** YOU ARE.

OH **DANG.**

I'M REALLY GLAD SOMEONE WENT **OUT OF THEIR WAY** TO DESIGN YOU AS YOU ARE!

ME TOO... I HAVE **FULFILLED** MY PURPOSE, FUNCTIONAL AND AESTHETIC...

MUST BE **NICE** TO KNOW WHAT YOU ARE **MEANT** TO DO...

YES, IN **ESSENTIALIST** TERMS, MY **ESSENCE** PRECEDED MY **EXISTENCE.** I CAME INTO BEING IN ORDER TO FULFILL A PREORDAINED PURPOSE -- THAT IS THE CORE OF MY **ESSENTIAL** NATURE AND THE **MEANING** OF MY EXISTENCE (AND OF **ALL** DESIGNED THINGS). YOU HUMANS SEEM MORE... **COMPLICATED.**

HAH. PERHAPS. **EXISTENTIALISM** WOULD SAY OUR **EXISTENCE** PRECEDES OUR **ESSENCE** -- THAT IS, THERE IS NO PREORDAINED PURPOSE OR **MEANING** IN OUR COMING INTO BEING, WHICH I REALIZE CAN SEEM QUITE **SAD** AND **DEPRESSING.**

ON THE **CONTRARY**, I FIND IT EXCEEDINGLY **BEAUTIFUL.** TO BE BORN **WITHOUT** PREORDAINED PURPOSE IS TO BE **TRULY FREE.** I CAN'T IMAGINE ANYTHING **MORE** FREE. YOU GET TO **MAKE** YOUR PURPOSE FOR YOURSELF.

HMM. THAT **IS** BEAUTIFUL, THOUGH NOT SURE IF WE QUITE HAVE A CHOICE... **SARTRE** WOULD SAY WE ARE "**CONDEMNED** TO BE **FREE**,"

WELL... SARTRE DIDN'T HAVE A FUN PENCIL BAG LIKE ME!

WE DESIGN FROM AN *AESTHETIC IMPULSE*

ANOTHER *IMPETUS* TO DESIGN SEEMS TO ARISE FROM AN INNATE *AESTHETIC IMPULSE.* IT'S WHY WE *ARRANGE* OUR HOMES, OUR WARDROBES, AND ALL OUR THINGS -- IT GIVES US A SUBTLE JOY WHEN THINGS *FIT*, WHEN WE BRING THINGS INTO *ALIGNMENT* WITH OUR DEEPLY HELD *NOTIONS* AND *PRINCIPLES* OF "HOW THINGS OUGHT TO BE." IT MAY *ARISE* FROM *FUNCTION*, BUT IT IS *GROUNDED* IN *AESTHETICS.*

...I THINK IT'S *AWESOME* THAT WE FIND THINGS *BEAUTIFUL*...

YA KNOW...

...THAT WE *CAN* IS DEEPLY *BEAUTIFUL* IN ITSELF!

BUT *WHY* DO WE FIND *ANYTHING* BEAUTIFUL? WHY ARE WE *NOT CONTENT* WITH PURE FUNCTIONALITY? WHY ARE WE NOT SATISFIED TO LIVE IN IDENTICAL WAREHOUSE-LIKE DWELLINGS, TO WEAR THE SAME CLOTHES, TO BE WITHOUT MUSIC?

WHATEVER ITS NATURE OR ORIGIN, *AESTHETICS* SEEMS TO COME OUT OF SOME COMBINED CAPACITY FOR *PERCEPTION*, *EMOTION*, AND *REASON* -- AND LIKE *PLAY*, IT OCCUPIES A REALM OF ITS OWN, ARISING OUT OF AND SOMEHOW RISING *ABOVE* THE WORLD AROUND US.

THIS IS EITHER A *GROTESQUE*, BROKEN, FROZEN TREE TRUNK, OR IT'S A *DANCE* OF LIGHT AND SHADOWS, TEXTURES THAT YOU CAN NEARLY *FEEL*, *GOTHIC SHAPES* AND *CROOKED SPIRES* WAYWARDLY REACHING TO THE SKY, EVOKING AT ONCE WINTERY CALM AND SOMETHING *UNSETTLING.* THERE IS A *MUSIC* TO IT IN THE SAME WAY *SCULPTURES* HAVE MUSIC, FROZEN IN TIME, THAT WE CAN TACITLY *FEEL*, AS A CONSEQUENCE OF *EXPERIENCE* AND *AESTHETIC JUDGMENT.*

AS WE'VE EXPLORED THROUGHOUT THIS BOOK, **AESTHETICS** FINDS ITSELF IN THAT CATEGORY OF ENDS-IN-THEMSELVES, WITH THINGS LIKE **PLAY** AND **ART.** ON ONE HAND, IT SEEMS UTTERLY **SUPERFLUOUS** AND **ESSENTIALLY USELESS** OUTSIDE ITSELF, AND YET THIS **SUPERFLUITY** -- THE **LACK OF NECESSITY** -- IS A FOUNDATION OF THE PURSUIT'S **INTRINSIC VALUE** AND A PRECONDITION TO **BEING FREE.** AND THIS FREEDOM, IN TURN, IS A PRECONDITION FOR **PLAY**, **ART**, PURSUIT OF **HUMAN FLOURISHING**, AND THINGS WE CHERISH FOR THEIR **OWN SAKE.** SUCH AN APPARENT **CONTRADICTION** -- THAT THE "USELESS" THINGS ARE AMONG THE MOST **ESSENTIAL** IN LIFE -- HAS ALWAYS SEEMED, TO ME, SOMETHING OF A **MIRACLE**...

WHATEVER **HAPPINESS** MAY MEAN, ONE ROUGH MEASURE OF IT MIGHT BE FOUND IN THE **RATIO** OF OUR LIFE WE SPEND IN ENDS-IN-THEMSELVES (THINGS WE DO PRIMARILY FOR THEIR INTRINSIC WORTH AND THAT WE **CHOOSE** TO DO) VERSUS MEANS-TO-ENDS (WHAT WE MORE OR LESS **NEED** TO DO TO GET ANYWHERE) -- OR MORE PRECISELY, IN THEIR **BALANCE**...

ENDS IN THEMSELVES

BALANCE POINT?

MEANS TO ENDS

WHAT IS A **GOOD** BALANCE?

TOO MUCH OF EITHER SEEMS **UNDESIRABLE**: ALL **MEANS** AND NO **ENDS** ROBS US OF MEANING AND JOY; BUT WHILE THE OTHER EXTREME -- A 100 PERCENT, **PURELY** ENDS-IN-THEMSELVES TYPE OF EXISTENCE -- MAY SEEM SUPREMELY ATTRACTIVE, **CAN** WE, AS HUMANS, ACTUALLY **HANDLE** IT? WITHOUT THE **NEED** FOR MEANS-TO-ENDS, WILL THE ENDS-IN-THEMSELVES FEEL AS **MEANINGFUL** AND AS **SATISFYING**, WITHOUT HAVING TO, SAY, **WORK** AT THEM?

VS.

EVEN IF WE COULD ACHIEVE IT, I HAVE NO IDEA IF WE CAN **HANDLE** THIS ALMOST **GOD-LIKE** LIFE?

ENDS IN THEMSELVES

IN OTHER WORDS, MAYBE OUR **STRUGGLES** ARE WHAT, AT LEAST IN PART, GIVE MEANING TO THE AESTHETICS OF BEING HUMAN. AND THAT **MIXTURE** AND **SYNTHESIS** OF MEANS AND ENDS, TO ME, **IS** THE **HUMAN** CONDITION -- NOT A 100 PERCENT ENDS-IN-THEMSELVES TYPE OF LIFE (WHICH SOUNDS LIKE THE "LIFE" OF A **GOD**), NOR A 100 PERCENT MEANS-TO-ENDS TYPE OF EXISTENCE, WHERE ALL ENDS ARE JUST THEMSELVES MEANS TO YET GREATER ENDS AND NEVER ENDS IN THEMSELVES (WHICH SOUNDS LIKE THE LIFE OF AN UNQUESTIONING **AUTOMATON** OR WORSE, A **SLAVE**).

IN THIS SENSE, THE MEANS-TO-ENDS AND THE ENDS-IN-THEMSELVES ARE AS **DIFFERENT** AS **NIGHT** AND **DAY**, BUT LIKE NIGHT AND DAY, EACH GIVES **MEANING** AND **DEFINITION** TO THE OTHER. THERE IS SOMETHING BEAUTIFULLY **IRONIC** ABOUT THIS **INTER-DEPENDENCE**, A DEFINING CHARACTERISTIC OF WHO WE ARE.

WHATEVER THE BALANCE, OUR LIFE TAKES ON BOTH A **PRAGMATIC** AND AN **AESTHETIC** DIMENSION. IN A TIME OF EVOLVING TECHNOLOGY, THE PRAGMATICS ARE **NOT** WHAT IS **AT STAKE** -- WE ARE ADEPT AT **INNOVATION**, MORE SO THAN FIGURING OUT WHAT TO DO WITH ITS RESULTS.

IT IS THE **AESTHETIC DIMENSION** THAT WE, IN OUR SO-CALLED **MODERNITY**, ARE EVER AT RISK OF **LOSING**, AS WE TOSS AND TURN IN THE TIDES OF TECHNOLOGICAL, CULTURAL, ENVIRONMENTAL, AND SOCIO-POLITICAL CHANGE THAT LEAVE NO PART OF LIFE UNTOUCHED. THE PROJECT OF **ARTFUL DESIGN** IS **NOT** ONE OF **SALVATION**, BUT **RECLAMATION** OF OUR **AUTHENTICITY**, A **RECOGNITION** AND **RESTORATION** OF THE VALUE OF LIFE'S ENDS-IN-THEMSELVES, IN **INTERPLAY** WITH THE INCREASINGLY TECHNOLOGY-MEDIATED MEANS-TO-ENDS THAT SHAPE OUR LIVES. LIKE ALL EUDAIMONIC UNDERTAKINGS, THIS IS NOT A ONE-TIME GOAL, BUT SOMETHING WE MUST **CONFRONT** AND **RECONQUER** EACH DAY.

RECLAIM!

RESTORE!

RECOGNIZE!

WHY SHOULD THE **DESIGN** OF HUMAN THINGS...

...BE ANYTHING **OTHER** THAN HUMAN?

⚘ PRINCIPLE 8.21

AESTHETICS AS SELF-EMANCIPATION!

NO MATTER WHAT WE DESIGN -- TOOLS, TOYS, GAMES, POLICY, LAW -- WE AIM TO MAKE THINGS THAT FULFILL THEIR **FUNCTION**. AT THE SAME TIME, WE DESIGN TO MAKE LIFE **INTERESTING, AUTHENTIC,** AND **MEANINGFUL**. ULTIMATELY, WE STRIVE TO DESIGN **USEFUL** THINGS THAT ARE INTRINSICALLY **BEAUTIFUL** -- THAT BRING THE WORLD INTO **ALIGNMENT** WITH OUR IDEA OF THE **GOOD**, THE **AUTHENTIC**, AND THE **JUST** -- AS AN **END IN ITSELF**. DESIGN SHOULD EXPRESS OUR HUMANITY, ACKNOWLEDGING THAT WE VALUE **PLAY**, FIND **BEAUTY** IN AUTHENTICITY, SEEK TO CONNECT WITH ONE ANOTHER, AND DESIRE TO BE **FREE**.

PRINCIPLE 8.22

DESIGN IS BORN OF LIFE, INCORPORATES IT, AND IS INSEPARABLE FROM IT

DESIGN IS ALL AROUND US, WITH ALL THE JOY AND SORROW, DIRT, GRIME, BEAUTY, AND IMPERFECTIONS OF LIFE.

WE DESIGN AS A *SYNTHESIS* OF LIFE'S MEANS AND ENDS, AS *LIFE ITSELF* IS A SYNTHESIS. TO DESIGN IS TO SEE ITS NEEDS THROUGH THE EYES OF AN ENGINEER, BUT ALSO TO KNOW ITS *PULSE*, ITS *DREAMS*, *FEARS*, AND *YEARNINGS* THROUGH THE LENS OF A *HUMAN BEING* TAKING IN THE WORLD... TO EARNESTLY TRY TO FOREGROUND *BEAUTY* OUT OF *CHAOS*...

FOR THERE IS **BEAUTY** IN PEOPLE, IN WHAT THEY **DO**, IN THE **KINDNESS** THEY SHOW OUT OF **CHOICE**, NOT NECESSITY; IN **SIMPLICITY**, IN **LESS** RATHER THAN **MORE**, AND IN THE **SHEER WONDER** AND STILLNESS OF MOMENTS IN WHICH WE FIND OURSELVES -- IN NATURE, HUMAN-MADE THINGS, IN THE VERY PHENOMENON THAT IS **LIFE** ITSELF.

⚙ PRINCIPLE 8.23   AS PHOTOGRAPHER **KATHERINE EMERY** URGES US

PAUSE. FOR THE SMALL GOOD THINGS.

THERE IS SOMETHING SUBLIME **NESTLED** IN THE EVERYDAY, BEYOND THE **FACADE** OF THE SEEMINGLY MUNDANE, THAT GIVES LIFE MORE MEANING AND **RICHNESS** THAN WE MIGHT **ALLOW** OURSELVES TO BELIEVE.

THE **SUBLIME** EXISTS AT THE **RAZOR'S EDGE** BETWEEN LIFE AND THE ETHEREAL BEAUTY HIDDEN BENEATH ITS SURFACE, LIKE **INVISIBLE CITIES** IN AN ALL-TOO-VISIBLE WORLD. WE LIVE IN IT, BUT ONLY FROM TIME TO TIME DO WE REALIZE ITS **MEANING** AND **BEAUTY.**

AND **BEYOND**, THERE ARE THINGS FAR GREATER THAN OUR **TRIUMPHS** AND OUR **FAILURES**...

THEY ARE **OUT THERE**, BUT THEY ARE ALSO **HERE** IN THE STUFF OF OUR HOPES, FEARS, AND DESIRES -- WHERE WE FEEL THE SUBTLE CONTOURS OF **INVISIBLE TRUTHS**, TIMELESS LIKE PHOTOGRAPHS IN THE MIND, AND JUST BEYOND THE VEIL OF OUR PERCEPTION.

WE ARE **REMINDED** OF THESE THINGS BEYOND THE WORRIES OF OUR DAILY LIVES, THROUGH THE ENDS-IN-THEMSELVES...

THE BONDS OF FAMILY...

THE KINDNESS IN FRIENDSHIP...

BEAUTY AND TRUTH NOT AS **SEPARATE** FROM THE **MUNDANITY** OF EVERYDAY LIFE, BUT ARISING OUT OF IT, CUT FROM ITS VERY **FABRIC**.

BEAUTY IN THE **STILLNESS** OF THE MOMENT

IN THE **ARTFUL**...

...AND THE **SUBLIME**,

WHERE **TIME** CAN STAND **STILL**...

443

SOME MOMENTS **STAY** WITH US, TIMELESS EVEN AS TIME MARCHES ON.

FROM TIME TO TIME, I DREAM OF MY **GRANDFATHER.** WE EAT LUNCH, AND HE LOOKS UP TO SAY HELLO, AND I AM SURPRISED AS ALWAYS TO SEE HIM IN MY DREAMS, AMID ORDINARY LIFE. I ALWAYS WANT TO GET MY CAMERA TO REMEMBER THE OCCASION, BUT I NEVER FIND ONE. SO WE SIMPLY ENJOY THE MOMENT, BEFORE MY MIND CARRIES ME AWAY TO THINGS I USUALLY DREAM ABOUT (WEIRD WORLDS OF CRAZY CATS, SWAT TEAMS, SORCERERS, AND FALLING HAMBURGERS).

HELLO, WANG GE!

TENNESSEE WILLIAMS WROTE: "IN MEMORY EVERYTHING SEEMS TO HAPPEN TO MUSIC." I DON'T HEAR MUSIC IN MEMORIES OR DREAMS, BUT THERE IS A TRUTH TO THIS. NOT ALL MUSIC IS **HEARD.**

THERE ARE **PEOPLE**, **EXPERIENCES**, AND **MOMENTS** IN EACH OF OUR LIVES THAT STAY WITH US. THE DETAILS ARE DIFFERENT BUT UNIMPORTANT COMPARED TO THEIR **MEANING**, BOTH PERSONAL BUT ALSO **UNIVERSAL.**

BE A **GOOD** PERSON. BE HAPPY FOR THE SUCCESS OF **OTHERS.** DO THE BEST YOU CAN, AND BE GLAD WHEN YOUR FRIENDS AND COLLEAGUES DO WELL, BETTER THAN YOU -- AND ESPECIALLY YOUR **STUDENTS**, TO WHOM YOU HAVE THIS RESPONSIBILITY.

SEE YOU NEXT TIME YOU COME HOME!

IT IS THE **AESTHETICS** OF THESE ENCOUNTERS IN EVERYDAY LIFE -- MOMENTS OF **SUBLIMITY**, IN SIMPLICITY, KINDNESS, AND HUMANITY -- THAT I HAVE BEEN **SEARCHING FOR**, THAT I HAVE TRIED TO **IMBUE** INTO THE THINGS I **DESIGN**, TO MAKE THE TECHNOLOGICAL WORLD MORE LIKE THIS **HUMAN WORLD.** BECAUSE, AT THE END OF THE DAY, THE TECHNOLOGY WORLD **IS** PART OF THE HUMAN WORLD. AS **STEWART BRAND** WROTE ON THE BACK COVER OF THE **WHOLE EARTH CATALOG**, "WE CANNOT PUT IT TOGETHER. IT **IS** TOGETHER."

ALTHOUGH THESE ARE **MY** GRANDFATHER AND GRANDMOTHER, THERE IS SOMEONE LIKE THIS IN **ALL** OF OUR LIVES, WHOSE WORDS AND DEEDS ECHO THROUGH US.

THE BITTERSWEET **EVERYDAY SUBLIME**

It is in the moments of everyday life that we are reminded of something greater. We squint at the contours of the transcendent, and we are both moved and made still, seized by the beauty of the moment, and saddened to know we will never be here again, as we are.

Like the picture of Dorian Gray in reverse, these are moments frozen in time, out of time—a photograph of a meaningful occasion or a perfect day, while we, the participants, move on and grow older. But having experienced the moment, we are changed by it.

Perhaps innately we know that all happiness is fleeting, but that beauty is eternal—and it exists both in the immensity of things, and also in the cracks of everyday life, in the small good things, which seem no less miraculous.

This is the sublimity we search for, in design as in life, for these are the moments we live for. They arise from life itself, while reaching for the transcendent that all truly artful things reach toward—to bear witness to a kind of **truth** about things as they really are.

ARTFUL CREATION, THOUGH ARISING OUT OF INDIVIDUALS, SOMEHOW SEEMS **UNIVERSAL.** SUCH DESIGNS SPEAK TO SOMETHING **INVISIBLY SHARED** IN PEOPLE, AND **TRANSPORT** US, **ELEVATE** US WITH THEIR SUBLIMITY. BEAUTY IS NOT ABOUT WHO MADE IT, BUT MORE THAT IT **WAS** MADE, THAT IT COULD HAVE EXISTED AT ALL...

THEY REMIND US WE ARE **ALIVE**, THAT WE **BREATHE**...

...**MARVEL** AND **FEEL.**

THEY TAP INTO OUR SECRET DESIRE TO **EXPRESS**, TO SING OUT LOUD TO FRIENDS AND STRANGERS, TO **PLAY**, WITHOUT WORRY OR NECESSITY...

...TO **HOLD ON** TO THE CHILD-LIKE WONDER WE HAVE FOR THE STRANGE, UNCANNY, AND WHIMSICAL!

THIS IS THE **SUBLIMITY** THAT PERMEATES THE **MUNDANITY** OF EVERYDAY LIFE, CONTEXTUALIZED THROUGH THE GRIDWORK OF TIME, SPACE, AND THE **FRAGILITY** OF OUR MEMORY...

...IN OUR YEARNING TO **REMEMBER.**

TO DESIGN ARTFULLY IS TO KNOW THAT THE MEDIUM EVER **MODULATES** THE MESSAGE, THAT TOOLS AND INSTRUMENTS CARRY WITH THEM DEEP **NUANCES** THAT SUGGEST THEIR **TRUE AFFORDANCES**, AND MAKE US **FEEL** WHEN WE USE THEM...

...TO VALUE **PLAY** AS PART OF OUR HUMANITY, AS AN EXPRESSION OF BEING **TRULY FREE**, AND TO UNDERSTAND THE **VALUE** OF DOING THINGS BECAUSE THEY ARE **GOOD IN THEMSELVES**...

...TO REALIZE **WE** ARE BOTH OUR OWN ORGANISM AND PART OF A GREATER **COMMUNITY** -- AND THAT THROUGH THIS MUTUALITY WE ARE FULLY DEFINED AS **INDIVIDUALS** AND AS A **SOCIETY**.

TOTAL STRANGERS

FAMILIAR STRANGERS

ACQUAINTANCES (E.G., CO-WORKERS)

FRIENDS

LOVED ONES (FAMILY AND CLOSE FRIENDS)

THE SELF?

...AND TO KNOW THAT THE THINGS WE DESIGN ARE NOT **NEUTRAL**, BEAUTY, AUTHENTICITY, MORALITY ARE, AT THEIR CORE, **CHOICES**, THEY **SHOW THROUGH** IN WHAT WE **MAKE**, NO LESS THAN THE ACTIONS WE TAKE, DESIGN **DEFINES** WHO WE ARE.

⚙ PRINCIPLE 8.24 — *DESIGN* TOWARD THE *SUBLIME*

A THING IS *USEFUL* IF IT FULFILLS A PURPOSE -- AND *BEAUTIFUL* FOR BEING JUST THE WAY IT IS. DESIGN IS FOREVER THAT *RADICAL SYNTHESIS* BETWEEN TWO *ENGINES* OF HUMANITY: THE MEANS-TO-ENDS AND THE ENDS-IN-THEMSELVES, THE IMPERATIVE OF EXTERNAL *PURPOSE* AND A RECOGNITION THAT THE MOST *AUTHENTIC* REASON TO DO ANYTHING IS, OFTEN, BECAUSE IT IS *WORTHWHILE IN ITSELF.* IN *EVERY* PROCESS OF MAKING SOMETHING *USEFUL* IS THE OPPORTUNITY TO *AUTHENTICATE* WHO WE ARE -- AND TO REACH FOR SOMETHING MORE THAN WE ARE...

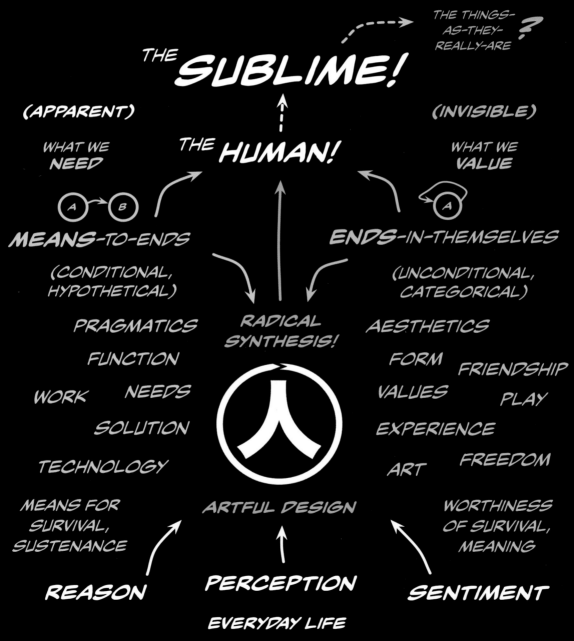

THE THINGS-AS-THEY-REALLY-ARE **?**

THE **SUBLIME!**

THE **HUMAN!**

(APPARENT)

WHAT WE **NEED**

Ⓐ→Ⓑ

MEANS-TO-ENDS

(CONDITIONAL, HYPOTHETICAL)

PRAGMATICS

FUNCTION

WORK   NEEDS

SOLUTION

TECHNOLOGY

MEANS FOR SURVIVAL, SUSTENANCE

(INVISIBLE)

WHAT WE **VALUE**

Ⓐ

ENDS-IN-THEMSELVES

(UNCONDITIONAL, CATEGORICAL)

AESTHETICS

FORM   FRIENDSHIP

VALUES   PLAY

EXPERIENCE

ART   FREEDOM

WORTHINESS OF SURVIVAL, MEANING

RADICAL SYNTHESIS!

ARTFUL DESIGN

REASON   **PERCEPTION**   **SENTIMENT**

EVERYDAY LIFE

IF THE TRANSCENDENT *THINGS-IN-THEMSELVES* RESIDE IN A MORE *IDEAL* PLANE OF EXISTENCE, THEN THE *SUBLIME* IS LIKE A *STAIRWAY* IN THE FOG, LEADING UP TOWARDS IT -- A FLEETING BRUSH AGAINST TRUE PERFECTION, A GLIMPSE BEYOND THE VEIL OF OUR EXPERIENCE, INTO THE INVISIBLE AND THE INFINITE. AND YET, THE *SUBLIME* IS TO BE *EXPERIENCED* ALL AROUND US...

THE *RHETORICAL* SUBLIME

TURN OF A PHRASE, TO INDUCE SUBLIMITY IN THE LISTENER OR READER. THE STUFF OF POETS AND ORATORS. THE DESIGN OF WORDS IN SEQUENCE TO CONVEY TRUTH.

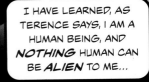

MAYA ANGELOU

I HAVE LEARNED, AS TERENCE SAYS, I AM A HUMAN BEING, AND *NOTHING* HUMAN CAN BE *ALIEN* TO ME...

THE *NATURAL* SUBLIME

AWE AND TERROR INDUCED BY THE IMMENSITY OF THINGS IN NATURE, EXPERIENCED IN THE FULLNESS OF OUR SENSES AND OUR REASON. BY DEFINITION, NOT OF HUMAN DESIGN.

MORAL SUBLIME

A CONSCIOUS, INTENTIONED MORAL CHOICE (E.G., AN ACT OF SACRIFICE). PRINCIPLED ACTION, NOT DESIGNED TO SERVE AN EXTERNAL MOTIVE. IN THE WORDS OF SYDNEY CARTON (FROM CHARLES DICKENS'S *A TALE OF TWO CITIES*): "IT IS A FAR, FAR BETTER THING THAT I DO, THAN I HAVE EVER DONE; IT'S A FAR, FAR BETTER REST THAT I GO TO, THAN I HAVE EVER KNOWN."

THE BITTERSWEET *EVERYDAY* SUBLIME

THE SUBLIMITY HIDDEN IN PLAIN SIGHT OF THE EVERYDAY, THE INVISIBLE NESTLED IN THE ALL-TOO-VISIBLE, THE MUNDANE THAT WE WOULD MISS DEARLY ONLY WHEN IT'S GONE (OR WE ARE GONE FROM IT). AN AESTHETIC "ENDGAME" OF ARTFUL DESIGN, TO REFLECT AND INDUCE POIGNANCY OF THE BITTER AND THE SWEET, OF LIFE AS WE LIVE IT.

I'VE... *SEEN* THINGS, YOU PEOPLE WOULDN'T BELIEVE...

BLADE RUNNER (1982)
(DIRECTOR: RIDLEY SCOTT)

THE *ARTFUL* SUBLIME

IMPERFECT ART MOVES YOU; *PERFECT ART* MAKES YOU *STILL*. SUBLIME *ART* IS ART WHICH IS MOST *INDISTINGUISHABLE* FROM *TRUTH*: IT DOESN'T IMPEL US INTO *ACTION*, BUT MAKES US *STILL*, SITUATING US IN ITS OWN *MENTAL REALM*, HELPING US TO RECOGNIZE SOMETHING ETERNAL AND AUTHENTIC AMID LIFE'S UNCERTAINTY, TRANSIENCE, HORRORS, AND UTTER IMPERFECTION. ART BECOMES *SUBLIME* NOT WHEN WE'VE UNDERSTOOD IT, BUT WHEN WE FEEL IT HAS *UNDERSTOOD US*. THROUGH SUBLIME ART, WE LEARN SOMETHING MORE CLEARLY ABOUT *OURSELVES* -- A *SUBTLE* BUT *AUTHENTIC* REWARD, TO KNOW WHAT IT MEANS TO BE HUMAN.

EACH IN ITS OWN WAY, **ART** OF VARIOUS MEDIUMS UNDERGOES ITS OWN **EVOLUTION** -- FIRST SERVING ONE OR MORE **PRACTICAL, SOCIETAL FUNCTIONS**, BUT EVENTUALLY ASCENDING INTO SOMETHING VALUABLE FOR **ITS OWN SAKE.** SUCH EVOLUTION ENTAILS BREAKING THROUGH A **TRANSCENDING POINT**, WHEN SOCIETY RECOGNIZES THE ART'S CAPACITY FOR THE **DEEPLY HUMAN** AND THE **SUBLIME.** THIS POINT OF **ARRIVAL** IS WHEN A CRAFT **TRANSCENDS** ITS PRESCRIBED **FUNCTIONS** AND BECOMES AN **END IN ITSELF.**

VISUAL ART
WAS: REPRESENTATIONAL

FOR MUCH OF WESTERN HISTORY, VISUAL ART WAS HELD TO BE **REPRESENTATIONAL**, DECORATIVE. IT WASN'T UNTIL THE **AGE OF ENLIGHTENMENT**, LEADING INTO THE **ROMANTIC ERA**, THAT VISUAL ART BECAME RECOGNIZED AS A **LANGUAGE** CAPABLE OF ADDRESSING THE DEEPEST HUMAN STATES OF MIND AND CONDITION, AND A **WINDOW** INTO THE SUBLIME. THERE IS **MORE** TO THE PICTURE THAN **MEETS THE EYE.**

TRANSCEND.

EVOLUTION!

EVOLUTION!

THE

SUBLIME!

EVOLUTION?

MUSIC
WAS: CLEVER CRAFT

WESTERN MUSIC CAN BE SAID TO HAVE UNDERGONE A SIMILAR EVOLUTION, SERVING SOCIETAL FUNCTIONS IN THE **COURTS OF ROYALTY**, IN **DIVINE WORSHIP** OF THE CHURCH, AND AS A MODE OF **ENTERTAINMENT** IN EVERYDAY CONTEXTS. AND WHILE WORKS OF GREAT SUBLIMITY WERE **CREATED** (BY THE LIKES OF BACH, MOZART, AND MANY OTHERS), MUSIC WAS UNDERSTOOD AND **SEEN** AS **CLEVER CRAFT** CAPABLE OF **INGENUITY** AND SOPHISTICATION. BUT IT WASN'T UNTIL THE TIME OF **ROMANTICISM** AND RULE-BENDING COMPOSERS LIKE **BEETHOVEN** THAT MUSIC WAS MORE WIDELY RECOGNIZED TO BE CAPABLE OF THE PSYCHOLOGICAL AND THE SUBLIME. IT WAS AS IF MUSIC **ALWAYS** HAD THIS POWER, BUT WE DIDN'T REALLY ALLOW OURSELVES TO BELIEVE IT.

TRANSCEND.

DESIGN?
WAS/IS: FUNCTIONAL TECHNOLOGY

IN THIS AGE OF **TECHNOLOGY**, WHERE TECHNOLOGY CAN BE BROADLY DEFINED AS "ANY INTENTIONALLY FASHIONED TOOL OR TECHNIQUE," DESIGN **HAS** THE **POWER** TO **TRANSCEND** ITS PREVIOUSLY UNDERSTOOD ROLES IN HUMAN LIFE, TO BE EVER-MORE CAPABLE OF ADDRESSING THE DEEPLY **HUMAN**... EVEN AS IT SPEAKS TO OUR EVERYDAY NEEDS (AND PERHAPS BECAUSE IT DOES).

IT IS _TIME_...

...FOR DESIGN TO _TRANSCEND_ INTO THE TRULY _ARTFUL!_ INTO THE _SUBLIME!_

ⓧ PRINCIPLE 8.25

TRANSCEND...

AND _YET_... _ARTFUL DESIGN_ IS _NOT_ ART BECAUSE DESIGN _MUST_ FULFILL A _FUNCTION_, SERVE A _PURPOSE_, IT THEREFORE CAN _NEVER_ BE A PURE END-IN-ITSELF LIKE _ART_. THAT IS THE _FUNDAMENTAL DISTINCTION_ BETWEEN DESIGN AND PURE ART, AND THEREIN LIES THE _UNIQUE NATURE_ OF _DESIGN_. IT IS _ITS OWN THING_, NEITHER PURE ART NOR ENGINEERING, BUT A _THIRD_ SPECIES, WHICH IS THE _ART_ OF MAKING _USEFUL_ THINGS -- THAT MAKE YOU _FEEL_, AND _FEEL HUMAN_... _DESIGN_ IS THE ART OF _HUMANIZING_ TECHNOLOGY.

AS _ART_ HOLDS A _MIRROR_ TO HUMAN LIFE, DESIGN CAN ENABLE, ENLIGHTEN, AND EMBODY THE HUMAN CREATURE -- A _HYBRID_ THAT CANNOT _LIVE_ WITHOUT THE MEANS-TO-ENDS OR _LIVE FULLY_ WITHOUT THE ENDS-IN-THEMSELVES.

DESIGN MUST _TRANSCEND_ TECHNOLOGY -- AND THE NOTION OF TECHNOLOGY AS "SIMPLY" A PROBLEM-SOLVER, OR AN AGENT OF CHANGE, SURVIVAL, HAPPINESS. THROUGH WHAT WE DO WITH IT, DESIGN IS -- AND CANNOT HELP BUT BE -- A _MIRROR_ TO DEFINE OUR _HUMANNESS_. WHAT WE MAKE, _MAKES US_.

...AND THAT, AS MUCH AS I AM ABLE TO TELL, _IS_ THE NATURE, MEANING, AND PURPOSE OF _DESIGN_.

"Man defines himself by his deeds—and what kind of image of man do we see in the mirror of our present times?"

—**Friedrich Schiller** (1795)
Letters on the Aesthetic Education of Man

IN SEARCH OF THE SUBLIME

We can design for the sublime no more than we can design for beauty, for these are not features of products but consequences of experience, manifested through the gridwork of sense and cognition. Yet, when we design with intention, as we do in art, we can create things that invoke the sublime, that bring into focus a truth and a beauty despite our limitations and chaos. Design cannot forsake the practical needs of humanity, but it—no less than art—can transcend them, seeking beauty in the authenticity of things, reaching for something more than we are, while speaking to precisely what we are. Sublime design is design that understands us.

"Man defines himself by his deeds—and what kind of image of man do we see in the mirror of our present times?"

—**Friedrich Schiller** (1795)
Letters on the Aesthetic Education of Man

IN SEARCH OF THE **SUBLIME**

We can design for the sublime no more than we can design for beauty, for these are not features of products but consequences of experience, manifested through the gridwork of sense and cognition. Yet, when we design with intention, as we do in art, we can create things that invoke the sublime, that bring into focus a truth and a beauty despite our limitations and chaos. Design cannot forsake the practical needs of humanity, but it—no less than art—can transcend them, seeking beauty in the authenticity of things, reaching for something more than we are, while speaking to precisely what we are. Sublime design is design that understands us.

ONE MORE thing...

...ONE MORE **PLACE**, ACTUALLY, I WANT TO TAKE US TO...

CODA
IN SEARCH OF
THE SUBLIME

COPENHAGEN, DENMARK.

A CITY, COUNTRY, AND REGION OF THE WORLD FAMOUS FOR ITS DESIGN SENSE...

ROTATING, SELF-RAISING, WIRE CHEESE CUTTER!

AN AESTHETIC SEEMINGLY ECHOED IN ITS PEACEFUL, HARMONIOUS WAY OF LIFE...

EVIDENT IN THE **SIMPLER** THINGS IN EVERYDAY LIFE...

8-HOUSE
DESIGNED BY BJARKE INGELS GROUP

INVENTIVE MODERN **ARCHITECTURAL DESIGN** NESTLES COMFORTABLY IN OPEN NATURAL SURROUNDINGS, IN AN EFFORT TO BE PRACTICALLY, SOCIALLY, AND ECOLOGICALLY PROFITABLE.

THINKIN' ABOUT THAT CLEVER CHEESE CUTTER UP THERE IS MAKIN' ME **HUNGRY**...

BJARKE INGELS GROUP AIMS FOR WHAT THEY CALL **PRAGMATIC UTOPIANISM**, WHICH IS VERY MUCH WHAT DESIGN STRIVES TO BE, I THINK.

454

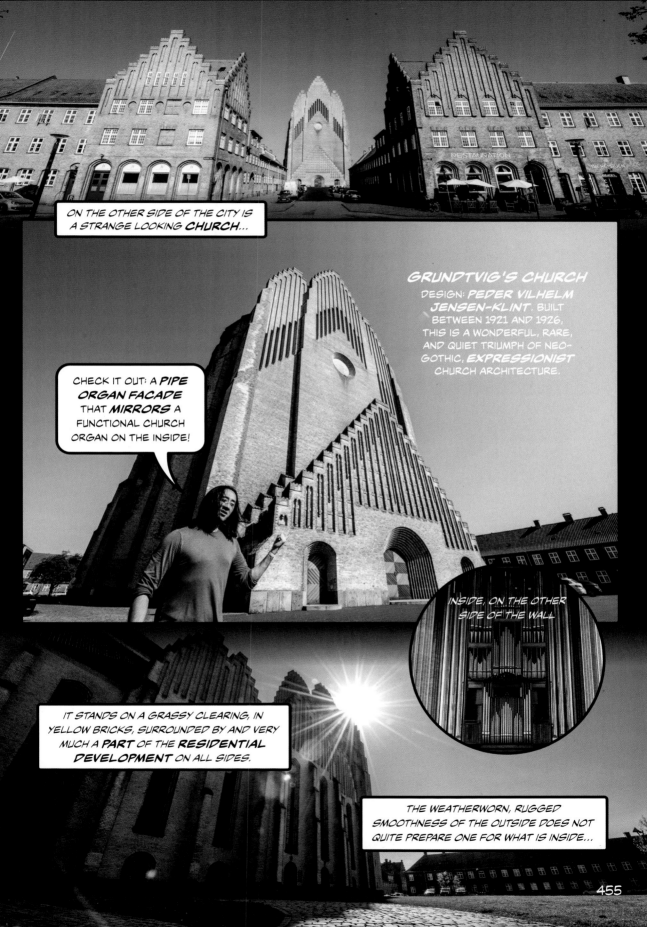

ON THE OTHER SIDE OF THE CITY IS A STRANGE LOOKING **CHURCH**...

GRUNDTVIG'S CHURCH
DESIGN: *PEDER VILHELM JENSEN-KLINT*. BUILT BETWEEN 1921 AND 1926, THIS IS A WONDERFUL, RARE, AND QUIET TRIUMPH OF NEO-GOTHIC, *EXPRESSIONIST* CHURCH ARCHITECTURE.

CHECK IT OUT: A **PIPE ORGAN FACADE** THAT **MIRRORS** A FUNCTIONAL CHURCH ORGAN ON THE INSIDE!

INSIDE, ON THE OTHER SIDE OF THE WALL

IT STANDS ON A GRASSY CLEARING, IN YELLOW BRICKS, SURROUNDED BY AND VERY MUCH A **PART OF THE RESIDENTIAL DEVELOPMENT** ON ALL SIDES.

THE WEATHERWORN, RUGGED SMOOTHNESS OF THE OUTSIDE DOES NOT QUITE PREPARE ONE FOR WHAT IS INSIDE...

...A PLACE OF ELEVATED **SIMPLICITY** AND **REFLECTION**... WHERE BOTH THE **DEVOTED** AND **WORLDLY** CAN FIND STILLNESS AND INSPIRATION.

"THE SHARP **EDGE** OF A **RAZOR** IS DIFFICULT TO TRAVERSE; THUS THE WISE SAY THE PATH TO ENLIGHTENMENT IS HARD."

-- THE **KATHA UPANISHAD**
(800-300 B.C.E.)

EACH IN OUR OWN WAY, WE SEARCH FOR THE **TRANSCENDENT**, FOR THE MEANING IN OUR LIVES, BEYOND THE MATERIALISTIC AND MECHANISTIC.

WHETHER IT'S BELIEF GROUNDED IN BEAUTY, HUMAN COMMUNION, SPIRITUALITY, OR THE IDEA OF HUMANITY'S FREEDOM TO DEFINE ITS OWN PURPOSE IN AN OTHERWISE PURPOSELESS EXISTENCE...

WE ALL SEEK TO PULL BACK THE **VEIL** OF **PERCEPTION** AND THE MUNDANITY OF EXISTENCE, TO REACH FOR **HIGHER** KNOWLEDGE AND BEAUTY -- THINGS AS THEY REALLY ARE, THAT WE NORMALLY CANNOT ACCESS...

AS WE ARE, WE MAY **NEVER GRASP** THE TRANSCENDENT, BUT IT IS THROUGH **THE SUBLIME** WE **REACH** FOR IT, AND CATCH A FLEETING GLIMPSE, A ROUGH CONTOUR, AND THE FEELING THAT THERE ARE THINGS INFINITELY MORE **IMMENSE** AND **BEAUTIFUL** THAN WE HAVE ALLOWED OURSELVES TO BELIEVE. WHAT'S MORE, **WE** -- HOWEVER INFINITESIMALLY -- ARE A **PART** OF IT IN THE MIRACLE OF EXISTENCE, IN THE **INVISIBLE POSSIBILITIES** OF EACH DAY.

THE SUBLIME **IS** THE **RAZOR'S EDGE**, BETWEEN THE TRANSCENDENT AND THE EVERYDAY...

LESS IS **SO MUCH** MORE...

ARTFUL THINGS HELP US TO TRAVERSE THE RAZOR'S EDGE, ALLOWING US TO SEE THE BEAUTY **HIDDEN** IN PLAIN SIGHT, IN THE CHAOS ALL AROUND US... SO THAT WE MAY **TRANSCEND** IT.

"AND IF YOU LISTEN VERY HARD
THE TUNE WILL COME TO YOU AT LAST
WHEN ALL ARE **ONE** AND ONE IS **ALL**
TO BE A **ROCK** AND **NOT TO ROLL...**"

THAT WHICH
MAKES US STILL

-- LED ZEPPELIN (1971)
"STAIRWAY TO HEAVEN"

BEAUTY   JUSTICE

THE
UNFATHOMABLY
IMMENSE

TRUTH

THE SMALL
GOOD THINGS

THE **SUBLIME!**

THAT WE **EXIST**
AT ALL!

AWE

TERROR

BEAUTY OF THE
MOMENT...

FREEDOM TO BE
OURSELVES

THE **INFINITE**

IF THE **SUBLIME**
IS THE **STAIRWAY** TO THE
TRANSCENDENT, THEN **DESIGN**
IS LIKE THE **MANUFACTURER**
OF THE STAIRWAY...

GENUINE ACTS
OF **KINDNESS**

STRANGE LOOPS!

STRANGE
PENCIL BAGS

THE **OMNI-**
BIOGRAPHICAL

THE **UNIVERSAL**

THE **ETERNAL**

THE **BITTERSWEET**

THE **ETHEREAL**

THE LAWS OF ARTFUL DESIGN

1. DESIGN IS THE ARTFUL SHAPING OF OUR WORLD INTO SOMETHING *USEFUL* AND *HUMAN.*

2. ANYTHING WORTH DESIGNING IS WORTH DESIGNING *BEAUTIFULLY.*

WHOA.

3. DESIGN IS AN ACT OF *ALIGNMENT* WITH OUR NOTIONS OF THE PURPOSEFUL AND THE GOOD.

4. DESIGN IS THE RADICAL SYNTHESIS OF *MEANS* AND *ENDS.*

5. DESIGN NOT ONLY FROM **NEEDS** -- BUT FROM THE **VALUES** BEHIND THEM.

6. DESIGN IS THE EMBODIED **CONSCIENCE** OF TECHNOLOGY.

7. DESIGN SHOULD **UNDERSTAND US.**

8. WHAT WE **MAKE, MAKES US.**

AND IF THE ARCHITECTURE OF THIS CHURCH REPRESENTS SOMETHING OF OUR DESIRE TO TOUCH THE **TRANSCENDENT**, THEN...

...DESIGN IS THE **COFFEE SHOP** NEXT TO IT, A **BRIDGE** BETWEEN THE **SUBLIME** AND THE **EVERYDAY.**

WHILE DESIGN MIGHT REACH FOR THE **TRANSCENDENT,** ITS FEET SHOULD **NEVER** LEAVE THE GROUND.

THE WAY I SEE IT, **ANYTHING** AND **EVERYTHING** WE DO CAN BE DONE IN ONE OF **THREE** GENERAL DIRECTIONS...

...THAT ECHO SOME INNATE **CAPACITIES** WITHIN US.

CAPACITY FOR **BEAUTY, GOODNESS, AND AUTHENTICITY**

HUMANS
WITH TECHNOLOGY

CAPACITY FOR THE **BANAL** AND MIND-NUMBINGLY **BORING**

CAPACITY FOR **HARM**

UHH... THIS MAY SEEM LIKE AN **OBVIOUS** CHOICE, BUT **IT'S NOT.** THE **CATCH** IS THAT THERE ARE **COMPELLING REASONS, HYPOTHETICAL IMPERATIVES,** THAT CONSTANTLY STEER EACH OF US TOWARD THE **BANAL** AND THE **HARMFUL,** WHEREAS BEAUTY, TRUTH, AND MORALITY HAVE TO BE CONSCIOUS **CHOICES** THAT CAN ONLY COME FROM **WITHIN** -- FOR THEY ARE **VALUABLE,** PRIMARILY, ONLY **IN THEMSELVES.**

IF DESTINY IS NOT A **CHANCE,** BUT A **CHOICE...** WHAT WOULD YOU **CHOOSE?**

"WE SHALL NOT CEASE FROM EXPLORATION
AND THE END OF ALL OUR EXPLORING
WILL BE TO ARRIVE WHERE WE STARTED
AND KNOW THE PLACE FOR THE FIRST TIME." -- T. S. ELIOT (1942)
                                          "LITTLE GIDDING"

AFTER ALL THAT HAS BEEN SAID -- AND WITH THE HOPE TO MORE FULLY UNDERSTAND DESIGN'S **NATURE**, **PURPOSE**, AND **MEANING** -- I STILL **CANNOT** SAY WHAT DESIGN TRULY **IS.**

PERHAPS DESIGN, LIKE **MUSIC,** IS **IRREDUCIBLE** TO A SINGLE DEFINITION, FOREVER **DEFYING** FORMALIZATION. FOR EVERY STATED **NOTION** OF WHAT DESIGN (OR MUSIC) IS, THERE ALWAYS LIES SOMETHING **BEYOND** IT, THAT **CHALLENGES** IT... MEANWHILE, IT'S **ALL TOO EASY** TO CONSTRUCT THINGS THAT TRIVIALLY FULFILL A DEFINITION BUT ARE UTTERLY **UNINTERESTING.** THIS IS THE NATURE OF DESIGN, AND OF ART: WE CANNOT SAY FOR CERTAIN WHAT IT IS, BUT WE **KNOW** IT WHEN WE ENCOUNTER IT, WHEN IT **SPEAKS** TO US.

IF ANYTHING, DESIGN IS, SIMPLY, WHAT WE **DO**: AN ACT OF **ALIGNMENT**, A **PURSUIT**, A **SEARCH**, A RADICAL **SYNTHESIS** OF THE USEFUL, AUTHENTIC, FREE, PLAYFUL, AND THE DESIRE TO REACH FOR THE **SUBLIME** -- AS PART AND PARCEL OF THE **TRANSFORMATION** OF THE PURELY **TECHNOLOGICAL** INTO THE **HUMAN.**

PERHAPS WE DON'T NEED TO DEFINE DESIGN PRECISELY, BUT RATHER TO KNOW **HOW** AND **FOR WHAT** WE DESIGN -- AND TO **EMBODY** THESE **VALUES** INTO THE THINGS WE DESIGN. ARTFUL DESIGN MAY BE **NOTHING MORE** AND **NOTHING LESS** THAN THIS.

THAT'S A GOOD **SMOOTHIE**...

461

A FINAL *DESIGN ETUDE*

THE FINAL DESIGN ETUDE IS ONE OF **SELF-REFLECTION.** TAKE A MOMENT TO THINK ABOUT THE FOLLOWING:

1. THINGS THAT **MOVE** YOU
2. THINGS THAT MAKE YOU **STILL**

THEY COULD INCLUDE AN OBJECT, A PIECE OF MUSIC, A PLACE, A MOMENT, OR AN ACT IN EVERYDAY LIFE THAT STAYS WITH YOU. WHAT IS THE **DIFFERENCE** BETWEEN THAT WHICH MOVES VS. THAT WHICH MAKES ONE STILL? **IS** THERE A DIFFERENCE TO YOU?

IN CHAPTER 1, YOU WERE ASKED TO PERFORM A **MEANS VS. ENDS ANALYSIS** OF AN OBJECT OF DESIGN. NOW DO A MEANS VS. ENDS ANALYSIS, BUT OF **YOURSELF** AND YOUR EVERYDAY LIFE, OF THINGS THAT **MAKE UP** YOUR LIFE...

1. WHAT ARE THE THINGS YOU DO FOR ANOTHER PURPOSE? (MEANS-TO-ENDS)
2. WHAT ARE THE THINGS YOU DO FOR THEIR OWN SAKE? (ENDS-IN-THEMSELVES)

DO YOU **LIKE** THE **BALANCE** BETWEEN THEM?

LASTLY, THINK OF **ALL** THE THINGS YOU **DESIGN** IN YOUR LIFE -- FOR **WORK**, FOR **SCHOOL**, FOR **LIFE**, OR FOR ANY OTHER PURPOSE. **HOW** DO YOU DESIGN? CAN YOU SAY **FOR WHAT** YOU ARE DESIGNING, AT THE END OF THE CAUSAL CHAIN? WHAT WOULD MAKE THESE DESIGNS SOMETHING THAT YOU'D **LOVE** AND FIND **MEANINGFUL**, THAT MOVE YOU, AND PERHAPS EVEN MAKE YOU PAUSE IN WONDER?

THAT'S IT FOR NOW.

SEE YOU AROUND, EVEN
IF IN ANOTHER TIME...

LIST OF PRINCIPLES

CHAPTER 3

PERRY'S PRINCIPLES

CHAPTER 6

WHOA. ARE WE, LIKE, IN A
NEW PLANE OF REALITY?

POSSIBLY.

HELLO VIRTUAL
CLEVELAND! PREPARE
TO BE ROCKED!

469

CHAPTER 7

CHAPTER 8

CONCEPTION OF *IDEAL* DONUT!

ACTUAL DONUTS!

IN *THEORY*, THERE IS *NO* DIFFERENCE BETWEEN THEORY AND PRACTICE...

...BUT NOT SO IN *PRACTICE.*

UH OH, WE MIGHT BE ONE OF GE'S *MIND WANDERINGS*, WHICH WOULD BE (1) VERY *META* AND (2) UNABASHEDLY *FOOD-MEDIATED*...

ALSO, WHAT'S A *DONUT?*

ACKNOWLEDGMENTS

MY DEEPEST THANKS TO MY EDITOR (AND S.U.P. EDITOR-IN-CHIEF) **KATE WAHL**, FOR CHAMPIONING SUCH AN UNCONVENTIONAL BOOK (A "PHOTO-REALISTIC ANTI-COMIC BOOK") AND GENTLY GUIDING IT AT EVERY TURN • TO S.U.P. DIRECTOR **ALAN HARVEY** FOR HIS UNFAILING SUPPORT AND ENTHUSIASM • TO **JENNIFER GORDON** FOR HER INSPIRING COPY EDITING ON A CRAZY BOOK.

HEARTFELT GRATITUDE TO THE **GUGGENHEIM FOUNDATION** FOR THEIR CURIOSITY AND FAITH IN THIS WORK, AND FOR THE FELLOWSHIP THAT ALLOWED ME TO COMPLETE IT • TO **STANFORD UNIVERSITY**, BOTH THE SCHOOL OF **HUMANITIES AND SCIENCES** AND THE SCHOOL OF **ENGINEERING**, FOR SUPPORTING THIS WORK -- DEANS **DEBRA SATZ**, **RICHARD SALLER**, **JENNIFER WIDOM**, AND PROVOST **PERSIS DRELL** • THANKS TO **ROBERTA DENNING** FOR SUPPORTING THIS BOOK AND COUNTLESS OTHER ARTISTIC ENDEAVORS AT STANFORD AND IN THE BAY AREA • TO PROVOST **JOHN ETCHEMENDY** FOR HIS TIRELESS SERVICE TO STANFORD -- AND FOR KEEPING A MEASURED AND OPEN MIND AT THE HIGHEST LEVEL OF OUR ACADEMIC INSTITUTION; I WOULD BE OUT OF A JOB OTHERWISE.

THIS BOOK OWES THANKS TO MANY WHO HAVE HELPED, INSPIRED, AND CONTRIBUTED DIRECTLY OR INDIRECTLY • **RANIA SANFORD** IN THE OFFICE FOR FACULTY DEVELOPMENT AND DIVERSITY (FOR INTRODUCING KATE TO FACULTY CRAZY ENOUGH TO WANT TO WRITE A BOOK AND FOR HOSTING THE FACULTY WRITING RETREATS, AT WHICH I HAVE BECOME A REGULAR) • TO **WENDY GOLDBERG** (FOR HELPING ME WEEK IN AND WEEK OUT TO APPRECIATE THE DESIGN OF WORDS IN SEQUENCE).

SPECIAL THANKS TO THE EARLY READERS OF THE MANUSCRIPT FOR THEIR CRITICAL FEEDBACK: JACK ATHERTON, MADELINE HUBERTH, JASON RIGGS, ERICA FEARON, DAN TRUEMAN, ROMAIN MICHON, REBECCA FIEBRINK, JONATHAN BERGER, DYLAN FREEDMAN, DAVE KERR, AND TRIJEET MUKHOPADHYAY.

TO MY **MUSIC•COMPUTING•DESIGN** RESEARCH GROUP AT STANFORD: JIEUN OH, N. J. B., SPENCER SALAZAR (FOR WORKING TOGETHER SINCE 2003), JORGE HERRERA, JACK ATHERTON, KITTY SHI • TO **FRIENDS** AND **COLLEAGUES** AT STANFORD UNIVERSITY, **CCRMA**, MUSIC, COMPUTER SCIENCE, PHILOSOPHY, COMMUNICATION, ART AND ART HISTORY, ENGLISH, HISTORY, STANFORD HUMANITIES CENTER • CHRIS CHAFE, JONATHAN BERGER, JULIUS SMITH, JAREK KAPUSCINSKI, MARK APPLEBAUM, JOHN CHOWNING (FOR INSPIRING ALL OF US), STEVE SANO (FOR AUTHENTICITY AND BELIEF IN PEOPLE), MAX MATHEWS (THE FIRST COMPUTER MUSICIAN), NETTE WORTHEY (FOR COUNTLESS CLUTCH SAVES), DEBBIE BARNEY, MATT WRIGHT, ROB HAMILTON, CHRYSSIE NANOU, MILTOS HAMILTON, TIM O'BRIEN, HANA SHIN, ALEX CHECHILE, JOHN GRANZOW, GINA GU, CHET GNEGY, VICTORIA GRACE, NICK GANG, AIDAN MEACHAM, SANJAY KANNAN, CARLOS SANCHEZ, FERNANDO LOPEZ-LEZCANO, JAY KADIS, SASHA LEITMAN, CHRISTOPHER JETTE, MARIO CHAMPAGNE, VELDA WILLIAMS, JERRY MCBRIDE, JONATHAN ABEL, TAKAKO FUJIOKA, ELEANOR SELFRIDGE-FIELD, CRAIG SAPP, MALCOLM SLANEY, JAMES ANDY MOORER (THANKS FOR THE ORIGINAL INFO ON THE DEEP NOTE!).

JAMES LANDAY, MANEESH AGRAWALA, MICHAEL BERNSTEIN, PAT HANRAHAN, KEN TAYLOR, JULIANA BIDADANURE, PAUL DEMARINIS, TERRY BERLIER, SCOTT BUKATMAN, CAMILLE UTTERBACK, NICK JENKINS, ALETA HAYES, FRED TURNER, JEREMY BAILENSON, BLAKEY VERMEULE, BERNIE ROTH, BILL BURNETT, LETICIA BRITOS CAVAGNARO, RYAN PHILLIPS, BILL VERPLANK, JANICE ROSS, KIM BEIL, JAMES STEICHEN, CAROLINE WINTERER, ELAINE TREHARNE, TINA SEELIG, RICH COX, MATTHEW TIEWS • FOLKS AT STANFORD CENTER AT PEKING UNIVERSITY (S-C, P-K-U!), JEAN OI, JOSH CHENG, LAP "LILI" LI, WENTWORTH, QIANG, FRANK HAWKE, PEGGY ZOU, 39, MARIAN WANG, LIU WANG, SUHUA HAO, CONNIE CHAO • TO THE 2016 SCPKU GRADUATE SEMINAR FOR LETTING ME RANT ON ABOUT AESTHETICS IN DESIGN: SARAH JIANG, TOMMY FANG, ANDREW MCCABE, MINNA XIAO, KYLE D'SOUZA, CAROLINE DEBS, JACKIE YANG, CINDY GUO, JING GAO, ALEX YANG, JAMES ZHU, YIQIN LU, SHUAISHUAI, DYLAN YUE, YINGQING XU, JANE E, ILENE E, EILEEN, ANDREW, AND TIMMY LANDAY • TO ANDREA LUNSFORD AND THE HUME CENTER FOR WRITING AND SPEAKING, JULIA BLEAKNEY, SHANLEY JACOBS, JANET KIM, KATIE FRITZ, SARAH PITTOCK, ZANDRA JORDAN.

TO **COLLABORATORS** AND **MENTORS** IN OTHER INSTITUTIONS AROUND THE WORLD: SENSEI **PERRY COOK** (FOR BEING A LIFE MENTOR), **DAN TRUEMAN** (FOR BEING A PARAGON OF RESEARCH, ART, AND CHARACTER), **REBECCA FIEBRINK** (CO-CONSPIRATOR AND COMRADE IN ARMS), PAUL LANSKY (FOR SHOWING ME THAT I CAN SIMPLY **LIKE** COMPUTER MUSIC, BEYOND CONCEPTUALLY LOVING IT), ROGER DANNENBERG, AJAY KAPUR, SCOTT SMALLWOOD, GEORG ESSL, GIORGOS TZANETAKIS, JON APPLETON, TED COFFEY, LUKE DAHL, ALEXANDER JENSENIUS, STEFANIA SERAFIN, DAN OVERHOLT, ATAU TANAKA, WOON SEUNG YEO, SID FELS, MICHAEL LYONS, THOR MAGNUSSON, THE NIME (NEW INTERFACES FOR MUSICAL EXPRESSION) COMMUNITY • **MENTORS** OWEN ASTRACHAN, SCOTT LINDROTH, JOHN MAEDA (FOR INSPIRING GENERATIONS OF DESIGNERS AND FOR HIS TIRELESS EFFORT TO TRANSITION S.T.E.M. EDUCATION INTO S.T.E.A.M. -- THE "A" IS FOR ART OR AESTHETICS!) • TO CASEY REAS, KATHERINE EMERY • FRANK GAO, JACKIE ZHANG • MICHAEL HAWLEY AND THE EG CREW, JANE ROSCH, CHRISTOPHER NEWELL • TO **TEDXSTANFORD**, **TED**, CCTV, TECH+, EL PAIS, AND OTHER PRESENTATION VENUES FOR HAVING ME PRESENT ABOUT DESIGN, MUSIC, AND TECHNOLOGY • TO SAKO FISHER (FOR LEADING WITH GRACE), JOHN GOLDMAN, BRENT ASSINK, CURT KIRSCHNER, SAN FRANCISCO SYMPHONY BOARD OF GOVERNORS • TO MY FELLOW FSP WRITING ACCOUNTABILITY BUDDIES: MARTA, STACEY, SARA, AND CHRISTEN.

DUDE!

TO THE CREW, PAST AND PRESENT, AT **SMULE**: JEFF SMITH (FOR STARTING AN ADVENTURE TOGETHER), DAVID ZHU, SPENCER SALAZAR (AGAIN), MATTIAS LJUNGSTROM, TRICIA HEATH, JODI ROPERT, MARK "THORRR" CERQUIERA, NICK KRUGE, SHAELYNE JOHNSON, TURNER "CAT IN THE HAT" KIRK, SUNIL PAREENJA, RIBBIT, MICHAEL WANG, STEFAN KOTES, MICHAEL BERGER, TOM LIEBER, ARI LAZIER, ALEX LI, JEANNIE YANG, PRERNA GUPTA, PARAG CHORDIA, NICK RUDOLFSKI, ELON BERGER, RENEE THOMAS, ERIKA SAN MIGUEL, SCOTT BOND, TOM RYAN, ARNAUD BERRY, JOSH WU, JOHN SHIMMIN, JENNIFER WU • EVERYONE WORKING TO MAKE THE PLANET MORE EXPRESSIVE AND SOCIAL THROUGH MUSIC • TO OUR BOARD AND INVESTORS (FOR HAVING GENUINE CURIOSITY): DAVID COWEN, CHRIS HOLLENBECK, TOM RYAN, BESSEMER VENTURE PARTNERS, GRANITE VENTURES, SHASTA VENTURES, ROTH CAPITAL PARTNERS, ADAM STREET PARTNERS, TENCENT • TO JAMES HANNON, ANDY GETSEY, MICHELLE SABOLICH AT ATOMIC PR • TO ALL **SMULE USERS** FOR MAKING MUSIC TOGETHER.

TO THE INTREPID FOLKS AND FRIENDS IN THE **PRESS** -- ESPECIALLY THOSE SEEKING **TRUTH** AND **BEAUTY** AMIDST THE CHAOS AND MADNESS OF THE WORLD. SPECIAL THANKS TO DAVID POGUE, ROB WALKER, CLAIRE CAIN MILLER, JEFFERSON GRAHAM, ARIELLE PARDES, AND TEKLA PERRY.

TO THE MEMBERS, PAST AND PRESENT, OF THE **STANFORD LAPTOP ORCHESTRA** AND THE **PRINCETON LAPTOP ORCHESTRA** • A SPECIAL CALL-OUT TO THE 2008 PRE-SLORK TEAM WHO LENT THEIR BLOOD, SWEAT, TEARS (MOSTLY SWEAT) TO BUILD SLORK • ADNAN MARQUEZ-BORBON, BAEK SAN CHANG, BRETT ASCARELLI, CHRIS WARREN, CHRYSSIE NANOU, DAVID BAO, DIANA SIWIAK, ETHAN HARTMAN, GINA GU, HAYDEN BURSK, JASON RIGGS, JEFF COOPER, JIEUN OH, JUAN CRISTOBAL CERRILLO, JUHAN NAM, KAYLA CORNALE, KYLE SPRATT, LAWRENCE FYFE, LUKE DAHL, MARK BRANSCOM, N.J.B., REED ANDERSON, ROB HAMILTON, STEINUNN ARNARDOTTIR, AND TURNER KIRK.

TO THE FANTASTIC PRODUCTION, MARKETING, SALES CREW AT **STANFORD UNIVERSITY PRESS** • MIKE SAGARA (COMIC SHERPA), PATTIE MYERS ("THORRY!"), STEPHANIE ADAMS, KATE TEMPLAR, EMILY SMITH, RYAN FURTKAMP, KENDRA ELIZABETH SCHYNER, ROB EHLE, BRUCE LUNDQUIST, LEAH PENNYWARK, JEAN KIM, JASMINE MULLIKEN, GRETA LINDQUIST, AND THE ENTIRE S.U.P. FAMILY.

GRANDMA   GRANDPA

MOM   DAD

RABBIT HAPPY

TO FRIENDS **TAO** AND **ROB** (FOR MANY LATE-NIGHT GAME SESSIONS AND ONGOING DISCUSSIONS OF GAMES AND GAME DESIGN) • TO **MANMAN** FOR HER KINDNESS AND FOR OPENING UP THE WORLD OF ARCHITECTURAL DESIGN TO ME • TO **PERRY** AND **STACIE** • TO **SHUCAO** (FOR A SPIRIT OF ADVENTURE AND EVER SEEKING THE SUBLIME) • TO **SPENCER** AND **MARK** FOR BEING GIANT VORGOMS TOGETHER IN STARCRAFT 2 AND LIFE • TO **FAMILY**, GRANDFATHERS AND GRANDMOTHERS: YUPING WANG (王裕平), RUXIAN WU (吴汝贤), TINGNAN ZHAO (赵廷楠), GUIMEI LI (李桂梅); MOM AND DAD, JIANMIN ZHAO (赵健民), JINCHANG WANG (王槿长); BIG WHITE RABBIT, HAPPY, MY AUNTS AND UNCLES • DEEP GRATITUDE TO **ONE** AND **ALL** FOR KEEPING ME NOURISHED IN MIND, BODY, AND SPIRIT IN WRITING THIS BOOK.

FOR MY TEACHERS AND MY STUDENTS...

...AND **ALL** WHO SEEK BEAUTY IN WHAT WE MAKE.

THERE ARE OVER **1,650** IMAGES IN THIS BOOK; SOME **1300** OF WHICH WERE -- ONE WAY OR ANOTHER -- CREATED BY ME; BELOW ARE THE **CREDITS** FOR THE IMAGES FROM EXTERNAL SOURCES. **PROFOUND THANKS** TO **WIKIMEDIA COMMONS**, WIKIMEDIA FOUNDATION, CREATIVE COMMONS, AND ALL THE **FOLKS** WHO HAVE SHARED THROUGH THESE.

color code: Creative Commons   by permission

IMAGE CREDITS

Prelude: Stanford Laptop Orchestra standing (2008) **and live coding** (2008) by Enrique Aguirre • **Ge programming laptop** (2015) by Cison Inversiones SL / El Pais SL.

Chapter 1: School of Athens (1509) by Raphael • **bone flute** (2013) by Jean-Pierre Dalbéra • **St. Stephen's Cathedral** (2007) by user:tobi_87 • **pipe organ console** by U.S. Navel Academy • **submarine console** (2010) by flickr:shiny things • **IBM 360** (late 1960) by NASA • **double spiral staircase at Vatican Museum** (2012) by user:colin • **original Utah Teapot** (2007) by Marshall Astor • **Sydney Opera House** by unknown • **momentary switch** (2009) by user:inductiveload • **door handle** (2007) by Kıvanç Nis • **radio set dial** (2005) by John Macomb • **Table of Mechanicks, Cyclopedia** (1728) by Fletcher Sculp • **piano keyboard** by user:takkk • **Lamborghini Gallardo** (2007) by Michal Pecyna • **Louvre Museum courtyard** (2010) by Benh Lieu Song • **plane wreckage at MOCA Los Angeles** (2010) by flickr:modernrockstar • **storage units** (2014) by flickr:Mike Mozart • **Three Disks in the Air** (1967) by Alexander Calder • **smoke** (2010) by Zeinab Abdel Meguid • **The Wanderer Above the Sea of Fog** (1817) by Caspar David Friedrich • **Opening Ceremony of Olympic Games** (2008) by U.S. Army • **Vitruvian Man** (1492) by Leonardo Da Vinci • **design for a flying machine** (1488) by Leonardo Da Vinci • **Mona Lisa** (1503-1506) by Leonardo Da Vinci • **David** (1501-1504) by Michelangelo; photograph by Rico Heil • **Sistine Chapel ceiling** (1508-1512) by Michelangelo • **Michelangelo's contribution to St. Peter's Basilica, reproduced in L'Architecture de la Renaissance** (1892) by Léon Palustre • **elevation plan of St. Peter's Basilica dome** by Michelangelo, engraved (1569) by Stefan du Pérac • **Richard Wagner** (1861) by Pierre Petit • **Walter Gropius** (1919) by Louis Held • **Bauhaus** (2007) by Janine Pohl • **crescent Earth** (2015) by flickr:Kevin Gill • **Apollo 11 launch** (1969) by NASA • **Verner Panton chair** (2007) by User:Holger.Ellgaard • **Dune 4.0 installation** (2011) by Studio Roosegaarde with Lotte Stekelenburg • **anthropomorphic corkscrew** (2007) by Kyle May • **"Gaoliang" Bridge at Summer Palace** (2006).

Chapter 2: Ge (2011) by EG Conference • **astronaut selfie** (2012) by NASA • **Take Along Telephones** (1973) by Popular Science • **Dynatac 8000x** (1983) by Motorola • **Mark Weiser portrait** • **Victoria Harbor, Hong Kong** (2013) by Diego Delso (delso.photo) • **R380** (2000) by Ericsson • **iPhone** (2007) by Apple • **2146** (1994) by Nokia • **3210** (1999) by Nokia • **Razr V3** (2005) by Motorola • **iPad** (2010) by Apple • **Galaxy S II** (2011) by Samsung • **iPhone 5** (2012) by Apple • **Isaac Asimov** by Phillip Leonian • **xun** (2008) by user:Badagnani • **Earth and horizon** (2009) by NASA • **ocarina users** (2008) • **clouds and horizon** by NASA • **Listening to the Qin** (c. 11th century) by Zhao Ji • **St. Cecilia with Two Angels** (1620-1625) by Antiveduto Grammatica • **family making music** (c. 1972) by Andy Sacks • **Edison GEM 1900** (2011) by user:AlejandroLinaresGarcia • **girl listening to radio** (1938-1945) by Franklin D. Roosevelt Library • **RCA 808 power vacuum tube** (2007) by Masaki Ikeda • **computer chips circuit boards** (2013) by Jon Sullivan • **Internet map** (2006) by the OPTE Project • **phonograph record** (2016) by user:Even-Amos • **cassette tape** by Paul Sherman • **Apple iPod classic** (2009) by Mathieu Riegler • **clouds** (2005) by user:PiccoloNamek • **State of the Union Address** (2010) by White House [Peter Souza] • **Barack Obama** (2009) by U.S. Navy • **I Am T-Pain users** (2010) Harrison Spiller, Laura Bermudez, Diane Morgan, Jacques Duvoisin, the Corvino brothers, Justin Fang, Hagler Chanthilath.

Chapter 3: Royal British Bowmen archery club (1823) after J. Townshend • **Drawing Hands** (1948) by M. C. Escher • **Penrose triangle** (2007) rendering by Tobias R. • **Impossible Stairs** (2005) rendering by user:sakurambo • **Impossible Triangle sculpture** (2008) photograph by Bjørn Christian Tørrissen • **Ascending and Descending** (1960) by M. C. Escher • **barber pole** (2009) by user:Corpse Reviver • **The Treachery of Images** (1928-1929) by René Magritte • **Firebirds** (2004) by Paul DeMarinis • **Infinite Cat Project** by Mike Stanfill and Internet users • **Ernest Hemingway** (c. 1927).

Chapter 4: Ada Lovelace (1836) by Margaret Sarah Carpenter • **DX7** (1983) by Yamaha • **George Forsythe** portrait (1971) • **Max Mathews and radio baton** • **Max Mathews presenting on his 80th birthday** (2007) by flickr:Kevin • **Paul Lansky's Homebrew album cover** (1992) artwork by P. Craig Russell • **Paul Lansky at Princeton** (c. 2005) by Kui Dong • **Paul Lansky at DA Lab** (1981) by John W. H. Simpson / Princeton University Library • **plates** (2006) by Egan Snow • **cooking pot** (2008) by George Shuklin • **copper pot** (2010) by user:JaneArt • **Joseph Fourier** (c. 1823) by Jules Boilly • **League of Automatic Music Composers** by Peter Abramowitsch • **The Hub performing at Mills College** by Jim Block • **Alex McLean hacking perl in nightclubs** (2004) • **logos of music programming languages**.

Chapter 5: sliders (2010) by Jeremy Keith • **tabla strike** (2009) by flickr:tiarescott • **hand on drum** (2007) by Randy Stern • **hand** (2015) by Golan Levin • **fist** (2009) by Mark Ramsay • **arm and trombone** (2011) by Paulo Philippidis • **singing** (2010) by flickr:Shopify • **the GUI's mental model of a user** (2004) by Dan O'Sullivan and Tom Igoe • **beige mouse** by Apple • **Theremin** (1927) by Corbis Bettmann • **playing modular synth** (2016) by Carsten Stiller • **minimoog** (2011) by Wolfgang Stief • **Max/MSP software** by Cycling '74 • **Max Mathews and radio baton** • **radio baton diagram** (1996) by Tim Stilson / Princeton SoundLab • **Michel Waisvisz playing The Hands** (1991) by Carla van Thijn • **Ben Knapp and bioinformatics** • **Atau Tanaka performing with Biomuse** (2009) by Nicholas Croft • **Rebecca Fiebrink** (2013) by Ricardo Barros (ricardobarros.com) • **Rebecca Fiebrink code + music** (2013) by Andrew Wilkinson • **Ajay Kapur** (2013) by Peter Stember • **Auraglyph** (2017) by Spencer Salazar • **Mouthesizer** (2001) by Michael Lyons, Michael Haehnel, and Nobuji Tetsutani • **Nic Collin's Sled Dog** (2006) by Marty Perez • **GuitarBot** (2002) by Eric Singer • **Tooka** (2003) by Sidney Fels • **Cook & Morrill Trumpet** (1989) by Perry Cook • **Roger Dannenberg and trumpet** • **Rebecca Fiebrink and glove interface** • **David Wessel and SLABS** • **Sensor Bass** by Curtis Bahn • **Pika Pika** by Tomie Hahn • **Mousketier** by Mark Applebaum • **Bo.S.S.A.** (1999) by Dan Trueman • **Sonic Banana** (2003) by Eric Singer • **Brofessors** (2016) by Jane E • **Princeton Laptop Orchestra** in Richardson • **Dan Trueman** portrait • **Scott Smallwood** portrait • **Stanford Laptop Orchestra** in Dinkelspiel (2008) by Enrique Aguirre • **night city** (2013) by flickr:Jo.

Interlude: Perry R. Cook in Humbug Studio (2013) by Peter Stember • **Sonic Banana** (2003) by Eric Singer • **Dan Levitin in Humbug Studio** (2011) by Perry Cook.

Chapter 6: William Shakespeare from the First Folio (1623) • **Friedrich Schiller** (1808-1809) by Gerhard von Kügelgen • **Starcraft 2** (2010) by Blizzard Entertainment • **Save the Date** (2013) by Chris Cornell • **Firewatch** (2016) by Campo Santo • **That Dragon, Cancer** (2016) by Numinous Games • **Super Mario Bros.** (1985) by Nintendo • **NES controller** (2016) by user:Evan-Amos • **Ein ermahnendes Wort** (trans. "A Word of Admonition", 19th century) • **military exercise** (2007) by user:Cloudaoc • **clockwork** (2011) by Arthur John Picton • **The Piano Lesson** (1785-1787) by Marguerite Gérard • **Rock Band series** (since 2007) by Harmonix • **Gone Home** (2013) by Fullbright Company.

Chapter 7: The Blue Marble (2007) by NASA / Reto Stöckli • **people dancing on the street** (2014) by Papak Sarkar • **festivity and dancing** (2012) by Rohit Jain • **park sing along in Beijing** by Jinchang Wang (thanks Dad) • **hands up in the air** (2009) by Whit Scott • **dance in purple** (2009) by USAID Africa Bureau • **cat on Magic Piano** (2010) • **We Feel Fine** (2008) by Jonathan Harris and Sep Kemvar • **diagram of The Turk** (1770) by Joseph Racknitz • **robot** (2015) by user:Akawikipic • **The Sheep Market** (2010) by Aaron Koblin • **hey ewe 2** (2017) by Rebecca Fiebrink • **leaf** (2011) by flickr:wholowhy • **people waiting on platform** (2013) by Nana B. Age • **train passengers** (2012) by Monik Markus • **passenger asleep** (2014) by Casey Hugelfink • **bus passengers** (2012) by flickr:prin_t • **live radio hour in Second Life** (2014) by user:HyacintheLuynes • **Paul Hindemith** (1923) • **singers in Beijing park** (2007) by Laurel F. • **Earth / North America** (2012) by NASA.

Chapter 8: Apollo 11 launch (1969) by NASA • **John F. Kennedy "We Choose to Go to the Moon"** (1962) by U.S. government • **Paris at night** (2011) by Vincent Ros • **Radio News magazine cover** (1928) • **Popular Science magazine cover "Perpetual Motion"** (1920) by Norman Rockwell • **KIM-1 computer** (2010) by user:Rama & Musée Bolo • **gas metal arc welding** (2004) by U.S. Air Force • **oil platform** (2009) by Agência Brasil • **soldering & tweezers** (2006) by user:Aisart • **Space Shuttle Atlantis, STS-79** (1996) by NASA / Kennedy Space Center • **windows** (2016) by user:W.carter • **rifling of cannon / lens** (2016) by Petar Miloševic • **race horse and rider** (1887) by Eadweard Muybridge • **race car drift** (2014) by Rowan Harrison • **screens in control room** (2014) by user:Nightscream • **Immanuel Kant** (c. 1790) • **Dubai at night** (2012) by Stefan Langmann • **Bogotá** (2015) by Fernando Garcia • **looking into night** (2017) by Pawel Maryanov • **Frankenstein's monster** (1935) by Universal Studios • **light trails** (2014) by Ronnie Khan • **Terminator exhibition, T-800** (2009) by Dick Thomas Johnson • **sad robot by ultra junk** (2014) by flickr:Jedimentat44 • **Space Shuttle Challenger in fog** (1982) by NASA • **Sistine Chapel ceiling** (1508-1512) by Michelangelo • **Francis Bacon** (c. 1617) by Paul van Somer I • **Leonardo, presumed self-portrait** (c. 1512) • **Ada Lovelace** (1840) by Alfred Edward Chalon • **John Keats drawing** (1896) from *Library of the World's Best Literature Vol. XV* • **Beijing: sea of buildings** (2014) by Jinchang Wang (thanks again Dad) • **light trails in car** (2009) by Craig Cormack • **striking night sky** (2011) by European Southern Observatory / Y. Beletsky • **Maya Angelou speaking** (2008) by Talbot Troy • **Blade Runner** (1982) Warner Bros. • **Wheatfield with Crows** (1890) by Vincent van Gogh • **Composition with Red, Blue, and Yellow** (1930) by Piet Mondrian • **Ludwig van Beethoven** (1820) by Joseph Karl Stieler • **ideal donut** (2014) by flickr:fdecomite • **actual donuts** (2012) by Ferry Sitompul.

ANNOTATED BIBLIOGRAPHY

HERE IS THE **BIB.** FOR **MORE** MATERIAL (ETUDE RESOURCES, IMAGES, AUDIO, VIDEOS, SOFTWARE, TALKS, ETC.), FIND YOUR WAY TO **THESE** WEBSITES BELOW... ASSUMING THE INTERNET HASN'T BEEN **REPLACED**, BY THE TIME YOU READ THIS, WITH SOME NEW-FANGLED ROBO-NANO-META-VIRTUAL-THINGYMAJIG, IMPLANTED IN YOUR BRAIN... BY THEY WAY, I'VE ORDERED THESE NOT ALPHABETICALLY, BUT BY SOME FLOW OF IDEAS AS RELEVANT TO EACH CHAPTER. GOOD LUCK.

BOOK HOMEPAGE
https://artful.design/

GE HOMEPAGE
https://www.gewang.com/

BITE-SIZED RANTS
@gewang

↑ ADDITIONAL CONTENT (VIDEOS, AUDIO, SOFTWARE, COMMENTARIES, CODE FOR DESIGN ETUDES)!

CHAPTER 1 DESIGN IS _____

MICHAEL POLANYI. **THE TACIT DIMENSION,** THE UNIVERSITY OF CHICAGO PRESS. 1966. "WE CAN KNOW MORE THAN WE CAN TELL." A SEMINAL BOOK THAT ARGUES TACIT KNOWLEDGE -- INHERITED PRACTICES, TRADITION, IMPLIED VALUES -- IS CRUCIAL TO SCIENTIFIC AND HUMAN KNOWLEDGE.

BRUNO MUNARI. **DESIGN AS ART.** PENGUIN BOOKS. 1966.
A MODERN TREATISE ON DESIGN BEING BEAUTIFUL, FUNCTIONAL, ACCESSIBLE.

PAUL RAND. **THOUGHTS ON DESIGN,** WITTENBORN AND COMPANY. 1947.
QUINTESSENTIAL TEXT ON GRAPHICS DESIGN AND THE SEAMLESS INTEGRATION OF FORM AND FUNCTION.

DON NORMAN. **THE DESIGN OF EVERYDAY THINGS,** BASIC BOOKS. 1988.
A CLASSIC BOOK ABOUT DESIGN IN EVERYDAY CONTEXTS: FROM DOORS TO TEAPOTS;
DESIGN AS A CONSCIOUS PRACTICE.

MARSHALL MCLUHAN. **UNDERSTANDING MEDIA: THE EXTENSIONS OF MAN,**
MIT PRESS. 1964. A PRESCIENT BOOK ABOUT THE POWER OF MEDIA AND TECHNOLOGY.
COINED "THE MEDIUM IS THE MESSAGE" AND THE CONCEPT OF A GLOBAL VILLAGE. A TRIP TO READ.

MARSHALL MCLUHAN AND QUENTIN FIORE. **THE MEDIUM IS THE MASSAGE: AN INVENTORY OF EFFECTS,** GINGKO PRESS, 1967. MCLUHAN AND FIORE BOLDLY EXPLORE
NATURE OF MEDIA AND TECHNOLOGY IN THIS KALEIDOSCOPE: IMAGES, IDEAS, UNIQUE FORMAT
(INCLUDING TEXT MEANT TO BE READ IN A MIRROR). EASIER READING THAN UNDERSTANDING MEDIA,
BUT NO LESS PROVOCATIVE. THE TITLE IS NOT A TYPO.

WILLIAM LIDWELL, KRITINA HOLDEN, AND JILL BUTLER. **UNIVERSAL PRINCIPLES OF DESIGN, REVISED AND UPDATED: 125 WAYS TO ENHANCE USABILITY, INFLUENCE PERCEPTION, INCREASE APPEAL, MAKE BETTER DESIGN DECISIONS, AND TEACH THROUGH DESIGN.** 2ND EDITION. ROCKPORT PUBLISHERS, 2010. DESIGN PRINCIPLES FROM HUMAN-COMPUTER INTERACTION, PSYCHOLOGY AND PERCEPTION, USABILITY.

PER GALLE. "PHILOSOPHY OF DESIGN: AN EDITORIAL INTRODUCTION," **DESIGN STUDIES** 23:(3). PP. 211-218 (2002). A CONCISE INTRODUCTION ON THE EMERGING ACADEMIC FIELD OF PHILOSOPHY OF DESIGN -- WHAT IT IS AND POSSIBLY WHAT IT'S GOOD FOR.

REM KOOLHAS AND BRUCE MAU. *S, M, L, XL.* 2ND EDITION. MONACELLI PRESS, 1997. I HAVE LONG BEEN INSPIRED BY THE QUEST FOR FUNCTIONAL-AESTHETIC UNITY OF CONTEMPORARY ARCHITECTURAL DESIGN.

JIMENEZ LAI. *CITIZENS OF NO PLACE: AN ARCHITECTURAL GRAPHIC NOVEL.* PRINCETON ARCHITECTURAL PRESS, 2012. A CEREBRAL EXPLORATION OF ARCHITECTURAL DESIGN, SHARING MANY COMMON VALUES WITH ARTFUL DESIGN.

BJARKE INGELS. *YES IS MORE: AN ARCHICOMIC ON ARCHITECTURAL EVOLUTION.* TASCHEN, 2009. A COMIC BOOK ON PROJECTS OF ONE OF THE BIGGEST ARCHITECTURE FIRMS WORLDWIDE.

SCOTT MCCLOUD. *UNDERSTANDING COMICS: THE INVISIBLE ART.* HARPERCOLLINS, 1993. A SUBLIME DECONSTRUCTION OF THE MEDIUM (AND DESIGN) OF COMICS, PRESENTED AS A COMIC.

CHAPTER 2 *DESIGNING EXPRESSIVE TOYS*

MARK WEISER. "THE COMPUTER FOR THE TWENTY-FIRST CENTURY," *SCIENTIFIC AMERICAN* 265:94-104 (1991). A LANDMARK PAPER ABOUT UBIQUITOUS COMPUTING AND CALM TECHNOLOGY; IT ARCHITECTS A VISION OF A FUTURE WHERE COMPUTING IS EMBEDDED IN EVERYDAY LIFE.

JOHN R. FREE. "NEW TAKE-ALONG TELEPHONES GIVE YOU PUSHBUTTON CALLING TO ANY NUMBER," *POPULAR SCIENCE* 203(1):60-62, 130 (1973). THIS 1973 ARTICLE HELPED TO INTRODUCE THE WORLD TO THE CONCEPT AND TECHNOLOGY OF MOBILE PHONES!

NICHOLAS COOK. "BETWEEN PROCESS AND PRODUCT: MUSIC AND/AS PERFORMANCE," *MUSIC THEORY ONLINE* 7(2) (2001). ON THINKING ABOUT MUSIC AS PERFORMANCE AND STUDYING ITS SOCIAL MEANING.

LEIGH LANDY. "THERE'S GOOD NEWS AND THERE'S BAD NEWS: THE IMPACT OF NEW TECHNOLOGIES ON MUSIC SINCE THE ARRIVAL OF HOUSEHOLD ELECTRICITY AND THE PHONOGRAPH INCLUDING POTENTIAL ADVENTURES TO LOOK FORWARD TO," *PROCEEDINGS OF IEEE CONFERENCE ON THE HISTORY OF ELECTRONICS.* 2004. A MEANINGFUL FOOD-FOR-THOUGHT LOOK AT TECHNOLOGY AND MUSIC; PART CELEBRATION, PART EULOGY.

DAVID POGUE. "SO MANY IPHONE APPS, SO LITTLE TIME," *NEW YORK TIMES.* FEBRUARY 5, 2009. AN ARTICLE ON APPS (AND A CASE STUDY OF OCARINA), FROM ONE OF THE INDUSTRY'S MOST INFLUENTIAL COLUMNISTS.

GE WANG. "DESIGNING THE IPHONE'S MAGIC FLUTE," *COMPUTER MUSIC JOURNAL* 38(2):8-21 (2014). AN ARTICLE ABOUT THE DESIGN AND IMPLEMENTATION OF OCARINA.

GE WANG, GEORG ESSL, AND HENRI PENTINNEN. "THE MOBILE PHONE ORCHESTRA," *OXFORD HANDBOOK OF MOBILE MUSIC STUDIES, VOLUME 2.* EDITED BY S. GOPINATH AND J. STANYEK. OXFORD UNIVERSITY PRESS, PP. 453-469. 2014. BOOK CHAPTER CHRONICLING THE FIRST MOBILE PHONE ORCHESTRA.

CHRISTOPHER SMALL. *MUSICKING: THE MEANING OF PERFORMING AND LISTENING.* WESLEYAN UNIVERSITY PRESS, 1998. A POWERFUL BOOK ON MUSIC AS ACTION, PROCESS, AND ACTIVITY; ENCOMPASSING MUCH MORE THAN THE RESULTING ARTIFACT OF SOUND.

GE WANG. "THE DIY ORCHESTRA OF THE FUTURE," *TEDxSTANFORD.* 2014.
http://www.ted.com/talks/ge_wang_the_diy_orchestra_of_the_future
A TALK (WITH LIVE DEMOS) ABOUT COMPUTER MUSIC, DESIGN, INSTRUMENTS -- AND WHY.
(IF YOU HAD ONLY 18 MINUTES FOR THIS CHAPTER)

CHAPTER 3 *VISUAL DESIGN*

EDWARD TUFTE. *THE VISUAL DISPLAY OF QUANTITATIVE INFORMATION.*
GRAPHICS PRESS, 1983. A CLASSIC (AND FUN) BOOK ON INFORMATION DESIGN.

VIRGINIA TUFTE. *ARTFUL SENTENCES: SYNTAX AS STYLE,* GRAPHICS PRESS, 2006.
BRILLIANT BOOK ON THE DESIGN OF SENTENCES, AND HOW SYNTAX IS STYLE -- AND STYLE, CONTENT.
(EDWARD IS VIRGINIA'S SON. WRITING INFLUENTIAL BOOKS SEEMS TO RUN IN THE FAMILY.)

SCOTT MCCLOUD. *MAKING COMICS,* HARPER, 2006.
AN INTIMATE EXPLORATION OF THE CRAFT OF STORYTELLING THROUGH VISUALS AND WORDS.

JOHN MAEDA. *DESIGN BY NUMBERS,* MIT PRESS, 1999.
A JOURNEY OF PROGRAMMABILITY AND VISUAL DESIGN. A TANGO OF CODE, IMAGE, AND NUMBERS.

MANEESH AGRAWALA, WILMOT LI, AND FLORAINE BERTHOUZOZ. "DESIGN PRINCIPLES FOR
VISUAL COMMUNICATION," *COMMUNICATIONS OF THE ACM* 54(4):60-69 (2011).
FASCINATING EXPLORATION OF IDENTIFYING, INSTANTIATING, AND EVALUATING DESIGN
PRINCIPLES FOR VISUAL COMMUNICATION, WITH CONNECTIONS TO THE DOMAINS OF PERCEPTION
AND COGNITION, AND AUTOMATED VISUAL DESIGN SYSTEMS.

GOLAN LEVIN. *PAINTERLY INTERFACES FOR AUDIOVISUAL PERFORMANCE.*
M.S. THESIS, MIT MEDIA LAB, 2000. INFLUENTIAL AND INSPIRING MASTER'S THESIS ON
DESIGNING EXPRESSIVE AUDIOVISUAL INTERFACES.

CASEY REAS AND BEN FRY. *PROCESSING: A PROGRAMMING HANDBOOK FOR
VISUAL DESIGNERS AND ARTISTS,* MIT PRESS, 2014. A HANDBOOK FOR
PROGRAMMING WITH THE PROCESSING VISUAL LANGUAGE, BY ITS CREATORS.

HARTMUT BOHNACKER, BENEDIKT GROSS, AND JULIA LAUB. *GENERATIVE DESIGN:
VISUALIZE, PROGRAM, AND CREATE WITH PROCESSING,* EDITED BY CLAUDIUS
LAZZERONI. PRINCETON ARCHITECTURAL PRESS, 2012. AN EXPLORATION OF ALGORITHMIC
GENERATIVE VISUAL DESIGN, ITS PRINCIPLES AND METHODS.

CASEY REAS, CHANDLER MCWILLIAMS, AND LUST. *FORM + CODE IN DESIGN,
ART, AND ARCHITECTURE,* PRINCETON ARCHITECTURAL PRESS. 2010. AN
ARTFUL EXPLORATION OF FORM THROUGH CODE AND SOFTWARE.

DOUGLAS HOFSTADTER. *GÖDEL, ESCHER, BACH: AN ETERNAL GOLDEN BRAID.*
BASIC BOOKS, 1979. CLASSIC TREATISE ON NATURE OF SELF-REFERENCE, FORMAL SYSTEMS,
INTELLIGENCE, AND STRANGE LOOPS.

GE WANG. "SOME PRINCIPLES OF VISUAL DESIGN FOR COMPUTER MUSIC,"
LEONARDO MUSIC JOURNAL 26 (2016). EARLIER PAPER ON VISUAL DESIGN
FOR MUSICAL EXPERIENCES. (ACADEMIC SELF-CITATION STRANGE LOOP.)

CHAPTER 4 *PROGRAMMABILITY & SOUND DESIGN*

MAX MATHEWS. *THE TECHNOLOGY OF COMPUTER MUSIC.* MIT PRESS, 1969. A SEMINAL BOOK ON USING DIGITAL COMPUTERS FOR MUSIC FROM THE FATHER OF COMPUTER MUSIC!

JEAN-CLAUDE RISSET. "SOME COMMENTS ABOUT FUTURE MUSIC MACHINES," *COMPUTER MUSIC JOURNAL* 15(4):32-36 (1991). AN INSIGHTFUL ARTICLE ABOUT COMPUTERS AS A TOOL FOR FUTURE MUSIC FROM A PROMINENT COMPOSER.

DONALD KNUTH. "GEORGE FORSYTHE AND THE DEVELOPMENT OF COMPUTER SCIENCE," *COMMUNICATIONS OF THE ACM* 15(8):721-727 (1972). KNUTH ON FORSYTHE, AND THE LATTER'S CONTRIBUTION TO THE ESTABLISHMENT OF COMPUTER SCIENCE AS A RECOGNIZED DISCIPLINE.

CURTIS ROADS. *THE COMPUTER MUSIC TUTORIAL.* MIT PRESS, 1996. A COMPREHENSIVE INTRODUCTORY REFERENCE ON MANY ASPECTS OF COMPUTER-BASED MUSIC.

DANIEL SHIFFMAN. *THE NATURE OF CODE: SIMULATING NATURAL SYSTEMS WITH PROCESSING.* D. SHIFFMAN, 2012. AN ENGAGING EXPLORATION OF PROGRAMMING!

MAX MATHEWS. "THE DIGITAL COMPUTER AS A MUSICAL INSTRUMENT," *SCIENCE* 142(3591): 553-557 (1963). PIVOTAL ARTICLE THAT ADDRESSED THE POTENTIAL OF COMPUTERS FOR MUSICAL EXPRESSION; IT INFLUENCED JOHN CHOWNING TO BEGIN EXPLORING COMPUTER MUSIC.

GE WANG. *THE CHUCK AUDIO PROGRAMMING LANGUAGE.* PH.D. THESIS, PRINCETON UNIVERSITY, 2008. A THESIS ON CHUCK AS A MUSIC PROGRAMMING LANGUAGE.

GE WANG, PERRY R. COOK, AND SPENCER SALAZAR. "CHUCK: A STRONGLY-TIMED COMPUTER MUSIC LANGUAGE," *COMPUTER MUSIC JOURNAL* 38(2):8-21 (2015). A JOURNAL ARTICLE DOCUMENTING THE CHUCK LANGUAGE AND ITS DESIGN.

AJAY KAPUR, PERRY R. COOK, SPENCER SALAZAR, AND GE WANG. *PROGRAMMING FOR MUSICIANS AND DIGITAL ARTISTS: CREATING MUSIC WITH CHUCK.* MANNING PRESS, 2015. A BOOK ON LEARNING TO PROGRAM MUSICAL THINGS WITH CHUCK.

SYDNEY PADUA. *THE THRILLING ADVENTURES OF LOVELACE AND BABBAGE.* PANTHEON BOOKS, 2015. A GRAPHIC NOVEL OF "THE (MOSTLY) TRUE STORY OF THE FIRST COMPUTER."

KEN STEIGLITZ. *A DIGITAL SIGNAL PROCESSING PRIMER: WITH APPLICATIONS TO DIGITAL AUDIO AND COMPUTER MUSIC.* PRENTICE HALL, 1996. A WONDERFUL BOOK ABOUT THE THEORY AND APPLICATIONS OF DSP FOR AUDIO AND MUSIC, FROM ONE OF THE FATHERS OF DIGITAL FILTERS.

JULIUS SMITH. *INTRODUCTION TO DIGITAL FILTERS (WITH AUDIO APPLICATIONS).* ONLINE: https://ccrma.stanford.edu/~jos/filters/ A MOST COMPREHENSIVE RESOURCE ON THE THEORY OF DIGITAL FILTERS FOR AUDIO FROM A FOREMOST EXPERT IN PHYSICAL MODELING FOR MUSIC.

PAUL LANSKY. *HOMEBREW.* BRIDGE RECORDS, 1992. A LANDMARK COMPUTER MUSIC WORK, INCLUDES "TABLE'S CLEAR" AND "NIGHT TRAFFIC"!

SOFTWARE: *THE CHUCK PROGRAMMING LANGUAGE* http://chuck.stanford.edu/ GET CHUCK HERE, AS WELL AS SOURCE CODE, LANGUAGE DOCUMENTATION, AND RESOURCES.

CHAPTER 5 INTERFACE DESIGN

PERRY R. COOK. "PRINCIPLES OF DESIGNING COMPUTER MUSIC CONTROLLER," **ACM SIGCHI, NEW INTERFACES FOR MUSICAL EXPRESSION (NIME) WORKSHOP.** 2001. AN HONEST, SIMPLE, AND DEEP DISCUSSION OF DESIGNING COMPUTER INSTRUMENTS, BY THE ZEN MASTER OF COMPUTER INSTRUMENT DESIGN.

DAVID WESSEL AND MATTHEW WRIGHT. "PROBLEMS AND PROSPECTS FOR INTIMATE MUSICAL CONTROL OF COMPUTERS," **COMPUTER MUSIC JOURNAL** 26(3):11-22 (2002). WHAT DOES IT MEANT TO DESIGN AND PLAY COMPUTER-BASED INSTRUMENTS? WHAT ARE THE CHALLENGES?

PERRY R. COOK. "REMUTUALIZING THE INSTRUMENT: CO-DESIGN OF SYNTHESIS ALGORITHMS AND CONTROLLERS," **STOCKHOLM MUSIC ACOUSTICS CONFERENCE (SMAC).** 2003. PERRY'S INSIGHTS INTO REGAINING THE INTIMACY OF COMPUTER INSTRUMENTS WE DESIGN.

DAN TRUEMAN. "WHY A LAPTOP ORCHESTRA?" **ORGANISED SOUND** 12(2):171-179 (2007). THE MANIFESTO OF THE MODERN LAPTOP ORCHESTRA.

 REBECCA FIEBRINK, GE WANG, AND PERRY R. COOK. "DON'T FORGET THE LAPTOP: USING NATIVE INPUT CAPABILITIES FOR EXPRESSIVE MUSICAL CONTROL," **NEW INTERFACES FOR MUSICAL EXPRESSION.** 2007. AN ARTICLE EXPLORING THE ETHOS OF DESIGNING INSTRUMENTS FROM THE PHYSICAL LAPTOP.

PERRY R. COOK. "RE-DESIGNING PRINCIPLES FOR COMPUTER MUSIC CONTROLLERS: A CASE STUDY OF SQUEEZEBOX MAGGIE," **NEW INTERFACES FOR MUSICAL EXPRESSION.** 2007. REVISITING AND EXTENDING THE PRINCIPLES.

SCOTT R. KLEMMER, BJÖRN HARTMANN, AND LEILA TAKAYAMA. "HOW BODIES MATTER: FIVE THEMES FOR INTERACTION DESIGN," **ACM DESIGNING INTERACTIVE SYSTEMS.** 2006. DEEPLY INSIGHTFUL ARTICLE ON THE ROLE OF THE HUMAN BODY IN HUMAN-COMPUTER INTERACTION DESIGN.

SCOTT SMALLWOOD, DAN TRUEMAN, PERRY R. COOK, AND GE WANG. "COMPOSING FOR LAPTOP ORCHESTRA," **COMPUTER MUSIC JOURNAL** 32(1):9-25 (2008). ON THE ISSUES AND CHALLENGES OF COMPOSING FOR AN EMERGING MEDIUM, DOCUMENTING ITS EARLY WORKS.

GE WANG, DAN TRUEMAN, PERRY R. COOK, AND SCOTT SMALLWOOD. "THE LAPTOP ORCHESTRA AS CLASSROOM," **COMPUTER MUSIC JOURNAL** 32(1):26-37 (2008). ON INTEGRATED CLASSROOMS FOR PROGRAMMING + SOUND & INSTRUMENT DESIGN + LIVE PERFORMANCE.

ATAU TANAKA. "BIOMUSIC TO BONDAGE: CORPOREAL INTERACTION IN PERFORMANCE AND EXHIBITION," IN **INTIMACY: ACROSS VISCERAL AND DIGITAL PERFORMANCE,** EDITED BY MARIA CHATZICHRISTODOULOU AND RACHEL ZERIHAN. PALGRAVE MACMILLAN, 1994. EXPLORATION OF BIO-SENSING TECHNOLOGIES AS VEHICLES FOR MUSICAL EXPRESSIVENESS.

DANIEL J. LEVITIN, STEPHEN MCADAMS, AND ROBERT L. ADAMS. "CONTROL PARAMETERS FOR MUSICAL INSTRUMENTS: A FOUNDATION FOR NEW MAPPINGS OF GESTURE TO SOUND," **ORGANISED SOUND** 7(2):171-189 (2002). A TREATISE ON MAPPING GESTURE TO MUSICAL PARAMETERS IN INSTRUMENT DESIGN.

ANDY HUNT, MARCELO M. WANDERLEY, AND MATTHEW PARADIS. "THE IMPORTANCE OF PARAMETER MAPPING IN ELECTRONIC INSTRUMENT DESIGN," **NEW INTERFACES FOR MUSICAL EXPRESSION.** 2002. AN ARGUMENT FOR THE SUBTLE, UNIQUE, AND ESSENTIAL PRINCIPLES OF DESIGNING MAPPING FOR COMPUTER INSTRUMENTS.

TINA BLAIN AND SIDNEY FELS. "CONTEXTS OF COLLABORATIVE MUSICAL EXPERIENCE," *NEW INTERFACES FOR MUSICAL EXPRESSION*, 2002. CREATIVELY EXTENDING THE TRADITIONAL ONE PERSON PER INSTRUMENT MODEL.

MICHAEL J. LYONS, MICHAEL HAEHNEL, AND NOBUJI TETSUTANI. "DESIGNING, PLAYING, AND PERFORMING WITH A VISION-BASED MOUTH INTERFACE," *NEW INTERFACES FOR MUSICAL EXPRESSION*, 2003. THE DESIGN OF THE UNCANNY AND UNIQUE MOUTHESIZER INSTRUMENT.

AJAY KAPUR, ARIEL J. LAZIER, PHILIP DAVIDSON, R. SCOTT WILSON, AND PERRY R. COOK. "THE ELECTRONIC SITAR CONTROLLER," *NEW INTERFACES FOR MUSICAL EXPRESSION*, 2004. A CASE STUDY OF AUGMENTING AND "DIGITIZING" NORTH INDIAN MUSICAL INSTRUMENTS.

SILE O'MODHRAIN AND GEORG ESSL. "PEBBLEBOX AND CRUMBLEBAG: TACTILE INTERFACES FOR GRANULAR SYNTHESIS," *NEW INTERFACES FOR MUSICAL EXPRESSION*, 2004. TWO TACTILE INTERFACES AND THE BROADER IMPORTANCE OF FEEL IN INSTRUMENT DESIGN.

BILL VERPLANK, MICHAEL GUREVICH, AND MAX MATHEWS. "THE PLANK: DESIGNING A SIMPLE HAPTIC CONTROLLER," *NEW INTERFACES FOR MUSICAL EXPRESSION*, 2002. CASE STUDY AND BROADER THINKING ON HAPTICS IN CONTROLLER DESIGN, FROM THREE GENERATIONS OF DESIGNERS.

BILL VERPLANK. *THE INTERACTION DESIGN SKETCHBOOK*, 2009. FRAMEWORKS FOR DOING INTERACTION DESIGN AS DESIGN FOR PEOPLE; CONCISE AND HUGELY INSIGHTFUL OBSERVATIONS FROM ONE OF THE FORERUNNERS OF INTERACTION DESIGN.

REBECCA FIEBRINK. *REAL-TIME HUMAN INTERACTIONS WITH SUPERVISED LEARNING ALGORITHMS FOR MUSIC COMPOSITION AND PERFORMANCE*, PH.D. THESIS. PRINCETON UNIVERSITY, 2011. PIONEERING WORK INTEGRATING MACHINE LEARNING, HUMAN-COMPUTER INTERACTION, AND MUSIC.

SALEEMA AMERSHI, MAYA CAKMAK, WILLIAM BRADLEY KNOX, AND TODD KULESZA. "POWER TO THE PEOPLE: THE ROLE OF HUMANS IN INTERACTIVE MACHINE LEARNING," *AI MAGAZINE* 35(4): 105-120 (2014). A POSITION PAPER ON THE VALUE OF HUMAN CURATION IN MACHINE INTELLIGENCE.

JOHN W. CAMPBELL. *TWILIGHT*. ASTOUNDING STORIES, 1934. A SHORT STORY FROM AN EARLIER GOLDEN AGE OF SCIENCE FICTION.

CHAPTER 6 GAME DESIGN

SEBASTIAN DETERDING, DAN DIXON, RILLA KHALED, AND LENNART NACKE. "FROM GAME DESIGN ELEMENTS TO GAMEFULNESS: DEFINING 'GAMIFICATION'," *MINDTREK*, 2011. AN EXCELLENT CLASSIFICATION AND DISTILLATION OF GAME DESIGN CONCEPTS.

ROBIN HUNICKE, MARC LEBLANC, AND ROBERT ZUBEK. "MDA: A FORMAL APPROACH TO GAME DESIGN AND GAME RESEARCH," IN *PROCEEDINGS OF THE AAAI WORKSHOP ON CHALLENGES IN GAME*, AAAI PRESS, 2004. A USEFUL FRAMEWORK FOR UNDERSTANDING THE DESIGN OF GAMES AND THE NATURE OF THE EXPERIENCE THEY PROVOKE.

ROGER CAILLOIS. *MAN, PLAY, AND GAMES*, UNIVERSITY OF ILLINOIS PRESS, 1961. INFLUENTIAL BOOK BREAKING DOWN "PLAY"; POWERFUL MODELS FOR THE CLASSIFICATION OF GAMES.

CLARKE C. ABT. *SERIOUS GAMES*, VIKING, 1970. FROM MILITARY TO EDUCATION TO BUSINESS, GAME-LIKE GOALS HAVE SHAPED HUMAN BEHAVIOR FOR MILLENNIA!

MIHALY CSIKSZENTMIHALYI. *FLOW: THE PSYCHOLOGY OF OPTIMAL EXPERIENCE.* HARPER AND ROW, 1990. BREAKING DOWN THE PHENOMENON AND PSYCHOLOGY OF BEING IN THE ZONE.

JONATHAN BLOW. "VIDEO GAMES AND THE HUMAN CONDITION." LECTURE, *COMPUTER SCIENCE COLLOQUIUM.* RICE UNIVERSITY. 2010. ON DESIGNING GAMES THAT REFLECT THE HUMAN CONDITION -- FROM THE DESIGNER OF *BRAID* AND *THE WITNESS*.

BLIZZARD ENTERTAINMENT. *STARCRAFT & STARCRAFT 2.* 1998-2015. THE MOST BALANCED AND NUANCED REAL-TIME STRATEGY GAMES OF ALL TIME. BEAUTIFULLY DESIGNED INSIDE AND OUT; ELEGANTLY SIMPLE MECHANICS THAT LEAD TO EXTREMELY VARIED AND COMPLEX DYNAMICS.

CHRIS CORNELL. *SAVE THE DATE.* 2013. FIGURING OUT WHERE TO GO FOR DINNER WITH FELICIA TURNS OUT BE ONLY ONE OF THE HARDEST THINGS IN THIS SUBLIME AND INGENIOUSLY DESIGNED GAME. STRANGE DESIGN LOOPS IN FULL MOTION!

RYAN AND AMY GREEN / NUMINOUS GAMES. *THAT DRAGON, CANCER.* 2016. AN AUTOBIOGRAPHICAL "SIMULATION" AND GAME OF RAISING AN INFANT WITH TERMINAL CANCER. DEVASTATING, BEAUTIFUL, AND REFLECTING HUMAN CHALLENGES IN THE MEDIUM OF GAMES.

STEVE GAYNOR / THE FULLBRIGHT COMPANY. *GONE HOME.* 2013. WHERE IS EVERYONE? YOU ARRIVE HOME (FROM SCHOOL IN EUROPE) IN THE MIDDLE OF THE NIGHT TO FIND YOUR PARENTS AND SISTER... GONE? A UNIQUELY ORIGINAL "STORY EXPLORATION VIDEO GAME" IN THE FORM OF A STORMY SHADOWY BIG HOUSE ADVENTURE GAME.

GE WANG. "GAME DESIGN FOR EXPRESSIVE MOBILE MUSIC," *NEW INTERFACES FOR MUSICAL EXPRESSION.* 2016. DESIGNING GAME ELEMENTS THAT PROMOTE MUSICAL EXPRESSION.

CHAPTER 7 SOCIAL DESIGN

THOMAS TURINO. *MUSIC AS SOCIAL LIFE: THE POLITICS OF PARTICIPATION.* UNIVERSITY OF CHICAGO PRESS, 2008. A GREAT EXPLORATION OF THE NATURE AND ROLE OF PARTICIPATION IN HUMAN LIFE.

LUIS VON AHN. *HUMAN COMPUTATION.* PH.D. THESIS. CARNEGIE MELLON UNIVERSITY, 2005. A PIONEERING WORK ABOUT USING HUMANS TO SOLVE HARD COMPUTATIONAL PROBLEMS, FROM ONE OF THE PIONEERS OF CROWDSOURCING.

LUIS VON AHN AND LAURA DABBISH. "LABELING IMAGES WITH A COMPUTER GAME," IN *PROCEEDINGS OF THE ACM SIGCHI CONFERENCE ON HUMAN FACTORS IN COMPUTING SYSTEMS* (PP. 319-326). 2004. KICKING OFF GAMES WITH A PURPOSE AND THE HARNESSING OF HUMAN COMPUTATION CYCLES.

JONATHAN HARRIS AND SEP KAMVAR. *WE FEEL FINE,* HTTP://WEFEELFINE.ORG/. 2008. A TECHNOLOGY-MEDIATED EMOTION BAROMETER OF THE INTERNET.

STANLEY MILGRAM. "THE FAMILIAR STRANGER: AN ASPECT OF URBAN ANONYMITY," IN *THE INDIVIDUAL IN A SOCIAL WORLD: ESSAYS AND EXPERIMENTS.* EDITED BY JOHN SABINI AND MAURY SILVER. MCGRAW-HILL, 1972. MILGRAM'S CLASSIC STUDY ON FAMILIAR STRANGERS.

JIM HOLLAN AND SCOTT STORNETTEA. "BEYOND BEING THERE," IN *PROCEEDINGS OF THE SIGCHI CONFERENCE ON HUMAN FACTORS IN COMPUTING SYSTEMS* (PP. 119-125). 1992. A PAPER OF KEEN INSIGHT ABOUT DESIGNING COMMUNICATION SYSTEMS.

GE WANG, SPENCER SALAZAR, JIEUN OH, AND ROBERT HAMILTON. "WORLD STAGE: CROWDSOURCING PARADIGM FOR EXPRESSIVE SOCIAL MOBILE MUSIC," *JOURNAL OF NEW MUSIC RESEARCH* 44(2):112-128 (2015). ON THE DESIGN AND IMPLEMENTATION OF A MASSIVELY-MULTIPLAYER MUSICAL ECOSYSTEM.

ROB HAMILTON, JEFF SMITH, AND GE WANG. "SOCIAL COMPOSITION: MUSICAL DATASYSTEMS FOR EXPRESSIVE MOBILE MUSIC," *LEONARDO MUSIC JOURNAL* 21:57-64 (2011). AN ARTICLE ON TOOLS, DATA, AND SYSTEMS FOR ENABLING MOBILE, SOCIAL MUSIC.

CHAPTER 8 MANIFESTO

ARISTOTLE. *NICOMACHEAN ETHICS.* 349, B.C.E. ARISTOTLE'S BEST-KNOWN WORK ON ETHICS, VIRTUE, THE GOOD, AND THE EUDAIMONIC. INTENDED TO BE A PRACTICAL GUIDE ON HOW ONE MIGHT LIVE ONE'S LIFE, THIS IS QUITE POSSIBLY THE FIRST SELF-HELP BOOK.

MARY SHELLEY. *FRANKENSTEIN: OR, THE MODERN PROMETHEUS.* 1818. POSSIBLY THE FIRST MODERN SCIENCE FICTION NOVEL, ONE THAT EXPLORES THE CONSCIENCE OF CREATION. A TECHNOLOGICAL AND MORAL MEDITATION THAT BRINGS INTO SHARP FOCUS THE PRIOR INDUSTRIAL REVOLUTION AND THE ROMANTIC ERA YET TO COME.

MELVIN KRANZBERG. "TECHNOLOGY AND HISTORY: 'KRANZBERG'S LAWS,'" *TECHNOLOGY AND CULTURE* 27(3):544-560 (1986). KRANZBERG'S LAWS OF TECHNOLOGY, THE FIRST OF WHICH STATES THAT TECHNOLOGY IS NEITHER GOOD NOR BAD, NOR IS IT NEUTRAL.

LONGINUS. *ON THE SUBLIME.* 1ST CENTURY C.E. THIS WORK, COMMONLY ATTRIBUTED TO LONGINUS, IS THE EARLIEST SURVIVING TREATISE ON THE SUBLIME. IT DEALS WITH RHETORIC AND THE DESIGN OF WORDS TO ACHIEVE A SUBLIMITY OF MIND AND EMOTION.

EDMUND BURKE. *A PHILOSOPHICAL ENQUIRY INTO THE ORIGIN OF OUR IDEAS OF THE SUBLIME AND THE BEAUTIFUL.* 1757. BURKE'S ONLY PURELY PHILOSOPHICAL WORK -- A DEEP EXPLORATION ON THE SUBLIME VS. THE BEAUTIFUL.

IMMANUEL KANT. *CRITIQUE OF PRACTICAL REASON.* 1788. KANT'S "SECOND CRITIQUE"; IT BRINGS INTO A CRISP FOCUS THE IDEA OF MORALITY AS AN END IN ITSELF. IT INCLUDES THE DELINEATION OF HYPOTHETICAL IMPERATIVES VS. CATEGORIAL IMPERATIVES.

IMMANUEL KANT. *CRITIQUE OF JUDGMENT.* 1790. KANT'S "THIRD CRITIQUE" WAS HIS TREATISE ON AESTHETIC JUDGMENT: THE AGREEABLE, THE BEAUTIFUL, THE SUBLIME, AND THE GOOD; INCLUDES AN EXPLORATION OF THE ROLE OF AESTHETIC EXPERIENCES IN HUMAN MORALITY, AND OF THE CONNECTION BETWEEN AESTHETICS AND ETHICS; THE MORAL SUBLIME.

ISAAC ASIMOV. *THE ROBOT SERIES, THE EMPIRE SERIES, AND THE FOUNDATION SERIES.* 1941-1992. THESE CLASSICS FROM THE GOOD DOCTOR ARE AS MUCH SCIENCE FICTION AS THEY ARE PHILOSOPHICAL RUMINATION ON POTENTIAL FUTURE HISTORIES FOR HUMANITY.

KOYAANISQATSI, DIRECTED BY GODFREY REGGIO. INSTITUTE FOR REGIONAL EDUCATION, 1982. AN EXPERIMENTAL FILM THAT CONFRONTS US WITH OUR LIFE OUT OF BALANCE, SCORED BY PHILIP GLASS.

E. M. FORSTER. "THE MACHINE STOPS," *THE OXFORD AND CAMBRIDGE REVIEW.* 1909. A CLASSIC SHORT STORY ABOUT THE ROLE OF TECHNOLOGY IN OUR LIVES, WRITTEN A CENTURY BEFORE THE INTERNET BECAME PART OF EVERYDAY LIFE.

KEN TAYLOR AND JOHN PERRY, CO-HOSTS. *PHILOSOPHY TALK.* FROM KALW AND ONLINE. http://philosophytalk.org/ 2004-PRESENT. TALK RADIO SHOW THAT "QUESTIONS EVERYTHING (EXCEPT YOUR INTELLIGENCE!)": FROM THE GOOD LIFE TO THINKING MACHINES, ANCIENT WISDOM TO MODERN CRISES, THIS IS FOOD FOR THE HUNGRY MIND.

ARTHUR SCHOPENHAUER. *WORLD AS WILL AND REPRESENTATION.* 1819. TO SCHOPENHAUER, LIFE IS PAINFUL. IT IS SUFFERING, DRIVEN BY A LESS-THAN-KIND INNER NATURE OF ALL THINGS HE CALLS WILL. THE WAY TO TRANSCEND SUCH SUFFERING IS THROUGH THE AESTHETIC (SEEING THE BEAUTY IN HUMANS) AND ETHICAL (COMPASSION AND RECOGNITION OF SELF IN OTHERS).

CAROL D. RYFF AND BURTON H. SINGER. "KNOW THYSELF AND BECOME WHAT YOU ARE: A EUDAIMONIC APPROACH TO PSYCHOLOGICAL WELL-BEING," *JOURNAL OF HAPPINESS STUDIES* 9(1):13-39 (2008). A MODERN REVISIT OF *NICOMACHEAN ETHICS* IN THE CONTEXT OF HAPPINESS, SELF, PURPOSE, FULFILLMENT, AND PSYCHOLOGICAL WELL-BEING.

TIMOTHY M. COSTELLOE, ED. *THE SUBLIME: FROM ANTIQUITY TO THE PRESENT.* CAMBRIDGE UNIVERSITY PRESS, 2012. A GREAT COLLECTION OF ESSAYS ON THE SUBLIME, FROM THE RHETORICAL SUBLIME OF THE ANCIENT GREEKS TO SUBLIMITY OF NATURE, MORALITY, AND FINE ART.

STEWARD BRAND. *THE LAST WHOLE EARTH CATALOG.* RANDOM HOUSE, 1968. ACCESS TO TOOLS THAT SUPPORT AN INDIVIDUAL'S SELF-EDUCATION AND SELF-FASHIONING. "STAY HUNGRY, STAY FOOLISH."

ALAN TURING. "ON COMPUTABLE NUMBERS, WITH AN APPLICATION TO THE ENTSCHEIDUNGSPROBLEM," *PROCEEDINGS OF THE LONDON MATHEMATICAL SOCIETY.* 1937. TURING'S LANDMARK PAPER THAT CONTAINED HIS PROOF OF THE "DECISION PROBLEM"; IT DEMONSTRATED A FUNDAMENTAL LIMIT ON THE NOTION OF COMPUTATION AS WE KNOW IT, WITH PROFOUND IMPLICATIONS FOR MATHEMATICS, LOGIC, AND THE PHILOSOPHY OF KNOWLEDGE AND THE AUTOMATION OF ITS PURSUIT.

ITALO CALVINO. *INVISIBLE CITIES.* HARVEST / HARCOURT, 1972. A SUBLIME BOOK OF DREAMS, AND VIVID IMAGERY WITHOUT PICTURES. A MEDITATION ON THE DESIGN OF CITIES OF THE MIND.

FRIEDRICH SCHILLER. *LETTERS ON THE AESTHETIC EDUCATION OF MAN.* 1794. "MAN IS NEVER SO AUTHENTICALLY HIMSELF, AS WHEN AT PLAY." A TREATISE (IN THE FORM OF A SERIES OF LETTERS) ON THE ROLE OF ART AND THE AESTHETIC DIMENSION IN HUMAN LIFE, SCHILLER ARGUES THAT IT IS THROUGH AESTHETIC EDUCATION THAT WE AS INDIVIDUALS AND SOCIETY CAN BE TRULY FREE.